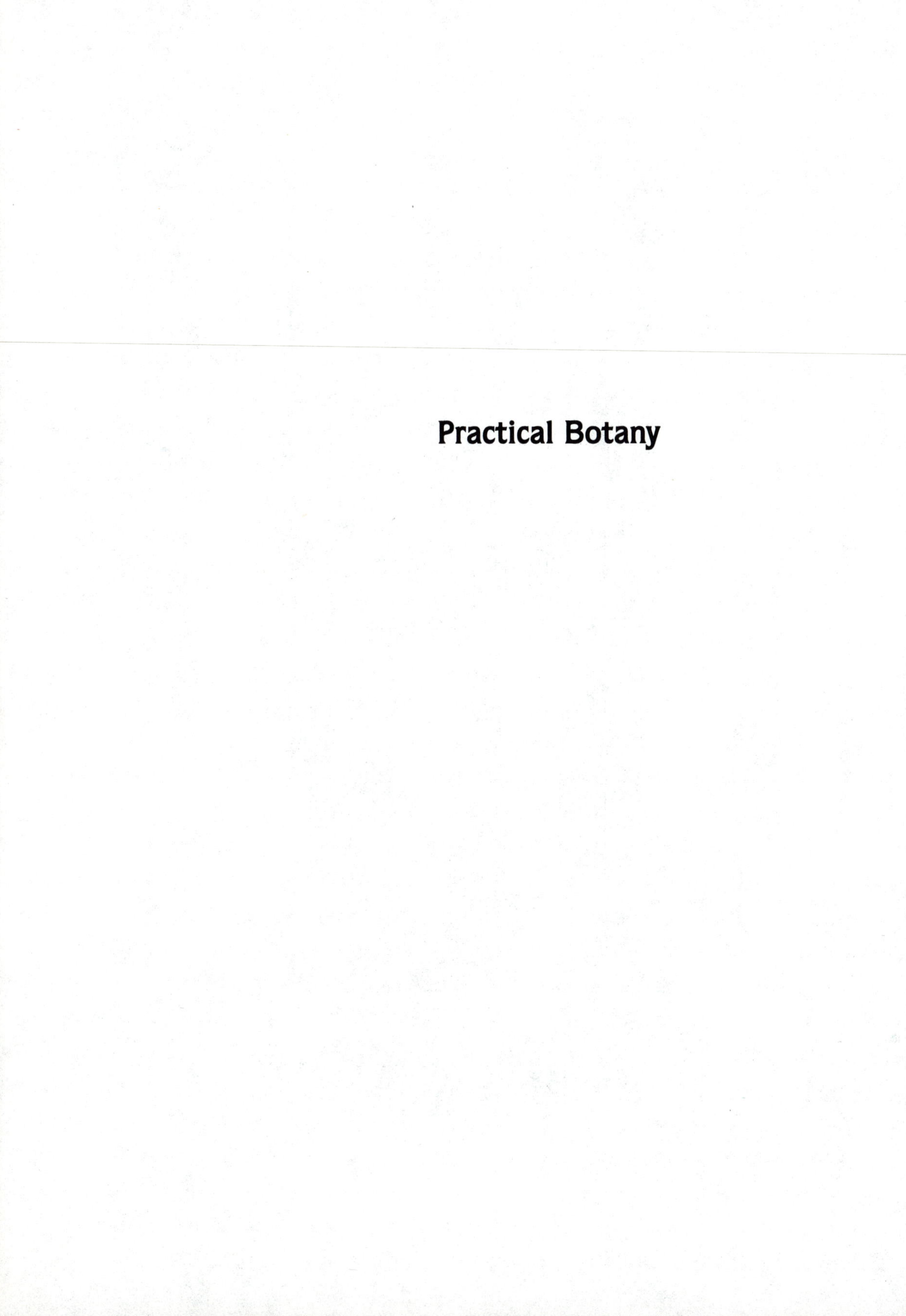

# Practical Botany

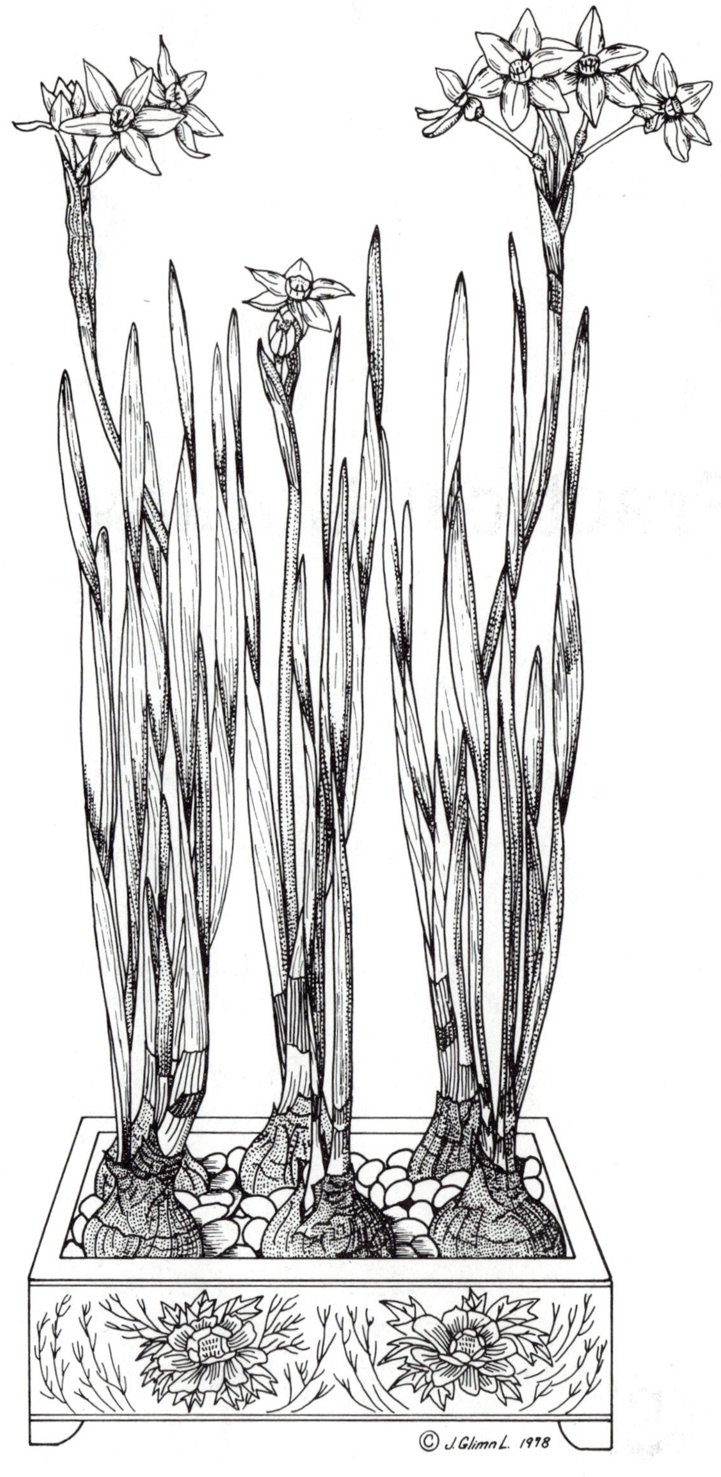
© J.Glimn L. 1978

QK
47
.P876
1983

WITHDRAWN

# Practical Botany

**Peter B. Kaufman**
*University of Michigan*

**T. Lawrence Mellichamp**
*University of North Carolina at Charlotte*

**Janice Glimn-Lacy**
*University of Michigan*

**J. Donald LaCroix**
*University of Detroit*

**RESTON PUBLISHING COMPANY, INC.**
*A Prentice-Hall Company*
Reston, Virginia

337081 Tennessee Tech. Library
Cookeville, Tenn.

*To our students, friends, and colleagues, who asked questions and stimulated our thinking on practical plant topics, we are most grateful, for it is for them that this book is written.*

*Library of Congress Cataloging in Publication Data*

Kaufman, Peter B.
Practical botany.

Includes bibliographies.
1. Botany. 2. Gardening. 3. Plants, Useful.
I. Title.
QK50.K38 1983 635.9 83-3333
ISBN 0-8359-5580-X

*Editorial/production supervision and interior design by Camelia Townsend*

©1983 by Reston Publishing Company, Inc.
*A Prentice-Hall Company*
Reston, Virginia 22090

All rights reserved. No part of this book may be reproduced in any way or by any means without permission in writing from the publisher.

10 9 8 7 6 5 4 3 2 1

PRINTED IN THE UNITED STATES OF AMERICA

# Contents

## 8
**Plant pests and their control, 235**



## 9
**Growing plants in controlled environments, 261**

# Preface

This book is meant to serve as an introduction and guide to the many diverse topics which can be covered in a practical botany course. It is designed for those interested in learning more about plants: how they grow and function, and how they may be manipulated, improved, and utilized by humans for the betterment of our society. A perusal of the table of contents will show the great variety of subjects covered. We hope that from reading this book, you will find new ideas about growing plants to suit your needs. Besides plant cultivation, additional subjects are natural plant communities, edible wild plants, preserving plants for crafts, uses of plant dyes, plant photography, and new uses of plants. We hope that you will be led to a greater understanding and appreciation of the many important ways plants touch our daily lives.

In the preparation of this book, each of us has contributed sections based on our individual areas of interest and expertise. Thus, plant structure, seeds, growth regulation, soils, cloning, controlled environments, edible wild plants, and natural dyes were written by P. B. Kaufman; introduction, landscaping, and photography, plus all photographic illustrations (except where noted) by T. L. Mellichamp; houseplants and natural communities by P. B. Kaufman and T. L. Mellichamp; preparing plant materials for crafts plus providing all the line drawings (except where noted), and editing the entire manuscript by J. Glimn-Lacy; plant pests, spice-herb-drug plants, and new uses of plants by J. D. LaCroix.

We are indebted to many people for help in the production of this book and would like to specifically acknowledge the following: David Bay and P. Dayanandan for converting the color slides into black and white prints; Patricia Pachuta and William Collins of the University of Michigan Matthaei Botanical Gardens for providing plant materials for the drawings; Kerry S. Walter for the line drawing plate of orchids; Imre Tamas of Ithaca College, Ithaca, New York, for many helpful editorial suggestions; Mary LaCroix and Audrey Mellichamp for their help in typing the manuscript and Camelia Townsend for her help as production editor. We would also like to gratefully acknowledge support from McMillan Horticulture fund of the University of North Carolina at Charlotte.

# Introduction

## PLANTS IN OUR LIVES

Everyone of us makes a daily encounter with plants, starting from the moment we get up in the morning. We wash our face with soap made from plant oils, using a cloth made from cotton, and brush our teeth with a mild abrasive paste containing microscopic diatoms (marine algae). We eat bread made from grains, use margarine produced from corn or soybean oil, drink coffee brewed from a tropical seed while sitting at a sturdy table made from oak. Walking out the door, we take for granted the beauty of the trees, enjoy the artistic decor of the neighborhood landscaping, marvel at the sparkle of colorful annuals which add color to the monotonous tones of green lawns and shrubs. We drive to work in cars powered by gasoline refined from the fossil remains of prehistoric plants. At dinnertime, we feast on the "fruits" of agriculture: meat (fattened with corn), potatoes, and vegetables seasoned with herbs and spices. We relax by puttering around the garden, leisurely watering the houseplants, or strenuously rebuilding a rock garden or rhododendron bed. We seek recreation by playing ball on a field of grass, waterskiing on a lake kept healthy by aquatic plants, or hiking in the high country where we can wonder at the tiny wildflowers or thousand-year-old redwood trees. And finally, we sink into our favorite chair with a glass of wine fermented from a pedigreed grape, sigh deeply as we breathe in the refreshing oxygen given off by green plants, and feel reassured that the sun (which affects all of life) will rise again tomorrow.

## NAMING PLANTS

Plants are given names to facilitate communication about them. It has always been important for people to talk about and refer to plants. Since earliest times, there have been efforts to learn which plants are useful for food, medicines, and aesthetics and to set up classification systems so the knowledge is organized and can be passed on and added to. In the beginning of organized botany, plants were given short phrase names in Latin, consisting of a series of descriptive words to briefly characterize them and indicate their uses. The short descriptive phrase was the *name* of the plant. As more and more plants were discovered, the phrase names had to become longer, with more descriptors, to distinguish one plant from the other. This became very cumbersome and was defeating the purpose of facilitating communication. Our present *binomial* system was devised by the famous Swedish botanist, Carl Linnaeus, whose *Species Plantarum* (or "species of plants") was published in 1753. In this book, he set forth the idea of a simple two-part name for each plant species. This two-part name, which replaced the multiparted phrase names, consisted of a *genus name* and a *specific epithet.* The generic name is a noun with the specific epithet modifying it as an adjective. This nomenclature system was then adopted under the International Code of Botanical Nomenclature.

Under the binomial system, for example, the name for white oak is *Quercus alba. Quercus* is the generic name or genus, and *alba* is the specific epithet modifying *Quercus.* In Latin *Quercus* means "oak" and *alba* means "white"; hence the common name, "white oak." You are probably familiar with many Latin generic names of common plants because they are also the common names: *Rhododendron, Chrysanthemum, Iris, Geranium, Hyacinthus, Lilium, Coleus, Citrus, Aster, Agave, Asparagus, Begonia, Canna, Caladium, Fuchsia, Gardenia, Hibiscus, Narcissus, Peperomia, Rosa, Sedum, Tulipa,* and *Yucca.*

A *genus* is an association of plants all of which share certain common characteristics which make them easily recognizable as a group. Oaks *(Quercus),* ashes *(Fraxinus),* maples *(Acer),* and birches *(Betula)* are familiar genera of trees. Within a genus, each specific type of plant is called a *species,* such as the white oak *(Quercus alba),* the red oak *(Q. rubra),* the black oak *(Q. velutina),* and the willow oak *(Q. phellos).* Species are capable of interbreeding within their kind, but rarely can two different species interbreed (orchids are a good exception). The specific epithet, then, is a Latin word, that goes along with the generic name to identify each separate species.

Many species, all of which have Latin names, may also have common names or English names. Sometimes, these are merely a translation of the Latin names (as "white oak" is a translation of the Latin, *Quercus alba*). In other cases, the common name is a colloquial name which means something to people only in a certain region. In this sense, common names can be confusing if you travel from one region to another. For example, in the northeastern United States, the plant known as *Caltha palustris,* an early spring marsh plant, is called marsh marigold or cowslip. In Europe, the common yellow primrose, *Primula verna,* is also called cowslip. The common white water lily has hundreds of common names in different countries around the world. Another good example of the confusion generated by common names is the aggregate of common houseplants known as "ivies." For example, there are German-ivy *(Senecio mikanioides),* Swedish-ivy *(Plectanthrus australis),* English ivy *(Hedera helix),* kenilworth-ivy *(Cymbalaria muralis),* Boston-ivy *(Parthenocissus tricuspidata),* and grape-ivy *(Cissus rhombifolia).* Each of the different ivies is in a totally different genus and may be totally unrelated. So what really does *ivy* mean? Apparently, it refers to any climbing or twining plant with lobed leaves.

How do we deal with the problem of common names? Most people seem to want to use common names since they seem to be "down to earth" and popularly understood. But we have to understand the value of the Latin scientific name for plants if we want to avoid confusion when talking about them. Latin has been used as the official language of botany for hundreds of years. It is now a dead language and does not change; hence, it can be understood around the world by anyone familiar with it. Even if you write or speak Chinese, you could write or speak the scientific names of plants in Latin. When professional botanists discover new plants, they use Latin to name and describe them, even if their native tongue is Russian or Japanese (which even have different alphabets). As a result, any botanist or informed layperson anywhere in the world can understand at least the briefest information about the new plant. Thus, these internationally accepted rules that govern the naming and describing of new plants help to minimize problems and confusion. If you use the accepted Latin name of a plant, you will stand a better chance of being understood when discussing a particular plant.

It might be noted that according to the rules of botanical nomenclature, there can only be one correct Latin name for a plant. Also, there can only be one correct cultivar name for a cultivated plant. Sometimes botanists and horticulturists get confused themselves or discover confusions between plants that have been named for many years; to correct this, they may have to change a plant name. This is annoying to the layperson! However, it is part of the process of scientific verification of plant identification and is an ongoing process. Thus, a plant name can change to a more correct older name that has been found for it, or a plant name may change if it is discovered that the plant has been confused with another plant through history. Also, taxonomists, those who name and classify plants and who constantly study large and difficult plant groups (families, genera, or species), occasionally come up with new information that they use to show better relationships of the plants to each other in their groups (taxa). Their way of expressing this new alignment is to change plants from one genus to another.

Another important category used in botany and horticulture is that of the plant *family*. Genera are grouped into families, which also have a certain set of defining characteristics. Some plant families are large, such as the orchid family with over 25,000 species in 750 genera; or small, with only a few genera and/or species. You are also familiar with some of the large and economically important plant families, such as the grass family (Poaceae), rose family (Rosaceae), pea family (Fabaceae), and aster family (Asteraceae). Family names written in Latin, as for example those indicated above, usually end in "-aceae," so they are readily identifiable. These large families have been around so long that some books still use the older family names that have been in use for centuries: Gramineae for Poaceae, Leguminosae for Fabaceae, and Compositae for Asteraceae. Other examples of large families where the family names have been changed include the carrot family (old name, Umbelliferae; new name, Apiaceae) and the mustard family (old name, Cruciferae; new name, Brassicaceae). These older family names are descriptive of the floral structures of the particular family and are very well known.

In this book, we use the new family names wherever appropriate. For further information on plant families, consult a reference such as L. H. Bailey's *Manual of Cultivated Plants* or his *Encyclopedia of Horticulture* or G. H. M. Lawrence's *Taxonomy of Flowering Plants.* Knowledge about plant families helps you to see relationships between genera and species and often tells you about the cultural requirements of the plants. For example, the cactus family (Cactaceae) consists of mostly desert dwellers, and members of the African violet family (Gesneriaceae) are mostly from the warm, humid tropics. The characteristics of the large families are easy to master and you will be amazed at how much you can learn by studying them.

Another important botanical term is *variety.* It is used when a species has more than one recognizable form in nature, and it may "vary." For example, the white flowering dogwood, *Cornus florida,* may occasionally produce red or pink bracts (floral leaves) instead of white ones. When this occurs, you may call the plant *Cornus florida* var. *rubra* to differentiate it from the typical white form. As you may know, the pink flowering dogwood is a horticultural favorite for home landscaping. Looking through a list of houseplants will also reveal several species that have recognized varieties. A variety, then, is simply a plant with a single character, or more rarely, two characters, differing from the typical form.

The trend recently is to give a special designation to cultivated varieties of plants when they exist *only* in cultivation; that is, new varieties of plants that are essentially man-made, or at least propagated and perpetuated through the care of man. These may be highly selected types of grain crops, outdoor woody landscape plants with unusual features, or hybrids of garden annuals. Whenever you are dealing with a selected type of plant that exists *only* in cultivation, never in the wild, you refer to the plant or strain as a cultivar, not a botanical variety. The word, *cultivar* is made up from *culti*vated *var*iety and means just what we have described, namely, a plant variation that exists only under cultivation. Thus, most houseplants used today with "fancy names" are cultivars, and they exist as clones of vegetatively propagated plants or specially produced hybrid strains of annuals.

A *clone* is a large number of identical specimens vegetatively propagated from one original plant. When a *sport* or *mutation* occurs, a keen-eyed horticulturist may spot it, like it for some outstanding characteristic, and propagate it. He or she has created a cultivar and may give it a fancy English name, not a Latin name. The English name appears in *single* quotes after the Latin generic name and species epithet (if the latter is known) and becomes the correct name of the plant. For example, an unusually good specimen of the pink flowering dogwood that has outstanding red flower bracts has been named, *Cornus florida* var. *rubra* cv. 'Cherokee Chief'. You may also have *Rhododendron* × 'English Roseum,' a garden hybrid of unknown species parentage. The "×" refers to a cross between two species, resulting in a hybrid. Many plants that have been in cultivation for years and/or are of varied hybrid origin are referred to only by their generic name followed by an accepted cultivar name. Most of the fancy hybrids of garden annuals and greenhouse plants have cultivar names. Many examples can be found in any current seed or nursery catalog.

In the above example of the red flowering dogwood, the words, *Cornus, florida,* and *rubra* are Latin, and because of this, are always italicized in print and underlined in writing or typing. The English cultivar name, 'Cherokee Chief' is never italicized, but is capitalized and set off in single quotes, with or without the abbreviation, "cv." (for cultivar).

A new cultivar may be found in nature, in a nursery somewhere, in a back yard, or in a greenhouse. All it is, is an unusually good specimen of an otherwise common plant that has at least one distinctive characteristic for which it is recognized as outstanding. Cultivars may be dwarf plants, weeping plants, unusual color forms, uncommonly large flowered forms, and so forth—anything that makes a plant worthy of recognition and propagation. Plant breeders grow thousands of specimens from seeds or cuttings, looking for new types that will have useful characteristics for further breeding or ornamental purposes.

In many scientific books on plants, you will find the scientific names followed by the name or initials of a person, such as *Quercus alba* L. The *L* stands for Linnaeus, the botanist who first correctly named the species. Cultivar names do *not* designate the namer, or "authority" as it is called. There may be some interesting historical facts associated with people who have named plants, but they are usually left out of common books on houseplants.

While a plant taxonomist names and classifies plants, it is often left up to the plant explorer to search the unknown reaches of the world for new findings. Even today, for example, much remains to be learned about the types of plants that grow in the tropics. It is well known that these plant explorers have been responsible for the introduction of the plants we now have in cultivation. The nineteenth century, in particular, was probably the most important time for new plant introductions into civilizations of northern countries. Still today, however, new plants are being brought into cultivation from all parts of the world as we seek to enrich our lives with more exotic specimens. Improvements are also being made on old favorites through hybridization and selection of outstanding cultivars.

The following categories of cultivated plants are recognized:

| *Categories* | *Example* |
|---|---|
| Family | Cornaceae |
| Genus | *Cornus* |
| Species | *florida* |
| Variety | *rubra* |
| Cultivar | 'Cherokee Chief' |

## REFERENCES

Bailey, L. H. 1963 (1933). *How plants get their names.* N.Y.: Dover Pub.

Dirr, Michael A. 1977. *Manual of woody landscape plants.* Champaign, Ill: Stipes.

# 1 Plant structure and development

It is important to begin our discussion with basic information on plant structure and development so that you will understand the various parts of the plant, how they function, and how they are used in different aspects of practical botany. Let us look first at the basic structure of the vegetative parts of seed plants. After that, we will examine fundamental aspects of sexual reproductive development and how this information can be put to use in plant breeding.

## Vegetative development

The major features of vegetative development of the plant include the production of roots, leaves, stem nodes and internodes, and lateral (axillary) buds (Figure 1–1). A knowledge of these vegetative plant structures is essential in nearly every aspect of practical botany, whether you are dealing with bonsai, rooting of cuttings, grafting, induction of flowering, pruning, or seed germination. It even enters landscaping because a plant's growth habit is a major consideration in deciding which plants to use near the house, on a fence, or as specimen plants. Thus, it is important to see how the vegetative plant is constructed and to be able to relate this structure to various practical uses.

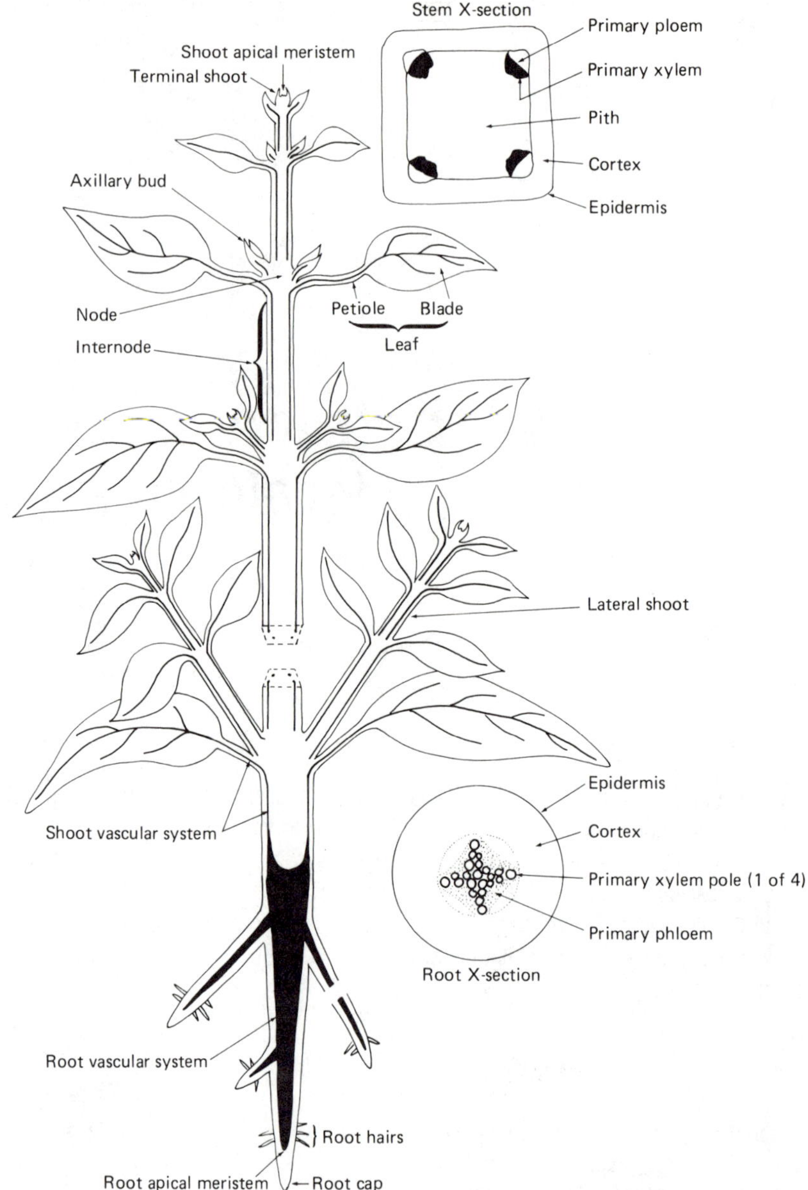


*FIGURE 1–1*

The plant body. A highly schematic diagram of the primary organs of the shoot and root systems of a vascular flowering plant. Black lines within the roots, stems, leaves, and buds represent the vascular system. The transverse sections of stem (top) and root (lower right) show the arrangement of vascular tissues that is characteristic of these organs. These sections are taken at levels of root and shoot, respectively, where the breaks (interruptions) occur.

## CELLS AND TISSUES: THE BASIC BUILDING BLOCKS OF THE PLANT

Plants are composed of living and dead cells (Figure 1–2). The living cells include the *parenchyma* and *collenchyma*. Parenchyma cells, which make up the major portion of the young plant, are *isodiametric* (same size in all directions) and generally, though not always, thin-walled. While at first unspecialized, they later develop into cells found in pith at the center of stems, into cortex just beneath the surfaces of roots and stems, into vein cells, and into photosynthetic cells (mesophyll) found in leaves. They also occur as food storage cells in seeds, bulbs, and tubers. Collenchyma cells have variously thickened walls which provide support of plant structures and are found

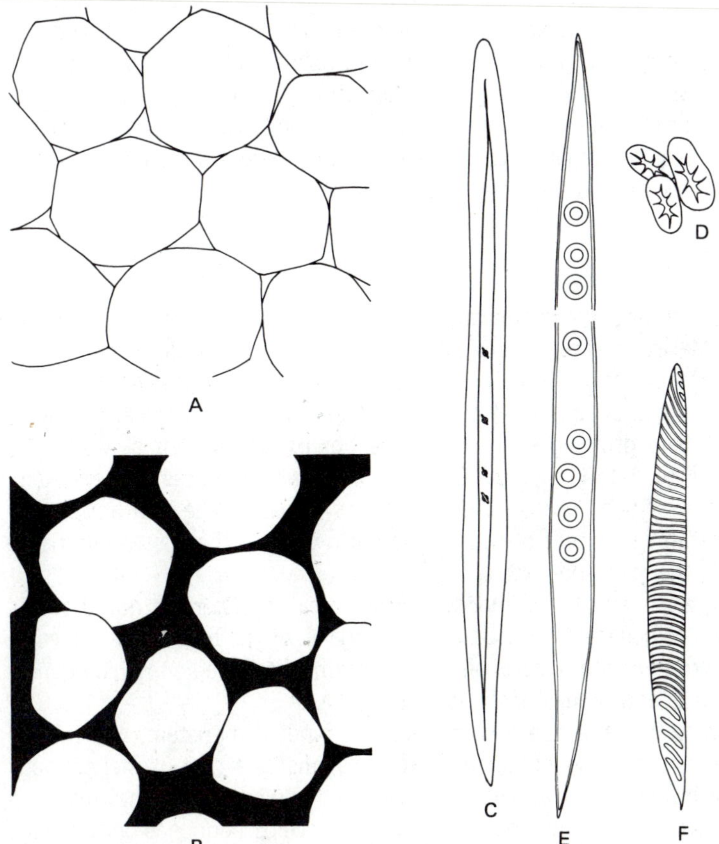


*FIGURE 1–2*
Cell types. Cells that make up the three basic types of tissue found in plants: (A) parenchyma, (B) collenchyma, and sclerenchyma. Sclerenchyma cell types shown are: (C) fiber, and (D) sclereid (stone cell). Examples of xylem cells are (E) tracheid and (F) vessel element.

near the surface around the veins of stems, in leaf petioles, and in swollen leaf bases of grasses which respond to gravity.

Nonliving cells include *tracheids* and *vessel elements* in the veins *(vascular bundles)*. They function in transport of water and mineral nutrients. Another nonliving type is the *fiber cell* which strengthens many tissues. Commercial hemp, jute, rattan, and cotton are composed of fiber cells. *Bark cells* and *sclereids* are also nonliving. Familiar plant forms composed of dense layers of sclereids are nutshells, fruit pits, and the gritty consistency of unripe pear fruits. Fibers and sclereids are examples of *sclerenchyma*, tissue composed of thick-walled cells which provide mechanical support.

Cells are arranged as *tissues*—such as *epidermis*, the tissue which covers plant surfaces, and *mesophyll*, the photosynthetic tissue of leaves. *Tissue systems* are made up of several tissues such as in vascular bundles with the xylem tissue system (involved in water and nutrient transport) and the phloem tissue system (involved in transport of organic solutes such as sugars). The *xylem* and *phloem* make up the major parts of the plant's vascular system. The tissue systems and the tissues are combined in various arrangements in the plant organs—as leaf, stem, root, flower, and fruit. Together, these organs make up the plant body. (Figure 1–1.)

## MERISTEMS

In practical botany, we continually talk about meristems. Why are they important? Meristems are regions of active cell division capable of generating new tissues that make up the primary plant body. There are several types of meristems in plants, one of which is the *apical meristem* located at the shoot tip of the main growing stem and at the tips of lateral buds as well as at root tips (Figure 1–1). The *shoot apical meristem* produces all new stem, leaf, and flower bud tissue as the stem grows upwards. The *root apical meristem* generates new cells of the root cap below it and the tissues and tissue systems of the root above it.

In grasses and other monocots, the apical meristem is not the only meristem that generates new tissues and organs. At the base of each sheathing leaf and each internode is an *intercalary meristem* region, which generates new cells and tissues of the leaf or internode.

In woody plants, there is still another type of meristem, namely, the *vascular cambium*. It is identified in woody stems as a ring of meristematic tissue just inside the bark. It is responsible for adding tissues that make up the secondary plant body that characterizes development in woody plants. Throughout the growing season, the vascular cambium produces secondary phloem to the outside of itself and secondary xylem to the inside. The secondary phloem conducts organic solutes such as sugars and amino acids; the secondary xylem conducts water and mineral nutrients. Another meristem, the *cork cambium* makes bark tissues.

*Practical uses of meristems:*

1. *Grafting* The vascular cambium is the primary meristem involved in the knitting together of the two parts grafted together; the top is *scion* and the base is *stock.*
2. *Flower initiation* The vegetative shoot apex with its apical meristem is converted to a reproductive (flowering) apex when we subject long- and short-day plants to appropriate day or night lengths (in other words, long or short days).
3. *Rooting stem cuttings* Roots are initiated from bases of stems to produce adventitious roots (roots out of place, so to speak) in stem cuttings.
4. *Mericloning* Shoot tips and associated embryonic leaves are used to vegetatively propagate orchids.
5. *Pruning* Axillary buds develop into leafy shoots when we prune terminal shoots. This is especially important in pruning Christmas trees, shaping bonsai plants, trimming hedges, training trees on trellises or walls, and stimulating more lateral shoots to develop on chrysanthemums, asters, or zinnias by pinching out the terminal portions of the shoots.
6. *Double harvesting* After harvesting rice grains from the first crop of seed heads (panicles), axillary buds (tillers) develop from the base of the plant to produce a second crop of panides with grains.
7. *Stem-piece propagation* Axillary buds form new shoots on stem cuttings of aroids such as *Philodendron, Scindapsus, Dieffenbachia,* and *Monstera.*
8. *Bud formation on cuttings* Adventitious buds (buds formed out of usual position) develop on leaf cuttings of *Begonia,* petiole cuttings of African violet, and gloxinia, root cuttings of chickory, horseradish, or salsify, and stems of plants that are layered to produce new plants. Embryonic plants form on the margins of leaves of the air plant *(Kalanchoë pinnata)* and the leaves of the piggyback plant *(Tolmiea menziesii).*

## Flower development

### INITIATION OF FLOWERS

Flowers, like leaves, branches, and buds, are initiated by the shoot apex. During the vegetative phase of plant development, the shoot apex initiates leaves laterally and forms internodes and nodes below. As it enters the reproductive phase, the apex enlarges immensely. It then switches from initiating leaf primordia to initiating an inflorescence with all its individual flower primordia (Figure 1–3). Each flower primordium has a shoot apex which initiates, in a very precise fashion, the various parts of the flower: sepals, petals, stamens, and pistils (Figure 1–4).

## FLOWER STRUCTURE

The parts of the flower include the following (Figure 1–4): *Sepals* generally serve to protect the flower bud from drying out during its development. Some sepals ward off predators due to the presence of spines or chemicals, especially in flowers where the sepals partly enclose the developing fruit after pollination. *Petals* serve to attract pollinators and are usually dropped shortly after pollination occurs. The *calyx* (the sepals collectively) and *corolla* (the petals collectively) can assume unusual, even bizarre, shapes, as in the orchid where the flower form, color, texture, position, and scent may be very specialized to attract specific animal pollinators. *Nectar,* containing varying amounts of sugar and proteins, may be secreted by any of the floral organs, and it usually collects inside near the base of the cup formed by the flower parts around the ovary.

The male (staminate) parts are called *stamens.* A stamen consists of a *filament* and *anther* where pollen is produced. The female (pistillate) part, the *pistil,* consists of *stigma, style,* and *ovary* containing ovules. After fertilization the ovules mature into seeds.

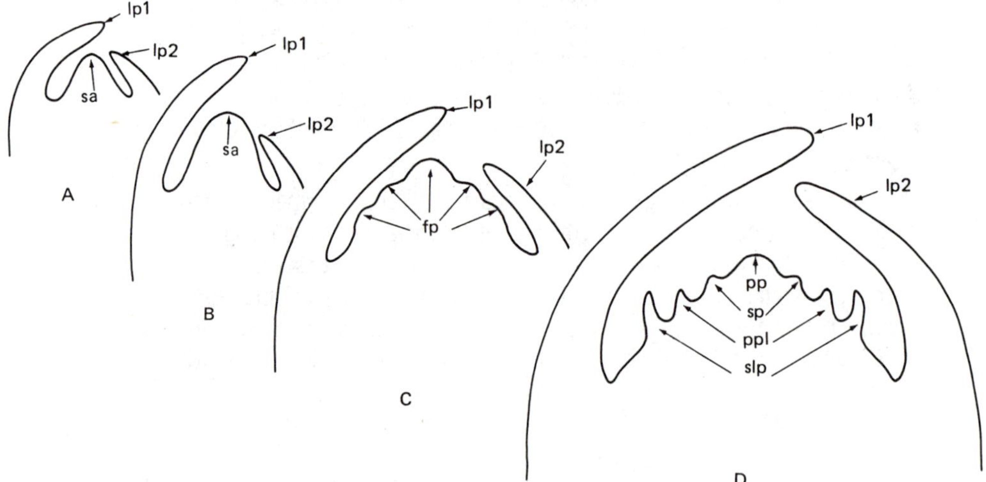


*FIGURE 1–3*
Stages in development of flower primordia in the shoot apex. Diagrams represent median longi-sections through a shoot tip showing transition from vegetative to reproductive phases of a single flower. (A) Vegetative shoot apex (sa) and leaf primordia 1 and 2 (lp). (B) Enlarging shoot apex (sa) at transition stage. (C) Flower primordia (fp) now appear as bumps on the enlarged shoot apex. (D) Flower primordia now begin to differentiate into what will become the separate flower organs: pistil primordium (pp), stamen primordia (sp), petal primordia (ppl), and sepal primordia (slp).

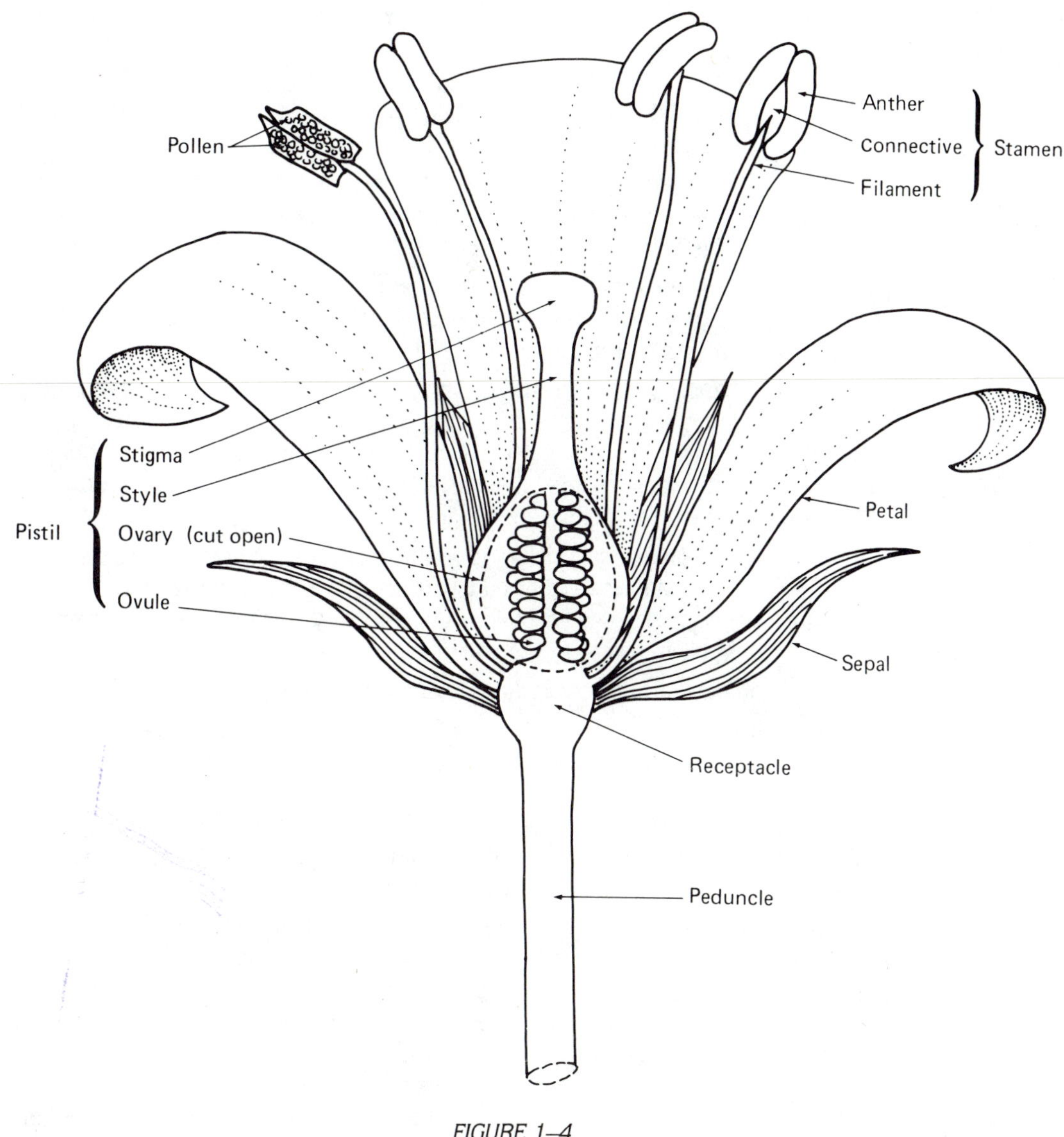


*FIGURE 1–4*
Parts of a flower.

## POLLINATION

Effective pollination occurs when viable pollen is transferred to the stigma. *Cross-pollination* is achieved by transfer of pollen from stamens of one flower to the stigma of another flower. *Self-pollination* occurs when pollen within a flower is transferred to its own stigma. The most desirable situation is cross-pollination, as it can increase the chances of new genetic material being introduced.

To insure cross-pollination, plants have evolved several ingenious pollination strategies:

1. Plants with perfect (bisexual) flowers, having both functional stamens and pistils in the same flower.
   a. *Self-incompatibility* Due to chemical and genetic factors many plants, including avocado, some tomatoes, some tulips, and some begonias, are self-sterile.
   b. *Heterostyly* The style position is not favorable for self-pollination. Some examples are loosestrife *(Lythrum)*, primrose, phlox, and water hyacinth.
   c. *Dichogamy* The anthers may mature and release pollen before the stigmas mature and emerge. Some examples are fireweed *(Epilobium)*, passion flower *(Passiflora)*, century plant *(Agave)*, Norway maple, members of the aster family (Asteraceae, formerly Compositae), peperomia, and geranium.
   d. *Pollinator specificity by flower structure* In the orchids, the pistil and stamens are fused into a structure called the *column*. Just the right size and shape insect is necessary to pick up and transfer the pollen masses at the end of the column.
2. Plants with imperfect (unisexual) flowers, having only stamens or only pistils.
   a. *Monoecious condition* (of one house) One plant has separate staminate-only *and* pistillate-only flowers; thus, pollen must be transferred between flowers. Examples are corn *(Zea mays)*, begonia, cucumber, squash, wind-pollinated nut trees, banana, wild rice, fig, and spurges *(Euphorbia)*.
   b. *Dioecious condition* (of two houses) One plant has only staminate *or* only pistillate flowers, thus pollen must be transferred between separate plants. Examples are papaya, willow, goat's beard *(Aruncus)*, black pepper, asparagus, smilax, *Nepenthes*, date palm, holly, aucuba, ash, silver maples, and aspen.

Many natural agents are involved in cross-pollination, such as wind, water, bees and wasps, butterflies and moths, beetles, flies, bats, birds, and in Australia, even some marsupials (Figures 1–5, 6, 7, 8). Table 1–1 indicates some of the pollination syndromes. We should emphasize the fact that pollinators are lured to flowers by odor, flower shape, and color patterns to obtain nectar and pollen for food. In general, insects do not perceive pure red color and are more often attracted to blue, purple, and yellow colors. In contrast to insects, birds can perceive red. Hence, many bird-pollinated flowers are brightly colored and have evolved red pigmentation, as seen in trumpet vine, many bromeliads, honeysuckle, cardinal flower, and many tropical plants.

*FIGURE 1–5*
Pollinator at work. A monarch butterfly gathering nectar from the blazing star, *Liatris aspera*. (Manistique, Michigan)

*FIGURE 1–6*
Pollinator at work. A hawk moth on a flower of *Mentzilia*. (Badlands, North Dakota, in August)

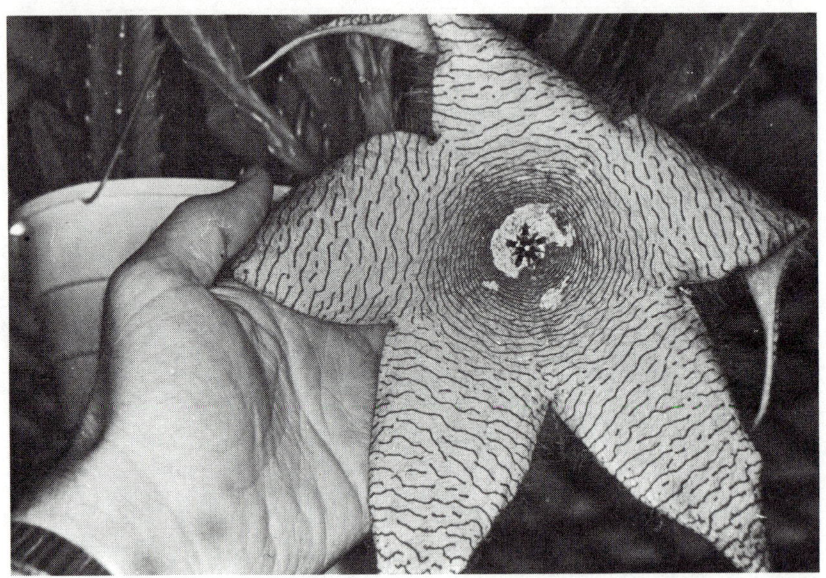

*FIGURE 1–7*
Flower of *Stapelia gigantea*. This succulent in the milkweed family *(Asclepiadaceae)* is native to Southern Africa. The white masses in the center of the flower are fly eggs.

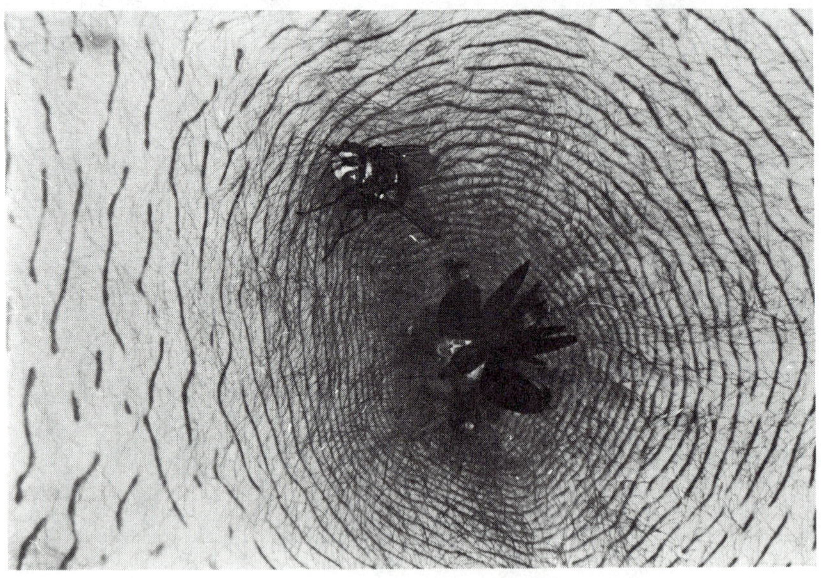

*FIGURE 1–8*
*Stapelia* flower. This enlargement of Figure 1–7 shows a fly pollinator.

*TABLE 1–1*
*Pollination syndromes*

| *Agent* | *Flower Structure* | *Odor* | *Color* | *Nectar* | *Pollen* | *Special Features and Examples* |
|---|---|---|---|---|---|---|
| Wind (*Anemophily*) | Simple, usually lacking petals; anthers usually dangling on long filaments; stigmas branched, feathery and freely exposed; often unisexual, with one ovule per flower; often arranged in catkins or aments | Little or none | Little or none | Usually none | Light, abundant, often round and smooth | Flowers usually appear before foliage in temperate forest trees; uncommon in tropical rain forests; common where insects are rare: arctic or mountain, prairies; plants usually gregarious; grasses, sedges, poplars, oaks, maples, hickories, walnuts, conifers |
| Beetle (*Cantharophily*) | Flowers with few visual attractants, generally large, flat, cylindric, and bowl-shaped; usually easy access; sexual organs exposed; flowers may be fleshy or have "food bodies"; typical primitive flower | Usually strong, fruity | Dull, frequently reddish, brownish, or white | May or may not be present, but if present, then exposed | Very abundant; often with its own odor | Most likely the most primitive of pollination syndromes; some trap flowers; magnolia, calycanthus, tulip tree, rose, some composites |
| Fly (*Sapromyophily*) | Flowers radially symmetrical, "deceitfully attractive," with great depth; often developed as guiding traps with light "windows"; often with motile hairs; sex organs hidden inside flower body | Usually rotten or putrid; can attract from great distances | Dull, brown, purple, green, or reddish | Often none, but if present, usually exposed | Sticky | There are some dramatic physiological mechanisms to generate odor; in some cases, flies try to lay eggs in what they perceive to be rotten meat; succulent stapelids, Dutchman's pipe, euonymus, aroids, skunk cabbage, Jack-in-the-pulpit, calla lily, anthurium, caladium, Queen Anne's lace, wild parsley, sassafras |

(Continued)

*TABLE 1–1 (Continued)*
*Pollination syndromes*

| *Agent* | *Flower Structure* | *Odor* | *Color* | *Nectar* | *Pollen* | *Special Features and Examples* |
|---|---|---|---|---|---|---|
| Bee (*Melittophily*) | Flowers usually zygomorphic (nonsymmetric), with some landing platform; flowers often semiclosed, intricate and mechanically strong; nectar guides present; stamens few; many ovules per ovary | Fresh, sweet | Yellow or blue, not true red | Moderately abundant, usually in base of flower | Abundant and sticky | The most variable of the syndromes; flowering may last over a long period of time; to some bees, pollen is collected for food just as nectar is for honey; often found in arid, warm or tropical areas; bean, pea, lupine, mints, many orchids, rose, rhododendron, clover, snapdragon, horsechestnut, penstemon, beebalm, larkspur, primrose, forget-me-not, iris |
| Butterfly (*Psychophily*) | Flowers erect, radial, rim generally flat, open during day, not closing at night; nectar well hidden in tubes or spurs that are narrow; simple nectar guides or mechanical tongue guides | Moderately strong, fresh and sweet | Vivid, including pure red and bright orange or pink | Ample supply, concealed | Sticky | Even though butterflies and moths are closely related, they have quite different behavioral patterns: butterflies are diurnal and alight; moths are nocturnal, and hover while feeding: daisies, phlox, azalea and rhododendron, pinks, sweet William, lilies, zinnia |

| | | | | | | |
|---|---|---|---|---|---|---|
| Moth *(Sphingophily)* | Flowers radial, larger than for butterflies, and horizontal or pendent; open late evening and night, nectar hidden in long narrow tubes or spurs; visual nectar guides absent, but mechanical tongue guides may be present | Strong, heavy to sweet, attracting moths from great distances | Usually white or pale (or green) | More abundant than bee or butterfly flowers, concealed | Sticky | Many of the "night-blooming" plants are moth-pollinated, especially in the deserts and tropics; white campion, many orchids, night-blooming cereus, Cactaceae, evening primrose, honeysuckle, tobacco, four o'clock, azalea, yucca, phlox, petunia, some cacti, columbine |
| Bird *(Ornithophily)* | Flowers open during day; corolla often tubular with a hard flower wall to protect the ovules from the bird's beak; nectar occasionally found in deep spurs; nectar guides absent | Usually faint or absent | Vivid, especially red and yellow | Abundant, with many essential nutrients | Sticky | There are some similarities between bird and insect flowers; over 2,000 species of birds from 50 families visit flowers; many bird-pollinated species are tropical; coral tree, fuchsia, coral honeysuckle, hibiscus, African tulip tree, trumpet vine, columbine, cardinal flower, century plant, some cacti, Christmas cactus, bromeliads |
| Bat *(Cheiropterophily)* | Flowers short-tubular, suspended from a strong pedicel away from the branches, or on the trunk; often lasting only one night; rather strong structurally | Strong "mousey" or musty odors (only at night) | Dull, often dark red or maroon | Ample supply | Abundant | Bats that have adapted to flower pollination as a way of life (as opposed to insect catching) have a poorly developed sonar system and well-developed eyes; many tropical trees; calabash, baobab, sausage tree, banana, cup-and-saucer vine, some cacti |

## FERTILIZATION

Once pollination is achieved, fertilization is necessary for embryo (and seed) formation. It occurs as follows. The pollen grain germinates on the stigma, producing a pollen tube which grows down through the style to the ovary (Figure 1–9). The pollen tube reaches an ovule and liberates two sperm into the embryo sac within the ovule. One sperm nucleus fuses with the egg nucleus, and the other fuses with two polar nuclei in the embryo sac. The product of the first fusion is a *zygote*. It will develop into the embryo of the seed, and thus, the next generation of the plant. The product of the second fusion is the *primary endosperm* nucleus. It will develop into the endosperm or food storage tissue surrounding the embryo in the seed.

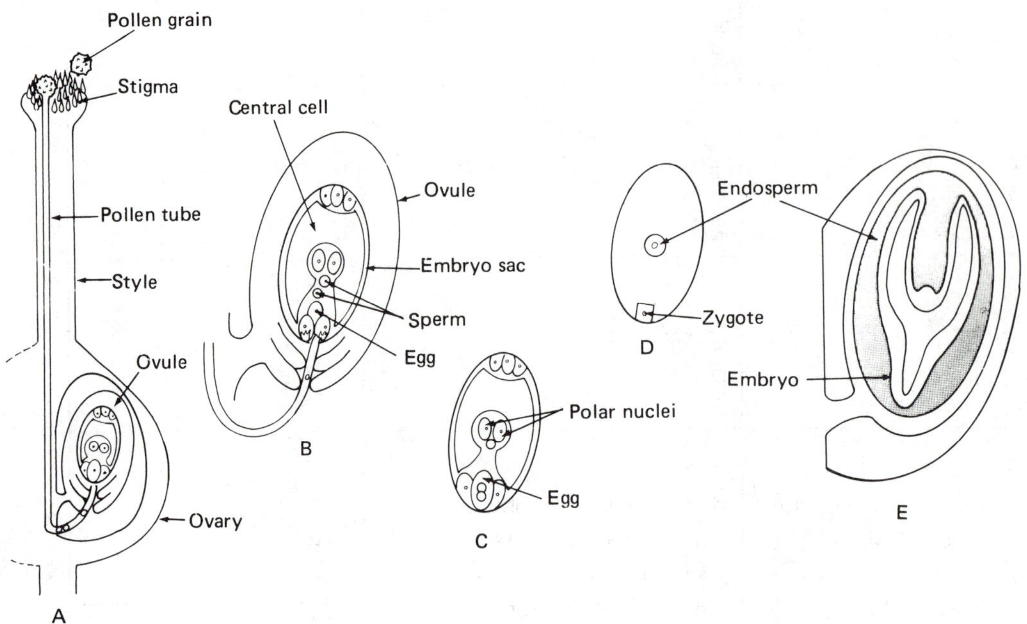


*FIGURE 1–9*
Double fertilization and seed development. (A) The female part of the flower consists of stigma, style, ovary. A pollen grain germinates on the stigma producing a pollen tube which grows down to the ovary. (B) Within the ovary is an ovule where double fertilization occurs. Two sperm are released from the pollen tube. (C) One sperm unites with the egg, and the other sperm unites with the two polar nuclei in the embryo sac. (D) The sperm-egg union is a 2N (diploid) zygote nucleus and the sperm-polar nuclei union is usually a 3N (triploid) primary endosperm nucleus. (E) The seed (mature ovule) consists of an embryo, derived from the zygote, with the surrounding food storage tissue (endosperm), derived from successive divisions of the endosperm nuclei.

## Breeding new plants

Many of the plants which affect human lives are wild plants, growing without the interference of man, but selectively collected by people for thousands of years. Most of our useful plants, however, are those that have been taken from the wild and modified to yield superior products. The art and science of selective breeding and plant improvement is as important today as it ever has been. We continue to increase food crop yields, produce new ornamental cultivars, discover new medicines from plants, and hope to provide energy through plant sources (see Chapter 15).

### OBJECTIVES OF PLANT BREEDING

There are several objectives in plant breeding:

1. By increasing crop yields, we can obtain more wood and cellulose from forest trees, more hay from forage crops, more kilograms per hectare (bushels per acre) from grain crops, fruits and vegetables in greater quantity, more oil in flax, soybeans, and peanuts, and more latex in rubber plants *(Hevea brasiliensis)*, guayule *(Parthenium argentatum)*, and milkbush or pencil tree *(Euphorbia tirucalli)*, a plant now grown to produce petroleum from its latex.
2. By improving the quality of plants and their products, we can increase the percentage of useful protein in grains such as rice, corn, and wheat and increase the sugar content and flavor in fruits.
3. By improving the ability of plants to withstand stress, we can increase frost resistance, obtain greater drought resistance, increase ability to withstand low winter temperatures, and increase resistance to lodging in grains by developing dwarf cultivars. Tall shoots fall over or "lodge" due to wind and rain, thus decreasing grain yields.
4. By increasing disease resistance, we can obtain blight-resistant American chestnuts, rust-resistant snapdragons, wilt-resistant tomatoes, and other fungal-resistant plants.
5. By increasing diversity in cultivars, we can obtain larger flowers and fruits, double flowers, more color diversity, exotic forms, and changes in plant habit. Because horticulturists are continually breeding new plants, we now have the normal single plus double hollyhocks, the normal yellow plus white marigolds, and the normal vine plus bush cucumbers (Figures 1–10, 11).
6. By increasing storage life, we can transport root crops, fruits, and vegetables over great distances while maintaining edibility.

These objectives are important. They help us to meet the goals of our energy requirements, improve the quality of the food we eat, make more food available for our increasing population, and improve the quality and diversity of our lives.

*FIGURE 1–10*
Field work involved in marigold breeding being carried out at Burpee's Santa Paula, California farm *(Photo courtesy of W. Atlee Burpee Seed Company, Inc., Doylestown, Pa.)*

*FIGURE 1–11*
Plant breeder in field working on tomato to produce new tomato hybrids. Barbara Gollimare is shown removing anthers from the blossoms (emasculation) in preparation for making a cross using pollen from another parent. *(Photo courtesy of W. Atlee Burpee Seed Company, Inc., Doylestown, Pa.)*

## POLLINATION PROBLEM

Once a plant flowers, it is important to see that it is pollinated properly. Plants, such as tomato and pepper, for example, must be hand-pollinated if grown indoors. This can be done by brushing the flowers with a small brush or with a gentle wand that vibrates at a specific frequency to "shake" the pollen out of the anthers. Outdoors, this is not necessary because the flowers are pollinated by insects. Bumblebees often buzz at a certain frequency while attracted to these flowers, thereby shaking out pollen. The flowers on these plants are self-compatible (self-pollinating). In many garden plants, self-sterile flowers are the rule. Plants, such as hazelnuts, filberts, and a number of fruit trees, must be cross-pollinated because on an individual shrub or tree, the flowers are self-incompatible (that is, not capable of self-fertilization). Insects, notably bees, are the primary pollinators for fruit trees, while wind carries pollen between nut trees. Self-incompatibility is a major problem in trying to breed plants sexually. It can be due to failure of the pollen to germinate on the stigma of the receptor plant, or the pollen tube bursts in the style. Thus, when efforts are made to cross-pollinate plants from the same clone (where all individuals in a group are genetically identical), no seeds result. One of the parents, either the pollen donor or the pollen receiver, must be genetically different for fertilization and seed production to take place. In most cases, the only way to perpetuate and multiply self-incompatible plants is by means of vegetative propagation (see Chapter 6). If one is interested in breeding new garden plants, one must learn to obtain fertile seeds.

## PARTS OF THE FLOWER INVOLVED IN PLANT BREEDING

To breed plants one must know the basic parts of a flower and their functions. Cross-pollination involves the transfer of pollen (male) from one parent plant flower to the stigma (female) of another parent plant flower. The seeds obtained are a result of sexual reproduction. Plant breeding is the selection of parent plants to be used, then manually cross-pollinating the plants. The procedure for selective cross-breeding involves several steps:

1. Remove the immature anthers (emasculation) with a pair of tweezers from the flower of the plant which is to become the female parent (Figure 1–12). It is important to remove all the anthers and bag the flower until the stigma emerges to prevent any chance of interference from unwanted pollen.
2. Then with a small brush or tweezers, transfer mature pollen from another plant (male parent) to the stigma of the first plant (female parent).

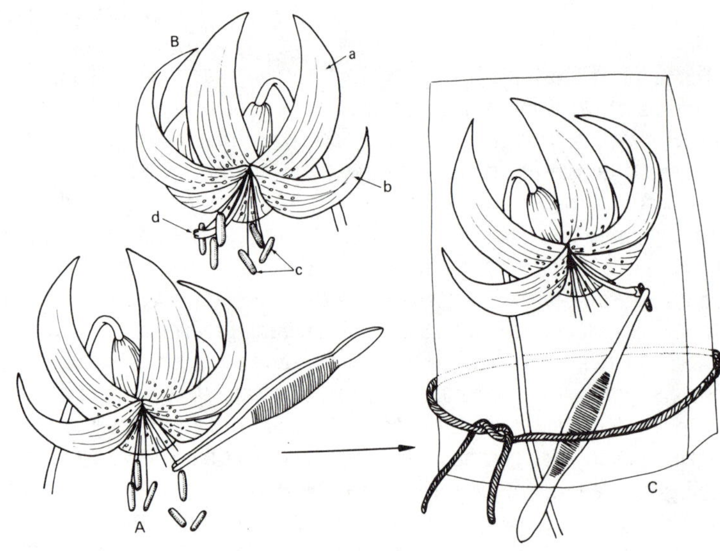


*FIGURE 1–12*
Cross-pollination in Lily *(Lilium)*. (A) Female parent flower, before emergence of style and stigma. Immature anthers (before pollen development) are removed. Bag flower to prevent random pollination when stigma emerges. (B) Male parent flower, showing sepals (a), petals (b), anthers (c), and stigma (d). (C) Pollination with transfer of pollen in mature anther from male parent to stigma of female parent. The bag over flower is then tied and labelled.

3. Cover the flower of the pollinated female parent with a glassine bag and tie with a labelled string tag at the base. Indicate the cross you have made on the tag (Figure 1–13).
4. Check the labelled flower periodically to see if "seed-set" has occurred. This is indicated by the swelling of the ovary(ies). In crosses that fail, the ovary aborts and falls off.
5. Harvest the mature fruits, remove the seeds, and plant (or store properly to satisfy dormancy requirements of the species). The plants from these seeds will be the first felial (F-1) generation. Hybrids will sometimes be indicated by F-1 plants having phenotypic (external appearance or characteristics) traits *intermediate* between those of the two parents. In many cases, however, the F-1 hybrids may resemble one or the other parent completely in certain characteristics.

## PRODUCTION OF HYBRID PLANTS

It is by combining the genetic characteristics of different strains of parent plants that we are able to obtain new combinations. Then we can select favorable plant characteristics. Plant breeding is like inventing: you try a lot of experiments to see if you get something that works. Essentially, plant breeding involves producing many thousands of seeds, planting them, growing them, and then seeing what characteristics the plants have. You select only the best for further breeding and discard the rest! Developing new types of plants is simply a matter of observing many plants and recognizing the best ones. Desired plants are those that are disease-resistant, have pleasing fruit

*FIGURE 1–13*
Bagged and tagged flowers after cross-pollination. (From a scientific study of breeding systems of *Senecio* at the University of Michigan's Matthaei Botanical Gardens)

and flower characteristics, or have new and unusual colors and forms. You never know when a single seed may give you the desired plant. Once this plant is obtained, then you may want to devise ways of maintaining that strain and multiplying it to produce many identical plants for a specific purpose such as food production or ornamentation. Then, you may have to use cloning methods or controlled cross-breeding to produce hybrid seed.

Following are some specific examples of hybridization which have produced some of our most important modern garden plants:

1. *Garden flowers* Iris, day-lilies, tulips, daffodils, pansies, impatiens, marigolds, and zinnias are among those plants in which hybridization between wild species, and then between cultivated strains, have yielded enormous diversity of strikingly attractive flowers. The hybrids exhibit larger sized flowers, a broader range of colors, and longer lasting flowers than the wild species ever could (Figure 1–14). Most of the material for producing the garden hybrids have come from collecting seeds from the wild, growing thousands of seedlings, and then selecting unusual mutations or strains that have appeared.

(A)

(B)

*FIGURE 1–14*
Cross-pollination in breeding petunias. (A) Petunia flower with corolla cut away to show anthers. (B) Transfer of pollen to the stigma with a brush. (C) Pollination achieved by holding a mature stamen with forceps to dust mature

(C)

(D)

pollen from the anther onto the stigma; note in both (B) and (C) that the flowers have been emasculated (anthers removed). (D) Tagged petunia flower after cross-pollination. *(Photos courtesy of W. Atlee Burpee Seed Company, Inc., Doylestown, Pa.)*

2. *Garden shrubs* Camellias, rhododendrons, azaleas, hollies, viburnums, roses, and lilacs occur in a large number of colors and forms due to many years of hybridization efforts. Rhododendrons, for example, have been bred repeatedly to obtain cold-hardy cultivars. Using the wild purple rhododendron *(R. catawbiense)* from the Appalachian Mountains as a parent imparts cold hardiness down to −32° C (−25° F) and has resulted in many hybrids known as *iron clads.*
3. *House and greenhouse plants* Orchids, bromeliads, gesneriads (especially African violets), fuchsias, and many of our nicest flowering plants have been developed by crossing wild species and then selecting from the resulting seedlings.
4. *Hybrid corn* One of the most famous examples of plant breeding is that of hybrid corn. Corn has separate male and female flowers on the same plant. The corn *tassel* is composed of male (staminate) flowers growing at the top of the stalk. *Ears* develop from aggregated female (pistillate) flowers. Corn *silk* on an ear of corn is the remains of styles. Each style, with a stigma at the end, grows from each ovary which becomes a *kernel,* or caryopsis type of fruit, after fertilization and maturity.
   a. *Inbreds* The first step in producing any kind of hybrid is to obtain pure parental lines called *inbred parents* (Figure 1–15). Hybrid vigor is attained by crossing two unrelated pure parental lines. Inbreeding is a method of obtaining pure lines of corn. Inbreds are produced either from field-pollinated varieties or from crosses between inbreds having desirable characteristics which could be beneficially combined, such as high-yield and fast-drying ears crossed with low-eared and short-stalked plants. This process of inbreeding and selection takes from five to seven generations. These inbreds must be tested to see if they will cross well with other inbreds. Out of every 1,000 inbreds produced in this manner, one is fortunate to find a parental selection that will prove to be adequate for use in producing commercial hybrids.
   b. *Single crosses* The second step involves crossing two unrelated inbred parents to produce seed, which when planted, results in a single cross plant. This is more difficult than it sounds because very few inbreds produce enough kernels to make a good female parent. And inbreds are, as a whole, poor pollen shedders, which makes them poor male parents as well. It takes two or three generations to build up the supply of kernels of the inbreds to have enough for planting commercial seed fields.
   c. *Double crosses* The straight double crosses or four-way hybrids are simply two unrelated single cross parents crossed to produce the seed which produces the commercial double cross.

Hybridization is rewarding when you have a diverse gene pool to work with and when cross-pollination is fairly easily accomplished. Hybridization is

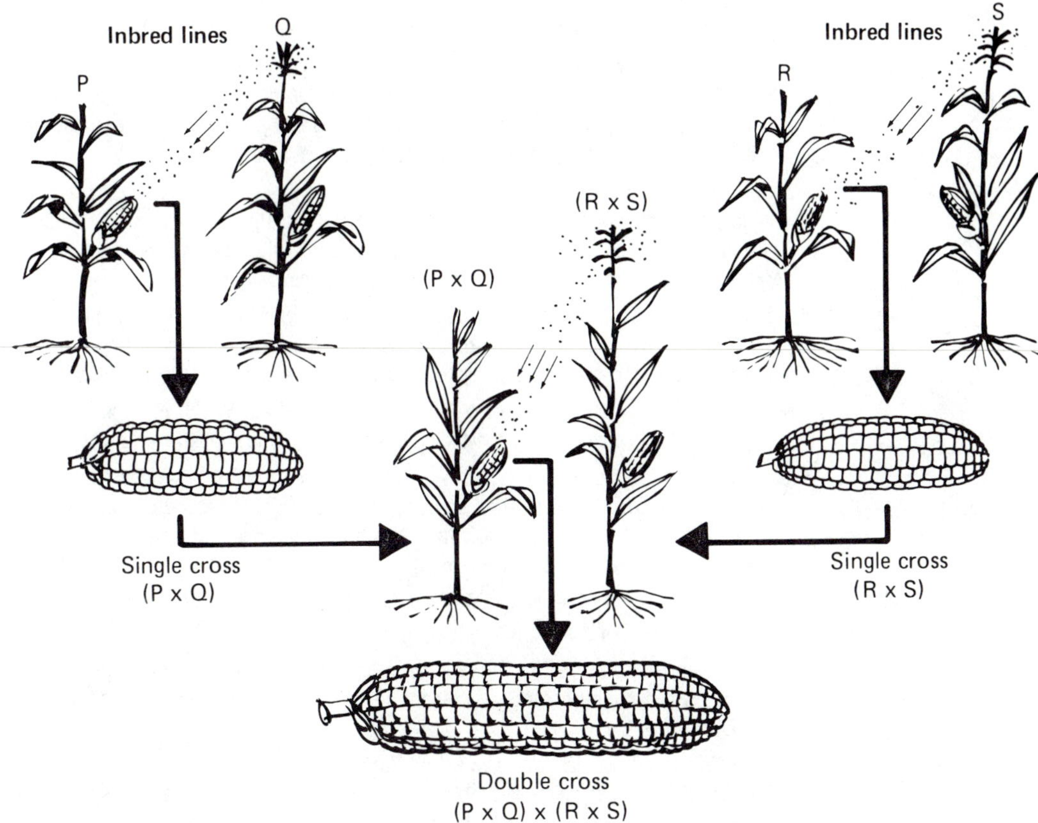


*FIGURE 1–15*
Double cross hybrid kernel production with corn or maize *(Zea mays)*. All female parents are hand-detasseled to prevent self-pollination. All kernels from male parents (self-pollinated) are saved for animal feed. *(Drawing prepared by Ross Koning)*

also most often used in plants where showy flowers are desired. Succulents and many foliage houseplants, for example, have not been popular subjects for hybridizers because they are more often enjoyed without their flowers.

## REFERENCES

Brooklyn Botanic Garden. *Breeding plants for home and garden.* Brooklyn Botanic Garden Handbook 75. (Basic handbook on principles and methods of breeding horticultural plants for the home garden.)

Esau, K. 1977. *Anatomy of seed plants.* 2d ed. New York: John Wiley and Sons. (One of the most complete and lucid treatments available on the structure and function of cell organelles, cells, tissues, and organs in plants.)

Fahn, A. 1974. *Plant anatomy.* 2d ed. New York: Pergamon Press. (Another good textbook on the structure and function of plant organelles, cells, tissues, and organs; ecological adaptations are thoroughly discussed and illustrated.)

Grant, Karen A. 1966. "A hypothesis concerning the prevalence of red coloration in California hummingbird flowers." *Amer. Naturalist* 100:85–97. (A specialized article on the pollination biology in hummingbird-pollinated flowers and the role of red flower pigments in attracting such pollinators.)

Hartmann, H. T., and Kester, D. E. 1975. *Plant propagation: principles and practices.* 3d ed. Englewood Cliffs, N. J.: Prentice-Hall, Inc. (One of the most complete and detailed texts available on the methods and "why" behind the methods for the propagation of plants—both sexually and asexually.)

Heiser, C. B., Jr. 1981. *Seed to civilization: the story of food.* 2d ed. San Francisco: W. H. Freeman and Co. (One of the best paperbacks available on the origins of agriculture and man's domestication of food, beverage, and spice plants.)

Janick, J. 1979. *Horticultural science.* 3d ed. San Francisco: W. H. Freeman and Co. (Covers basic concepts on the biology, technology, and industry of horticulture.)

Kaufman, P. B., and LaCroix, J. D. 1979. *Plants, people, and environment.* New York: Macmillan Co. (An up-to-date account of the ecological nature of our environment, what is happening to it, and solutions to prevent its destruction from the perspective of plants, primarily.)

Meeuse, B. J. D. 1961. *The story of pollination.* New York: Ronald Press. (An excellent overview of pollination biology in plants.)

Proctor, M., and Yeo, P. 1973. *The pollination of flowers.* London: Collins Publishers. (Covers the major aspects of pollination vectors, and the evolution of the pollination process in plants.)

Schery, R. W. 1972. *Plants for man.* 2d ed. Englewood Cliffs, N.J.: Prentice-Hall, Inc. (One of the best texts available on economic botany with emphasis on the products from cell walls of plants, from exudates and extractions, and from plant parts used for food and beverage.)

Walsh, J. R. 1981. *Fundamentals of plant genetics and breeding.* New York: John Wiley and Sons. (This text summarizes the major topics in plant genetics that are of interest to plant breeders: Mendelian genetics, genes, chromosomes and their numbers, plant reproduction, natural genetic variability, plant breeding objectives, hybridization, mutation breeding, backcross breeding, pedigree breeding, chromosome breeding, and the release and marketing of new cultivars.)

Wareing, P. F., and Phillips, I. D. J. 1981. *Growth and differentiation in plants.* 3d ed. New York: Pergamon Press. (Covers the major topics in plant development: structural and morphological aspects, internal controls, environmental controls, and molecular aspects of differentiation.)

Wendorf, F., Schild, R.; Hadidi, N. E.; Close, A. E.; Kobusiewicz, M.; Wiekowska, H.; Issawi, B.; and Haas, H. 1979. "Use of barley in the late paleolithic." *Science* 205:1341–48. (A technical account dating man's earliest use of barley as a domesticated crop in the Middle East.)

# 2 Seeds to seedlings

We have discussed the important features of plant structure, vegetative development, and reproductive development. Our next objective is to learn more about the products of reproductive development in flowering plants—seeds. Thus, we will focus on such salient topics as seed structure, how seeds are dispersed, how long seeds remain alive (viable), how they become dormant, what we can do to break seed dormancy, and practical aspects of germinating seeds.

## Structure of seeds

A *seed* is a mature, fertilized ovule that is born in a mature ovary, the fruit (Figure 2–1). The basic parts of a seed include the embryo, variable amounts of food storage tissue (endosperm), and the seed coat (testa). Size varies from just visible, dustlike orchid seeds to coconut seeds which weigh up to 6.8 kilograms (15 pounds). Some seeds have huge embryos that occupy most of the interior of the seed, while others have much smaller embryos and abundant endosperm tissue (Figure 2–2). The orchid is an extreme case in which we encounter an undeveloped (undifferentiated) embryo, no endosperm, and a very fragile seed coat.

## Seed dispersal

Seed dispersal is essential for plants to survive and to spread into new habitats. Plants that we know as weeds are especially good at seed dispersal, as witnessed by dandelions and thistles with their wind-blown fruits held aloft by "parachutes"; and cocklebur and burdock with spine-covered fruits that

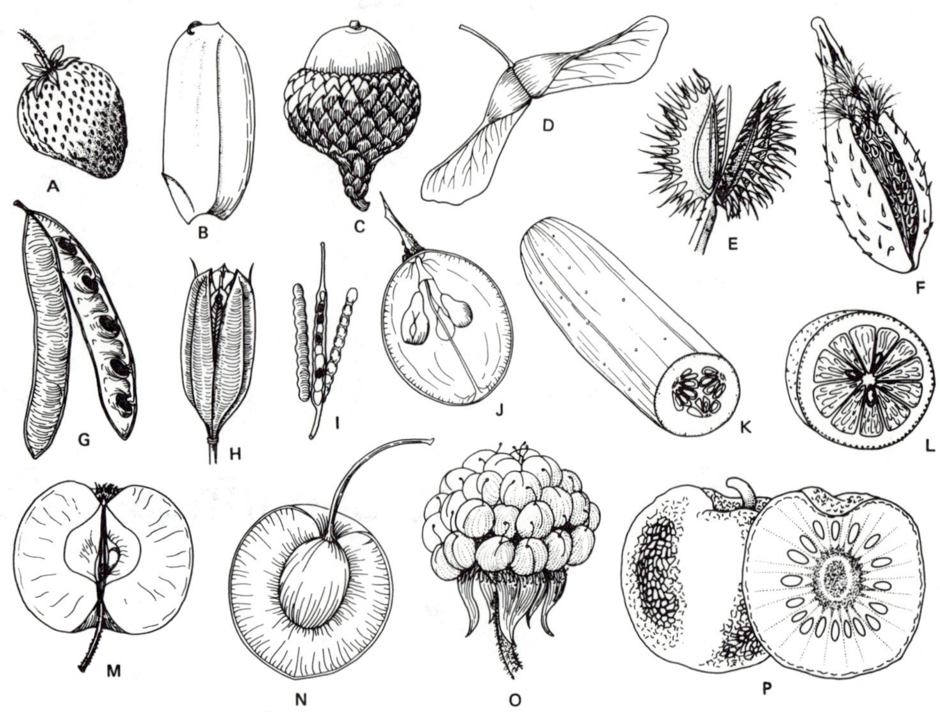


*FIGURE 2–1*

Fruit types. Fruits may be classified in various ways. One system includes simple, aggregate, and compound fruits. If a single flower produces one fruit, it is a simple fruit. If a single flower produces many fruits which stay together, the unit is called an aggregate fruit (A, O). If several flowers produce many fruits which stay together, the unit is called a multiple fruit (P). Examples shown are: (A) an *achene* is a dry, one-seeded fruit with a thin wall, the strawberry (×½) fruit is an aggregate of achenes; (B) *grain* or *caryopsis,* a dry achene-type fruit of the grass family (Poaceae), rice (×4½); (C) *nut,* a dry achene-type fruit, in oak (×1) called an acorn; (D) *samara,* a dry, winged achene, which is double in maple (×½); (E) *schizocarp,* mericarps break apart and function as achenes, wild carrot (×5); (F) *follicle,* a dry fruit which opens along one line, milkweed (×½); (G) *legume,* a dry fruit which opens along two lines and is restricted to the pea family (Fabaceae formerly Leguminosae), black locust (×½); (H) *capsule,* a dry fruit which opens by splitting or by pores, lily (×½); (I) *silique,* a dry fruit which consists of two narrow coverings with a partition between, mustard (×1); (J) *berry,* a fleshy covered fruit, grape (×1); (K) *pepo,* a type of berry with a hard rind, cucumber (×¼); (L) *hesperidium,* a type of berry with a leathery rind enclosing juice sacs, citrus, lemon (×3/8); (M) *pome,* a fleshy fruit covering derived from an expanded floral tube, over a cartilaginous seed container, apple (×½); (N) *drupe,* a fleshy wall enclosing a bony covering usually over one seed, cherry (×1); (O) aggregate of drupelets, black raspberry (×2½); (P) multiple fruit of achenes, osage orange (×¼).

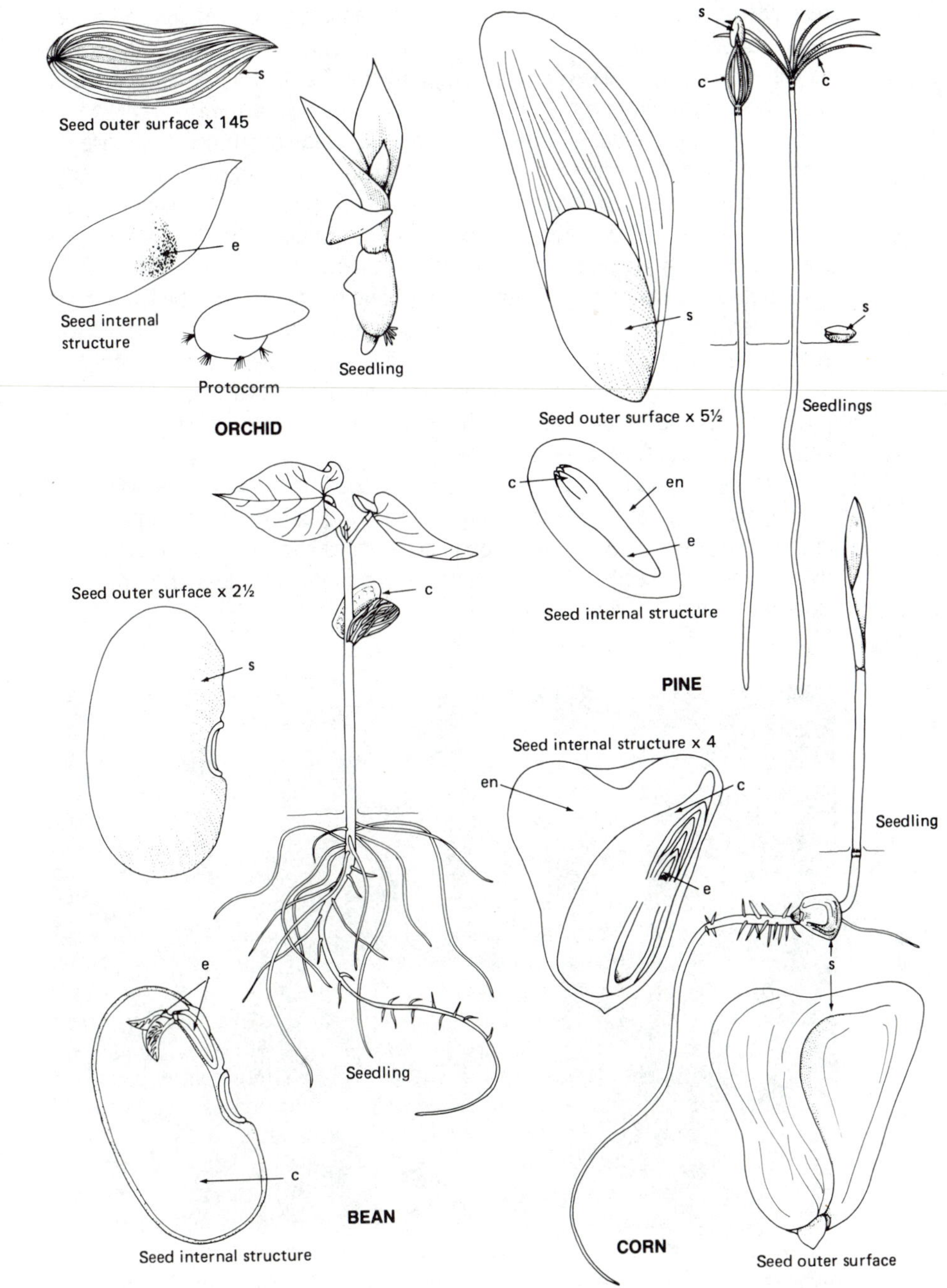


*FIGURE 2–2*
Basic seed and seedling structure. These economically important plants are: orchid *(Vanda tricolor)*, pine *(Pinus)*, corn *(Zea mays)*, and bean *(Phaseolus vulgaris)*. e = embryo, s = seed coat (testa), en = endosperm, c = cotyledon.

readily become attached to animal fur. How annoying it is for us to be seed dispersal agents because of "stick-tight" fruits that get caught in our clothes as we walk along a roadside or through a field! Plants have developed ingenious ways of facilitating the dispersal of their seeds or fruits.

Sometimes, seeds are the dispersal units, and sometimes fruits are the dispersal units. Examples include the following: winged seeds (catalpa, paulownia, pine, fir, spruce, trumpet vine, lily, yucca) (Figure 2–3); winged fruits (elm, ash, maple, birch, basswood, begonia); parachute seeds (milkweed, dogbane) (Figure 2–4); parachute fruits (thistle, dandelion, cattail) (Figure 2–5); stick-tights with barbed awns are fruits (many of the grasses, beggar's ticks, cocklebur, tick trefoil, enchanter's nightshade) (Figures 2–6 and 2–7); dry seeds (bean, pea, iris, lily, petunia, columbine, jimson weed); dry fruits (nuts and grains); fleshy seeds (pomegranate, magnolia, some cacti, yew, water lily, podocarpus, euonymus); and fleshy fruits (grape, honeysuckle, highbush cranberry, currant, gooseberry, apple).

Dispersal mechanisms include wind (winged or hairy covered seeds and fruits), water (floating seeds and fruits), ground animals (stick-tights, nuts, and berries), and birds (red-colored fruits such as cherries and strawberries). In addition, the seeds of some plants are explosively released from fruits (jewelweed, witch hazel) (Figure 2–8).

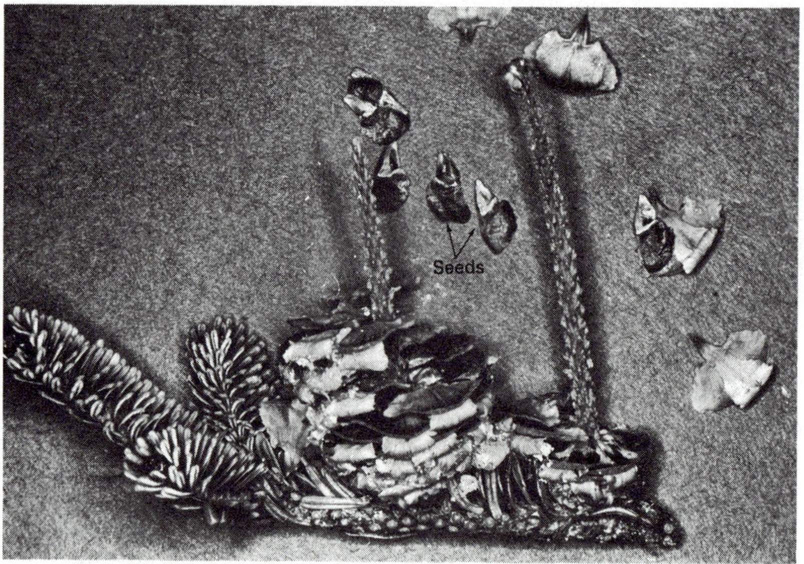


*FIGURE 2–3*
Winged seeds. Many conifers have winged seeds that are dispersed by wind. Shown here are the cone bracts (at right) and winged seeds (center) from a disintegrating cone of the Fraser fir *(Abies fraseri). (Photo from Great Smokey Mountain National Park, North Carolina-Tennessee)*

*FIGURE 2–4*
Parachute seeds. The fruit (follicles) of the milkweed *(Asclepias incarnata)* open, due to drying. Once liberated their parachute-type seeds are wind-dispersed.

*FIGURE 2–5*
Parachute fruits. The fruiting spike of a cattail shoot *(Typha)* has thousands of fruits with hairs which aid in wind dispersal.

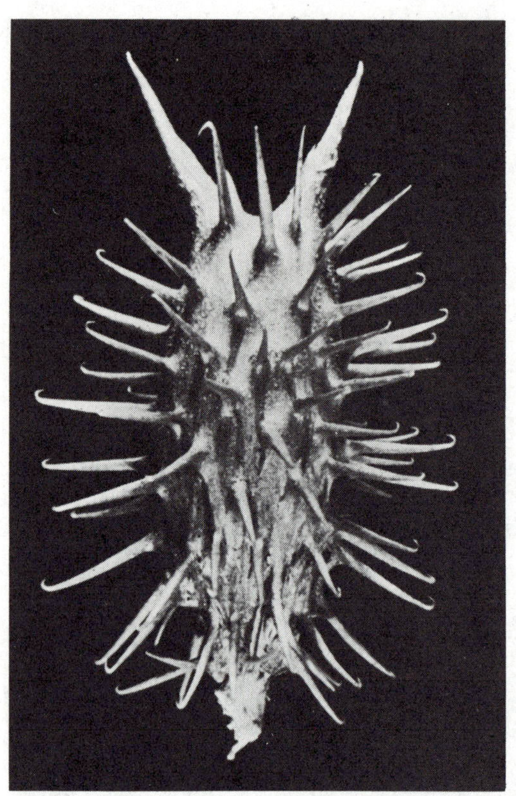

*FIGURE 2–6*
A stick-tight fruit. The fruit of the common weed, cocklebur *(Xanthium strumarium)*, in the aster family (Asteraceae, formerly Compositae), has barbs with apical hooks which aid in their dispersal by animals.

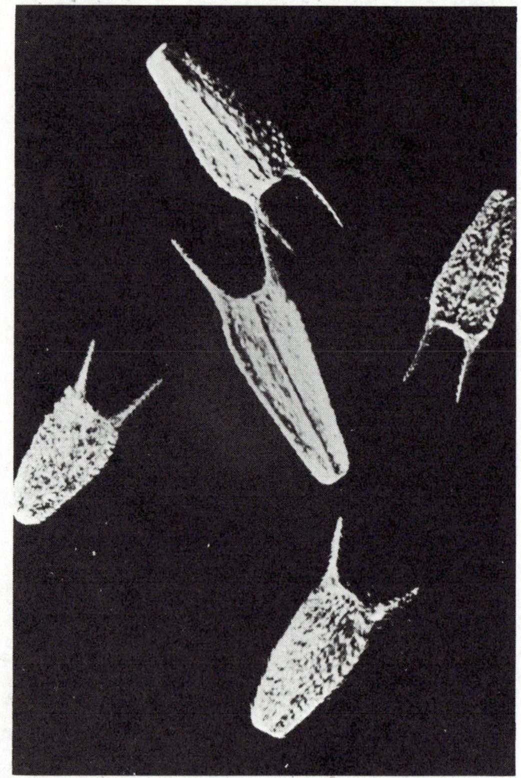

*FIGURE 2–7*
A stick-tight fruit. The barbed fruits of the weed, beggar's ticks *(Bidens)*, in the aster family (Asteraceae, formerly Compositae), have spines with downward pointing barbs. These spines and barbs are an adaptation to aid in animal dispersal of the seeds. *(Photo courtesy of Warren H. Wagner, Jr.)*

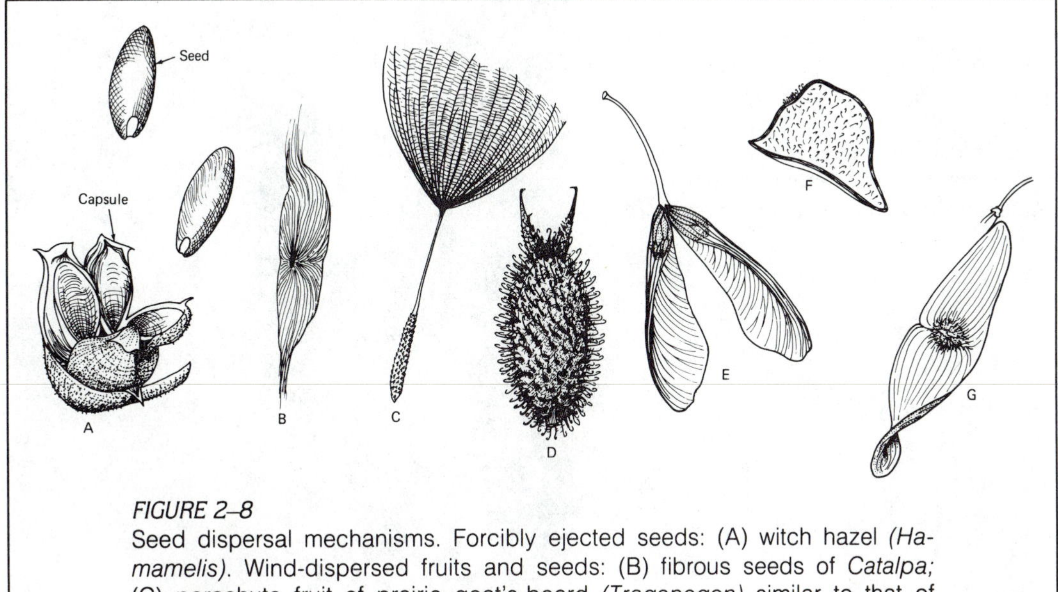


*FIGURE 2–8*
Seed dispersal mechanisms. Forcibly ejected seeds: (A) witch hazel *(Hamamelis)*. Wind-dispersed fruits and seeds: (B) fibrous seeds of *Catalpa;* (C) parachute fruit of prairie goat's-beard *(Tragopogon)* similar to that of dandelion; (E) "key" fruits of maple *(Acer);* and (G) twisted, membraneous fruit of tree of heaven *(Ailanthus)*. Animal-dispersed fruits: (D) prickly, fur-entangling fruit of cocklebur *(Xanthium)* and (F) stick-tight fruit of tick trefoil *(Desmodium)*.

## Seed viability or longevity

Once seeds are dispersed, many may lodge in the soil. When brought to the surface, they will germinate when conditions are favorable for growth and development of the embryo in the seed. Some seeds can lie dormant in the soil for years before germination. Thus, we refer to a seed's viability as the length of time over which it remains alive. Seeds which are viable for only a few weeks after they mature include willow, poplar, maple, magnolia, and buckeye. Most commercial flower and vegetable seeds remain viable for several years (two to three years or more) if stored properly. Seeds of desert plants, spruces *(Picea)*, true firs *(Abies)*, pines *(Pinus)*, and hemlocks *(Tsuga)* remain viable for 15 to 20 years. Several weeds have seeds which remain viable for much longer periods, such as 100 years for evening primrose *(Oenothera biennis)* (Figure 2–9) and curly dock *(Rumex crispus)* and 100 to 150 years for locoweed *(Astragalus* spp.*)*. Among cultivated legume seeds with hard seed coats are those of clover; its seeds are viable up to 90 years. The sensitive plant *(Mimosa pudica)*, also a legume, has seeds which are viable for periods of up to 220 years. One of the records for seed longevity is held by the Indian lotus *(Nelumbo nucifera)*. Seeds of this plant found in Manchu-

*FIGURE 2–9*
Long-lived seeds. A flowering and fruiting shoot of the common weed, evening primrose *(Oenothera biennis)*. The seeds, borne in the fruits of this plant, may remain alive (viable) in the soil for periods of up to 100 years. *(Robert Bandurski, Michigan State University, personal communication)*

rian peat beds have been shown to be 1040 years old, based on radioactive carbon dating, and are still capable of germinating! However, apparently the record is held by the Arctic tundra lupine *(Lupinus arcticus)* whose seeds had been buried in dry, frozen Arctic soil in small mammal burrows and which germinated readily in the laboratory. The age of these seeds, as estimated by geological methods and radioactive carbon dating, is 10,000 years!

To determine whether or not stored seeds remain relatively viable, simple seed germination tests can be run. Usually 100 seeds are counted out and placed in a "rag-doll." This involves placing the seeds in a wet, folded paper towel and inserting the latter upright in a jar, filled to one-quarter with water. The water moves up the towel by capillarity, thus wetting the seeds. This setup is then placed in the dark at a warm temperature and the progress of germination determined periodically. Alternatively, the seeds can be placed between moist blotters in a tray or on moist filter paper in a Petri dish, then placed in an incubator set at a given temperature regime that favors seed germination (Figure 2–10). If no germination occurs, the seeds are

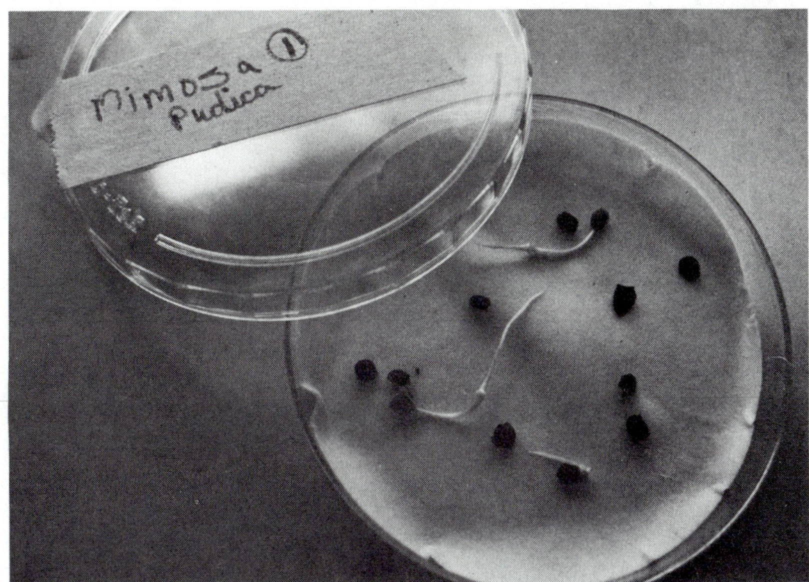


*FIGURE 2–10*
Standard seed germination test. Seeds of the sensitive plant *(Mimosa pudica)* in a Petri dish on moist filter paper. Three of the seedlings have germinated.

either nonviable, or they are dormant. If dormant, various treatments to break seed dormancy can be tried (see next section and Tables 2–2 and 2–3) before running another seed germination test.

Another method for testing seed viability is with TTC (2,3,5-triphenyltetrazolium chloride).[1] Presoak 10 to 12 seeds in water for 10 to 12 hours, cut in half, then soak in 0.1 percent TTC for 2 to 5 hours in the dark at room temperature. Development of a pink color in the embryo indicates a positive test for seed viability. The TTC test does not tell whether or not the seeds will germinate. It only indicates, when positive, that the seeds are respiring (absorbing oxygen and burning up stored foods) and hence, viable.

*PROLONGING SEED VIABILITY.* To prolong the viability or longevity of seeds, one must resort to methods which will reduce the rate of respiration. This can be achieved by placing the seeds in a package under vacuum (to reduce the amount of oxygen to near zero) and by lowering the temperature to 3° to 5° C (35° to 40° F). Some companies sell their seeds vacuum-packed to prolong seed viability. If one cannot vacuum-pack hand-harvested seeds, the next best thing is to dry them and then store them in jars or in airtight envelopes. Addition of powdered milk or silica gel in the bottom of the jar helps to keep the seeds dry during storage. Place them in a cool, dry

[1]TTC can be obtained from several chemical companies. One is Sigma Chemical Company, P.O. Box 14508, Saint Louis, Missouri 63178.

place but not where the seeds will freeze (below 0° C or 32° F) unless the seeds are proven to be winter-hardy. Seeds which remain viable at temperatures below freezing include pine, spruce, fir, most temperate zone species of deciduous trees and shrubs, perennials, and annuals.

The best time to collect seeds to store is when they are mature and ready to be shed. Seeds of most woody plants growing in temperate regions can be collected in late fall, winter, or even as late as early spring the following year if the seeds or fruits still persist on the plants. However, late autumn is best for seeds or fruits which mature in late summer to early autumn such as woody plants and herbaceous perennials and annuals. When fleshy seeds or fruits are mature, they should be collected immediately. If you wait until next year they will have disappeared due to removal by birds, other predators, wind, or by breaking apart due to freezing or microbial action. The fleshy portions should be removed, and the seeds either stored, treated if dormant, or planted.

## Seed dormancy

### MAJOR TYPES OF SEED DORMANCY

Seeds which do not germinate when provided favorable temperature, light, oxygen, and water necessary for germination, but are alive (viable), are said to be *dormant.* Seeds have developed dormancy mechanisms that allow them to survive environmental stresses such as dryness, high or low temperature, lack of oxygen, and unfavorable soil or water environments. Mechanisms of seed dormancy include: (1) hard seed coats, (2) chemical inhibitors in seed coats or other parts of the seed, (3) seed coats impermeable to water, carbon dioxide, or oxygen, and (4) the presence of immature embryos. Examples of these different types of seed dormancy are listed in Table 2–1.

*TABLE 2–1*
*Seed dormancy mechanisms and examples for each*

| *Dormancy Mechanism*[1] | *Examples* |
|---|---|
| Hard seed coat | Many legumes (family Fabaceae or formerly Leguminosae) including honey locust *(Gleditsia)*, black locust *(Robinia pseudoacacia)*, redbud *(Cercis canadensis)*, Kentucky coffee tree *(Gymnocladus dioica)*, acacia *(Acacia)*, clovers *(Trifolium)*, and alfalfa *(Medicago sativa)* |
| Chemical inhibitors in seed coats | Pines *(Pinus)*, spruces *(Picea)*, firs *(Abies)*, hemlocks *(Tsuga)*, and junipers *(Juniperus)*; members of the rose family (Rosaceae)—mountain ash *(Sorbus aucuparia)*, rose *(Rosa)*, apple *(Malus)*, pear *(Pyrus communis)*, and peach *(Prunus persica)*; viburnums *(Viburnum)*, and ashes *(Fraxinus)* |
| Seed coats impermeable to water or oxygen | Cocklebur *(Xanthium)* and cattail *(Typha)* |
| Seeds with immature embryos | Holly *(Ilex)*, ginkgo *(Ginkgo biloba)*, some viburnums *(Viburnum)*, and orchids |

[1]Lists of plants with these types of seed dormancy are well documented in the U.S. Forest Service's *Seeds of Woody Plants of the United States* and in Hartmann and Kester's *Plant Propagation, Principles and Practices.*

## NATURE'S WAYS OF BREAKING SEED DORMANCY

Nature has evolved various strategies for breaking dormancy in seeds. These include the following:

1. Action of alternate freezing and thawing which abrades or softens hard seed coats which are not permeable to water (many legumes), oxygen (cocklebur), and carbon dioxide (cattail).
2. Breaking down hard or impermeable seed coats by microorganisms.
3. Passage of seeds from fleshy fruits through the guts of birds or other animals and their subsequent excretion; thus, digestive acids help to break down hard or impervious seed coats.
4. Leaching out of water-soluble inhibitors from seed coats or other parts of the seeds by torrential rains. This is particularly important with desert plants such as Russian thistle and saltbushes (both are "tumble weeds").
5. Breakdown of chemical inhibitors present in various parts of the seed by low winter temperatures. This is especially important for members of the rose family, such as apple, rose, mountain ash, and cotoneaster.
6. Action of light in stimulating germination of seeds of many types of plants.[2] These include lettuce, some pines and spruces, and a number of bromeliads.

## ARTIFICIAL WAYS OF BREAKING SEED DORMANCY

We have developed various artificial ways to break dormancy in seeds. Many of these methods are faster or easier to perform than those in nature. The key in Table 2–2 indicates the different kinds of seed dormancy and specific methods one can use to break the respective types of seed dormancy. One of these methods is *scarification.* This refers to any process of breaking, scratching, or altering the seed coat to make it permeable to water and gases. For mechanical scarification, a metal, triangular file, grindstone, or sandpaper can be used. Take care to break through the seed coat without damaging the embryo. Acid scarification softens the seed coat, thus making it permeable to gases and water.

Chemical inhibitors, present in various parts of the seed, can be inactivated or rendered ineffective by treating the seeds (by soaking) in certain plant hormones.[3] (For a discussion of plant hormones, see Chapter 3.) Gibberellin is a hormone found in abundance in immature seeds. It is active in releasing enzymes that break down stored food reserves in the seed, making

---

[2]It is the red portion of sunlight which is most effective in promoting seed germination. The pigment, phytochrome, is responsible for absorbing the red light wavelengths that stimulate seed germination.

[3]A plant hormone is an organic substance produced in small amounts in one part of a plant, translocated to another part(s), where it causes a physiological response.

the products (sugars, amino acids, fatty acids) available for growth of the embryo. Seeds to be treated are placed in a cheesecloth bag and soaked for 24 hours in 1,000 ppm (parts per million) gibberellin. Use the potassium salt of gibberellic acid ($GA_3$). Kinetin is a synthetic material, related to the natural cytokinin hormones; it functions by stimulating cell division in the embryo and overriding the influence of chemical inhibitors in the seed. Treatment is the same as for gibberellin.

Cold stratification is necessary for seeds which require a cold period for germination to take place. This is true for many northern trees and shrubs. Stratification involves distributing seeds between layers of moisture-retaining material, separated by layers of cheesecloth (to prevent the seeds from being lost in the medium) and storing the seeds at low temperature for a specified time. The purpose is to after-ripen dormant embryos, to cause breakdown of chemical inhibitors and an increase in levels of growth-promoting hormones, and to make the seed coat more permeable to air and water. The stratification treatment involves placing the seeds in moist medium in containers and keeping at 0° to 5° C (32° to 40° F), usually for six to eight weeks. Examples of media are well-washed sand, peat moss, chopped sphagnum, vermiculite, or mixtures of these. Containers for stratification include plastic bags, cans, jars, or boxes.

In trying to decipher the type of dormancy in a seed and a suitable method for breaking it, there are a couple of clues which may help. If the

*TABLE 2–2*
*Key to known methods used to break various types of seed dormancy*

|  |  |
|---|---|
| 1. Seed coat hard or impermeable to water and/or gases ........ | Scarify by carefully cutting through seed coat in a few areas or soak in concentrated hydrochloric or nitric acid for 30 minutes, then wash with water |
| 1′. Seed coat not hard or impermeable to water and/or gases ........ | Go to no. 2 |
| 2. Seeds will not germinate but are viable, based on TTC test. Seed germination inhibitors are suspected ........ | Remove seed coat and treat seeds with gibberellic acid (or kinetin) or stratify seeds in moist media for 4 to 6 weeks at 0° to 5° C (32° to 40° F) |
| 2′. Seeds are viable but will not germinate after stratification ........ | Go to no. 3 |
| 3. Embryos in seeds are poorly developed ........ | Cut out embryos and culture in sterile potting soil or in sterile nutrient agar |
| 3.′ Embryos are well developed, but seeds require light to germinate ........ | While seeds soak in dish of water, illuminate with cool white fluorescent lamps or 15- to 20-watt red light bulb at a distance of 15 to 20 centimeters (6 to 8 inches) for 24 hours, then plant |

seed is from a plant which is a member of the legume family (Fabaceae), then you can predict that the dormancy is due to the presence of a hard seed coat (Figure 2–11). The treatment would be scarification. If the seed belongs to a plant in the rose family (Rosaceae), or it is one of the conifers such as hemlock, spruce, fir, or pine, then the dormancy is most likely due to the presence of chemical inhibitors in the seeds. In this case, one would use stratification. In the case of most garden vegetable and flower seeds no special treatments are required because these seeds are usually not dormant at maturity. Some wild flowers and native trees have a double dormancy and require two years to germinate. During the first stage the root is produced, and during the second stage, the shoot is produced. Breaking double dormancy may involve a warm, moist period followed by a cool, moist period. These seeds should be placed in half sand, half peat moss in a closed plastic bag, stored at room temperature for four to six months, then placed in a refrigerator (not a freezer) at 4° C (40° F) for three months before planting. Plants with double dormancy seeds include the following: fringe tree *(Chionanthus)*, cotoneaster *(Cotoneaster)*, hawthorn *(Crataegus)*, dove tree *(Davidia)*, witch hazel *(Hamamelis)*, silver bell tree *(Halesia)*, holly *(Ilex)*, juniper *(Juniperus)*, sumac *(Rhus)*, yew *(Taxus)*, viburnum *(Viburnum)*, lily *(Lilium)*, and peony *(Paeonia)*.

Table 2–3 is a listing of treatments commonly used to prolong viability or to break dormancy in seeds of woody plants.

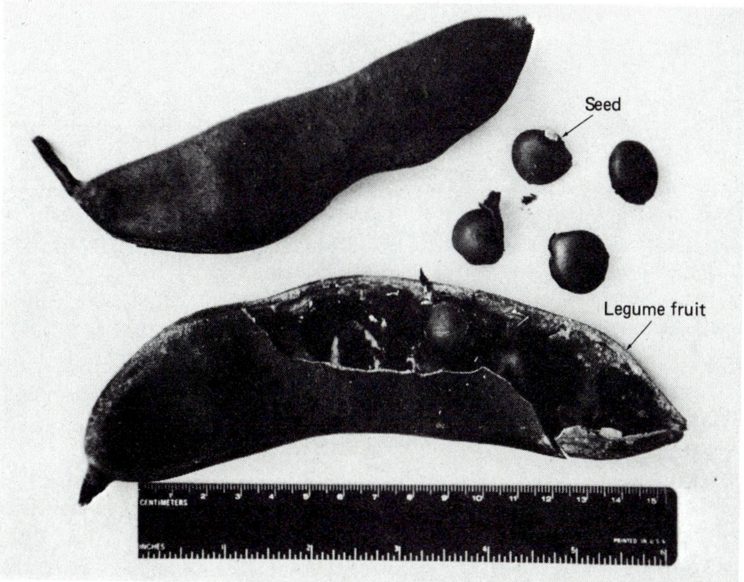


*FIGURE 2–11*
Legume fruit. Members of the legume family (*Fabaceae*, formerly *Leguminosae*), such as Kentucky coffee tree *(Gymnocladus dioica)* have fruits with hard seed coats.

## TABLE 2–3

### *Treatments commonly used to prolong viability or to break dormancy in seeds of woody plants*

*Seeds which have a short viability time (a few weeks to six months) should not be dried after harvesting but planted.*

| | |
|---|---|
| Alder *(Alnus)* | Mahonia *(Mahonia)* |
| Ampelopsis *(Ampelopsis)* | Nandina *(Nandina)* |
| Angelica tree *(Aralia)* | Oak *(Quercus)* |
| Bayberry *(Myrica)* | Pawpaw *(Asimina)* |
| Beech *(Fagus)* | Persimmon *(Diospyros)* |
| Buffalo berry *(Shepherdia)* | Poplar *(Populus)* |
| Cedar *(Cedrus)* | Sassafras *(Sassafras)* |
| Chestnut *(Castanea)* | Service berry *(Amelanchier)* |
| Cinquefoil *(Potentilla)* | Silver bell *(Halesia)* |
| Cryptomeria *(Cryptomeria)* | Snowbell *(Styrax)* |
| Dove tree *(Davidia)* | Sophora *(Sophora)* |
| Elm *(Ulmus)* | Spicebush *(Lindera)* |
| Franklinia *(Franklinia)* | Spirea *(Spiraea)* |
| Hickory nut *(Carya)* | Stewartia *(Stewartia)* |
| Hop hornbeam *(Ostrya)* | Sumac *(Rhus)* |
| Hornbeam *(Carpinus)* | Sweet gum *(Liquidambar)* |
| Horse chestnut *(Aesculus)* | Tulip poplar *(Liriodendron)* |
| Katsura tree *(Cercidiphyllum)* | Tupelo *(Nyssa)* |
| Maple *(Acer)* | Walnut *(Juglans)* |
| Magnolia *(Magnolia)* | Willow *(Salix)* |
| | Zelkova *(Zelkova)* |

*Seeds which have a long viability time may be dried after harvest but require a cold period for germination, with stratification times listed.*

| | |
|---|---|
| Fir *(Abies)* | 2 to 3 months |
| Maples *(Acer)*—most kinds | 3 months |
| Barberry *(Berberis)* | 2 to 3 months |
| Birch *(Betula)* | 2 to 3 months |
| Trumpet creeper *(Campsis)* | 2 months |
| Hornbeam *(Carpinus)* | 3 to 4 months |
| Hickory *(Carya)* | 3 to 4 months |
| Cedar *(Cedrus)* | 1 to 2 months |
| Bittersweet *(Celastrus)* | 3 months |
| False cypress *(Chamaecyparis)* | 2 months |
| Clematis *(Clematis)* | 3 months |
| Flowering dogwood *(Cornus florida)* | 3 months |
| Japanese dogwood *(Cornus kousa)* | 3 months |
| Beech *(Fagus)* | 3 months |
| Ash *(Fraxinus)* | 2 to 3 months |
| Privet *(Ligustrum)* | 3 months |
| Sweet gum *(Liquidambar)* | 3 months |
| Magnolia *(Magnolia)* | 3 to 4 months |

| | |
|---|---|
| Apple *(Malus)* | 1 to 3 months |
| Tupelo *(Nyssa)* | 3 months |
| Spruce *(Picea)*—most species | 1 to 3 months |
| Pine *(Pinus)*—most species | 2 months |
| Cherry *(Prunus)* | 3 to 4 months |
| Golden larch *(Pseudolarix)* | 1 month |
| Pear *(Pyrus)* | 3 months |
| Currant, gooseberry *(Ribes)* | 3 months |
| Mountain ash *(Sorbus)* | 3 months |
| Lilac *(Syringa)* | 2 to 3 months |
| Arborvitae *(Thuja)* | 2 months |
| Hemlock *(Tsuga)* | 3 months |
| Grape *(Vitis)* | 3 months |

## Seed germination

### THE PROCESS OF SEED GERMINATION

Seed germination is defined as the resumption of growth of the embryo, resulting in the protrusion of a root (also called the *radicle*) and a shoot (also called the *plumule*). Upon germination some seeds have their seed leaves *(cotyledons)* below ground. This is called a *hypogeous* condition. These subterranean cotyledons provide food reserves (fats, oils, carbohydrates, and proteins) that are broken down for development of the embryo into a seedling. They are nonphotosynthetic and eventually break down when all their food reserves have been utilized by the rest of the seedling. In contrast, other seeds, upon germination, have cotyledons which are elevated above ground. This is called an *epigeous* condition. These cotyledons not only provide food reserves for the developing seedling, but also become green and photosynthesize, thus providing additional food (sugars) for the developing seedling. Eventually, these cotyledons turn yellow, wither, and fall off (Figure 2–12).

### PRACTICAL ASPECTS OF GERMINATING SEEDS

The seed is basically a "care" package consisting of a young plant (the embryo), some stored food (in the endosperm and/or the cotyledons of the embryo), and a seed coat. If the seeds are planted too deeply, the food supply in the seed is used up before the light-requiring food manufacturing process (photosynthesis) can start. If seeds are planted too shallowly, many of the germinating seedlings dry out and die due to lack of water, or they get eaten by birds or other animals. The rule of thumb for planting seeds is to sow them at a depth of twice the diameter of the seeds. Minute seeds, such as those of begonia or petunia, are planted just below the soil surface, preferably in a pot in the house or greenhouse. Larger seeds, such as those of peas and corn, are planted 2 to 3 centimeters (3/4 to 1¼ inches) deep. In the vegetable garden, seeds are best covered with finely screened soil or sand to

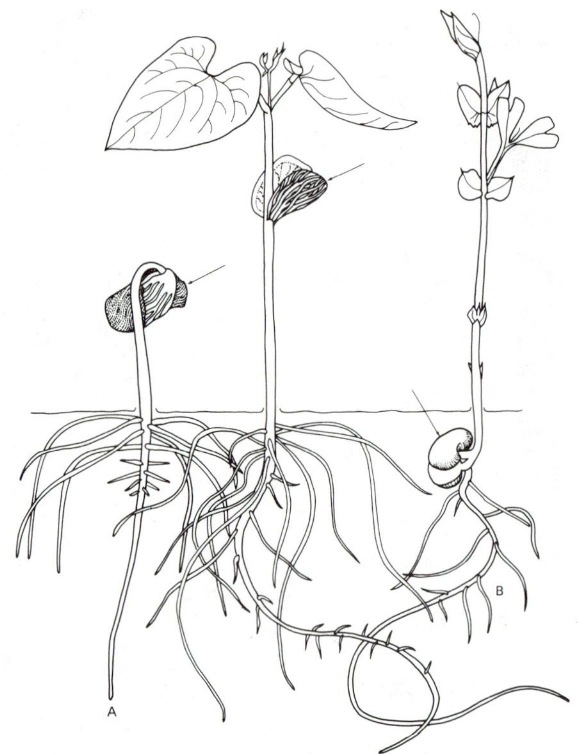


*FIGURE 2–12*
Seed leaf positions. Seedlings with above ground (epigeous) seed leaves (cotyledons), (A) bean, and below ground (hypogeous) cotyledons, (B) pea.

facilitate emergence of the seedling shoots. This is particularly necessary with clay soils which often form "hardpan," like a thin veneer of brick, after watering and drying. Finely sieved soil or sand mixed with milled sphagnum moss is also helpful with seeds planted in pots. Sphagnum moss contains antibiotic compounds which suppress growth of pathogenic fungi and bacteria. With any potting soil media, it is advisable to sterilize it before planting the seeds. Sterilize media in a 100° C (212° F) oven for 20 minutes or in a pressure cooker at 15 pounds per square inch pressure for 15 to 20 minutes. This soil "pasturization" treatment kills most insect eggs, weed seeds, bacteria, and fungi which cause *damping off* of seedlings. Seedlings damp off by falling over at ground level due to death of the hypocotyl (internode of the seedling below the cotyledons).

To start seeds of cacti and succulents, use at least one-half part fine sand. Generally, for seeds of other plants, use a 1:2:1 mixture of 1 part sand, 2 parts loam, and 1 part of leaf mold or peat moss, or 1 part of vermiculite and 1 part of perlite (a 1:1 ratio). Useful containers include plastic pans, shallow clay pots covered with glass or plastic-wrap, small plastic jelly dishes, paper-type egg cartons, and peat pots (Figures 2–13 and 2–14). Peat pots are especially good since they can be planted directly in the soil where they break down and contribute humus (decomposed organic matter) to the soil around the seedling(s).

*FIGURE 2–13*
Plastic pans of germinating seeds. Warm box with bottom heat in the greenhouse at the University of Michigan Matthaei Botanical Gardens.

*FIGURE 2–14*
Paper egg cartons and peat pots (lower right) of germinating seeds. The seedlings, peat pots and all, may be planted in the garden because the pot breaks down in the soil.

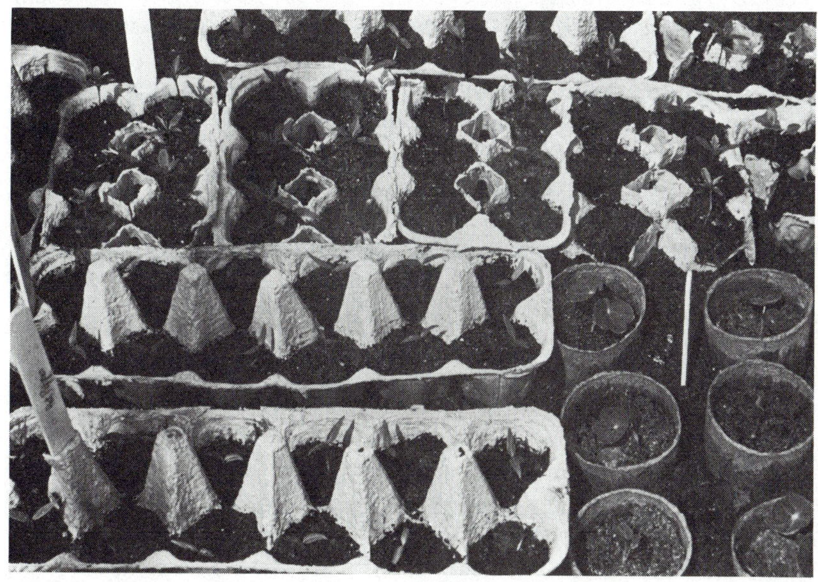

After the danger of frost is over in the spring, seeds of many annuals can be sown in the vegetable garden, preferably in rows. When vigorous seedlings have developed, they can be transplanted to beds in the garden. However, with seeds of perennials, it is best to plant them in July or August the year before you want plants for the garden. The plants can be overwintered in a cold frame (see Chapter 9). Or to save time, one can simply obtain divisions of perennials from friends or buy plants from a nursery. Perennials are very expensive to purchase, and if large numbers of plants are needed, it is far better to start them from seed. However, you must maintain optimal conditions and care for a longer period of time if you plant seeds and raise the plants yourself.

## GERMINATING ORCHID SEEDS

Orchid seeds are among the smallest seeds known. They are like fine dust! Inside the orchid seed is a small mass of cells which make up a globular, undifferentiated embryo. Surrounding this mass of cells of the embryo is a paperlike, flattened, winglike tissue, the seed coat, that aids in dispersal of the seeds. Because the embryo is undifferentiated, one must grow the seeds in a sterile, nutrient-rich culture in order for the embryos to complete their development and germinate. This artificial culture of orchid seeds also greatly speeds up both processes.

In nature, the development of the complete embryo may require a year or more. It occurs with the aid of a fungal partner (termed a *mycorrhizal fungus* partner) which provides vitamins and aids in water and nutrient uptake for the orchid seedlings and derives sugars and amino acids from the orchids. This is a *mutualistic association* or *symbiotic relationship.* The fungus is not needed in culture because the nutrient medium provides the needed metabolites.

To culture orchid seeds under sterile conditions, obtain ripe fruits (capsules), preferably green ones. Surface-sterilize the capsules in 10 percent household bleach (sodium hypochlorite or calcium hypochlorite) for 15 to 20 minutes. Cut open the capsules with a sterile scalpel or knife (sterilize by dipping in 95 percent ethyl alcohol and burning in a flame or by dipping in boiling water). Then remove a mass of seeds from the capsule with the scalpel or knife tip and transfer them to a nutrient agar medium especially developed for orchids. The one most commonly employed for germinating orchid seeds and growing the seedlings is *Knudsen's C* medium. The recipe for this medium is presented in Table 2–4. Sometimes, better seedling development is obtained with orchid seedlings in sterile culture if ripe banana (one per liter of medium) is included in the sterilized medium. This is called *banana mash* medium; it also includes the constituents of Knudsen's C medium.

Containers for media include baby food jars, serum bottles, flasks, milk bottles with appropriate stoppers or lids. The medium is placed in each container to a depth of two to three centimeters before both are sterilized. Sterilization can be achieved with an autoclave or a pressure cooker. In the pro-

*TABLE 2–4*
*Knudsen's C orchid medium (to germinate seeds and grow seedlings in sterile culture)*

| | |
|---|---|
| Monopotassium acid phosphate | 0.25 gm |
| Calcium nitrate | 1.00 |
| Ammonium sulfate | 0.50 |
| Magnesium sulfate | 0.25 |
| Ferrous sulfate | 0.025 |
| Manganese sulfate | 0.0075 |
| Sucrose | 20.00 |
| Agar | 17.50 |
| Water | 1000 ml |

(pH should be adjusted to between 5 and 5.2)

*Optional:* In blender, mash one ripe banana and add to above.

*Source:* Knudson, L. 1946. *Amer. Orchid Soc. Bull.* 15:214–17.

cess of transferring seeds to the culture medium in these containers, the mouths of the flasks or bottles should be flame-sterilized, and at the same time, the containers held horizontally so that contaminating fungal and bacterial spores in the air will not fall in. Once seeds are transferred, the containers are replugged with sterile cotton (bungs) or with special foamlike plugs, or lids with center holes containing foam plugs. This provides adequate air exchange without contaminating the sterile cultures with bacteria or fungi. Or simply use aluminium foil covers over the container mouths or tie heat-resistant plastic bag squares over the bottle/flask mouths (Figure 2–15).

Once inoculated with seeds, the cultures are placed under low light intensity bulbs such as cool white fluorescent lamps. One cannot be too careful to prevent microbial contamination of the seedlings in sterile culture! Otherwise, the seeds or seedlings will be overgrown by bacteria or fungi growing on the culture medium and will die. Often, a sterile transfer chamber, such as a laminar flow hood or sterile transfer "glove box," is used to reduce the possibility of contamination.

## Seedlings

### SEEDLING EMERGENCE FROM THE SOIL

As a seed germinates the root emerges first and anchors the seedling into the substratum. It is only after this occurs that the shoot portion emerges from the seed and starts to grow upward. The shoot portion, or *epicotyl,* forms a hook in most dicotyledon[4] seedlings such as beans (Figure 2–12). In grass seedlings, which are monocotyledons,[5] the modified leaf that develops up through the soil is called a *coleoptile* (Figure 2–16). Seedling hooks and co-

[4]*Dicotyledon* refers to flowering plants with two seed leaves or cotyledons.

[5]*Monocotyledon* refers to flowering plants with one seed leaf or cotyledon.

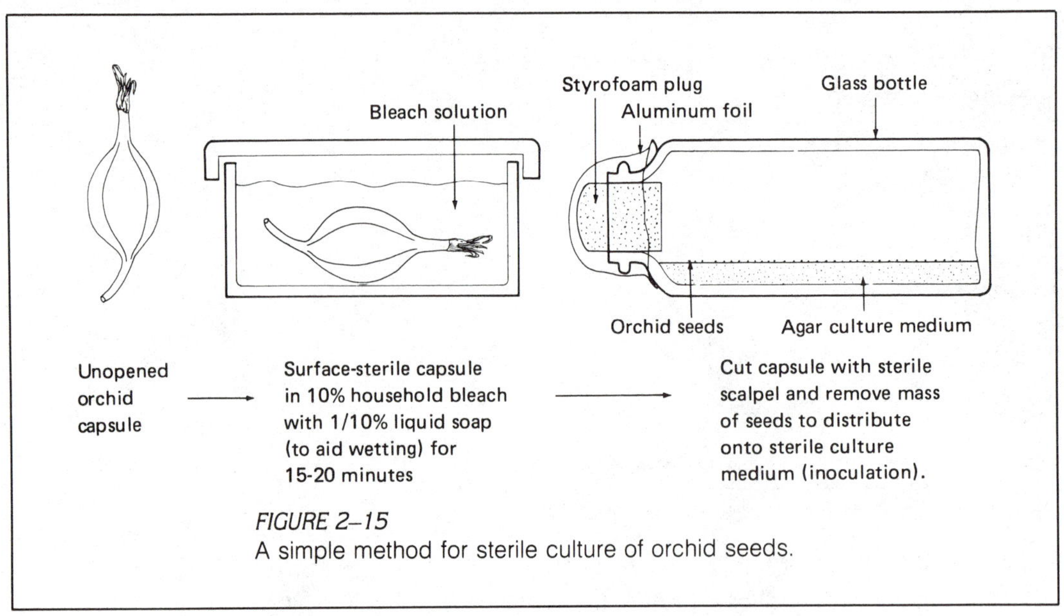


*FIGURE 2–15*
A simple method for sterile culture of orchid seeds.

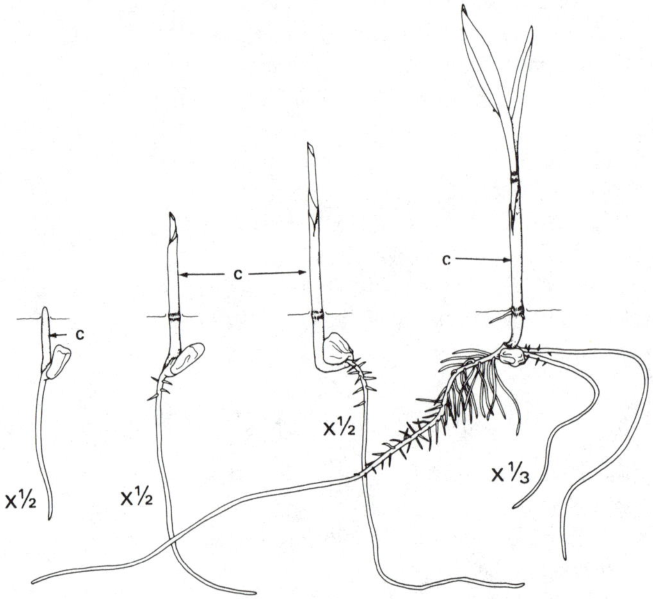

*FIGURE 2–16*
A grass seedling. Corn, a grass, in different stages of development during germination and emergence from the soil. c = coleoptile of seedling.

leoptiles have the function of protecting the young shoot tips. Once above the soil surface, the dicot seedling hook unbends, carrying the shoot tip 180° to the vertical position. In other words, the seedling shoot becomes "unhooked." Likewise, after the coleoptile of the grass seedling emerges from the soil, the coiled inner leaves protrude through its tip and unravel. In either case, the unhooking or unraveling is dependent upon sunlight, specifically, the red portion of the light spectrum. The plant pigment which absorbs the red light is called *phytochrome.* It plays an important role in many other processes including seed germination, leaf expansion, stem elongation, and flowering.

## REFERENCES

Hartmann, H. T., and Kester, D. E. 1975. *Plant propagation: principles and practices.* 3d ed. Englewood Cliffs, N.J.: Prentice-Hall.

Janick, J. 1979. Chapters 3, 4, and 5 in *Horticultural Science.* San Francisco: W. H. Freeman and Co.

Kramer, P. J., and Kozlowski, T. T. 1979. Physiology of seeds and seedlings. Chapter 14 in *Physiology of Woody Plants.* New York: Academic Press. (Covers, amongst other topics on the physiology of woody plants, the physiological aspects of the structure, dormancy, and germination of seeds of woody plants.)

North Carolina Wildflower Preservation Society. 1977. *North Carolina native plant propagation book.* Chapel Hill: North Carolina Wildflower Preservation Society. (A good practicum on how to propagate various types of wildflowers native to North Carolina by means of seeds and vegetatively.)

Porsild, A. E.; Harrington, C. R.; and Mulligan, G. A. 1967. *Lupinus arcticus* Wats. grown from seeds of pleistocene age. *Science* 158:114–15. (An article documenting by means of $^{14}$carbon radioisotope labelling the age of the the oldest known viable seeds of a plant—here the arctic lupine.)

Reilly, A. 1978. *Park's success with seeds.* Greenwood, S. C.: Park Seed Co. (Basic handbook on ways to successfully germinate seeds of many different kinds of horticultural plants, especially garden flowers and vegetables.)

Roberts, E. H., ed. 1972. *Viability of seeds.* London: Chapman and Hall. (A detailed discussion of how long seeds of various plant species remain alive and are capable of germinating; also discusses tests for seed viability and the methods for dating the age of seeds.)

U.S. Department of Agriculture Forest Service. 1974. *Seeds of woody plants of the United States.* USDA Agriculture Handbook No. 450. Washington, D.C.: U.S. Government Printing Office. (One of most detailed treatises available on collecting, storing, breaking dormancy, and germinating seeds of economically important species of woody plants—landscape, conservation, and timber species.)

Villiers, T. A. 1975. *Dormancy and survival of plants.* London: Arnold Publishers. (A good treatise on how plants are able to survive stress, such as low temperature or drought, through various dormancy mechanisms.)

Wareing, P. F., and Phillips, I. D. J. 1981. *Growth and differentiation in plants.* 3d ed. New York: Pergamon Press. (One of the most lucid, up-to-date paperbacks on how light, temperature, hormones, and genes control growth and development in plants. A classic!)

Wyman, D. 1977. Collecting and storing the seeds of woody plants. *American Horticulturist* 56:32–33, 37. (A fine article on how and when to collect seeds of woody plants and how to store the seeds properly for longest possible time to maintain viability.)

# Growth regulation by hormones, light, and temperature

# 3

Plant growth can be regulated through the use of hormones and by manipulating light and temperature. Anyone who has grown houseplants knows how important it is to provide the right kind of lighting conditions. The same goes for temperature. Cacti and succulents grow best in south-facing windows. African violets and their relatives prefer supplementary lighting in winter months. Chinese rubber tree plants and philodendrons thrive in shady locations. Some orchids prefer warm, moist surroundings while others prefer cool temperatures. We also know that temperature is very important in breaking seed and bud dormancy and in germinating seeds.

## Regulation of plant growth by hormones

### FUNCTIONS OF HORMONES IN PLANT DEVELOPMENT

Plants produce hormones, which are organic substances that are synthesized in small amounts and function to regulate growth and development throughout the life of a plant. A hormone can be produced at the *site of action* (also called the *target site*) or at some distance from it. Hormones coordinate the growth of different plant parts, acting in response to various environmental stimuli such as day-length, temperature, and gravity. Some of the important developmental processes regulated by plant hormones include the following:

1. Root initiation and elongation.
2. Leaf expansion and stem elongation.
3. The growth of lateral buds.

4. Curvature of roots toward the earth's center of gravity and stems and leaves away from it (gravitropism).
5. Curvature of plants toward light of higher intensities (phototropism).
6. Flower initiation.
7. Development of male (staminate) and female (pistillate) parts within the flower.
8. Fruit enlargement and ripening.
9. Leaf fall (abscission) and fruit drop.
10. Onset and breaking of bud dormancy.
11. Onset and breaking of seed dormancy.
12. Seed germination.

## THE MAJOR TYPES OF HORMONES IN SEED PLANTS

The major types of hormones native to seed plants include gibberellins, auxins, cytokinins, ethylene, and abscisic acid. Chemical structures of some of the primary plant hormones are shown in Figure 3–1. Many of the synthetic hor-

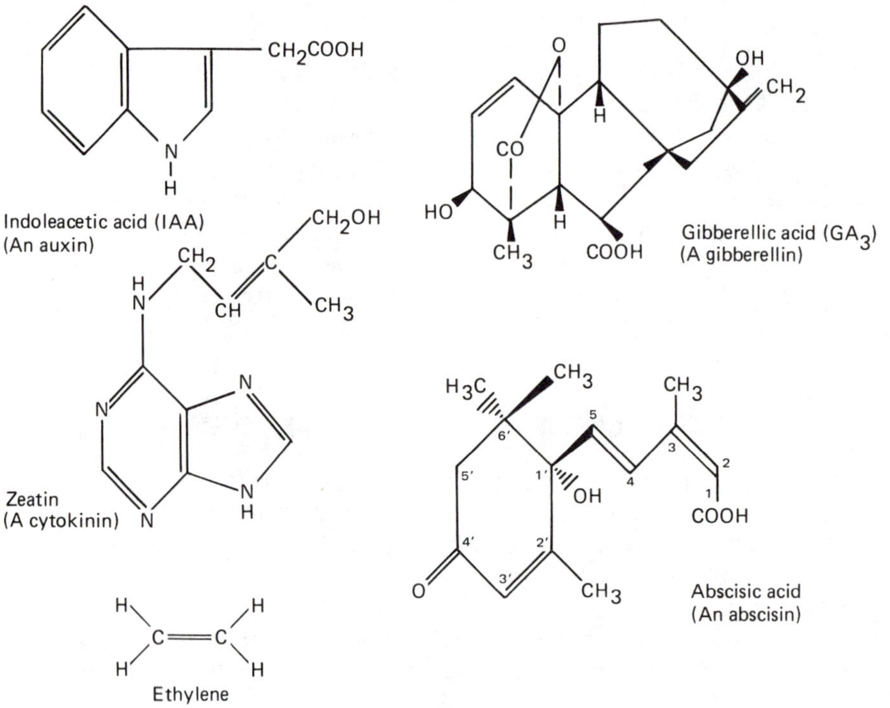

Indoleacetic acid (IAA)
(An auxin)

Gibberellic acid ($GA_3$)
(A gibberellin)

Zeatin
(A cytokinin)

Abscisic acid
(An abscisin)

Ethylene

*FIGURE 3–1*
Chemical structures of representatives of the major types of native (endogenous) hormones in plants.

mones, which act like the native hormones and which may have similar chemical structures or actions, include 2,4-D; 2,4,5-T; IBA; IPA; NAA—all synthetic auxins; kinetin and benzyl adenine—both synthetic cytokinins; Ethrel (also called Ethephon)—a producer of ethylene; and Gibrel—a source of gibberellic acid, $GA_3$ (one of 62 known gibberellins) (Figure 3–2).

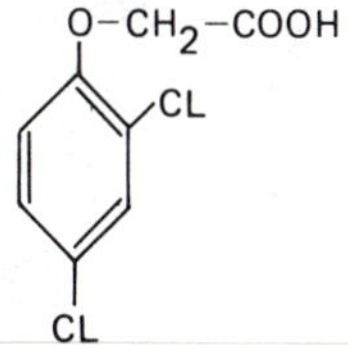

2, 4 - D (2, 4-dichlorophenoxyacetic acid)
(A synthetic auxin)

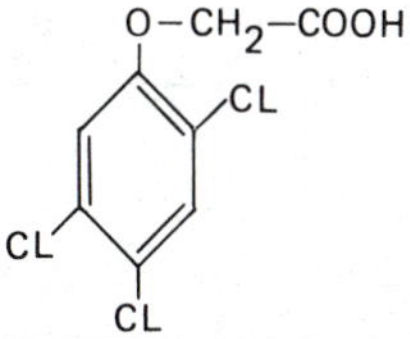

2, 4, 5 - T (2, 4, 5 - trichlorophenoxyacetic acid)
(A synthetic auxin)

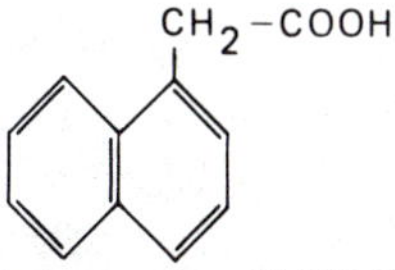

2- Naphthaleneacetic acid (NAA)
(A synthetic auxin)

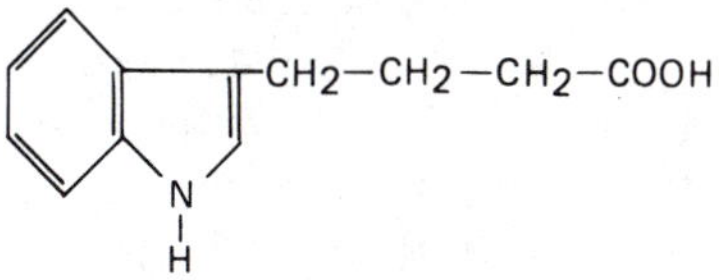

IBA (Indole-3-butyric acid)
(A synthetic auxin)

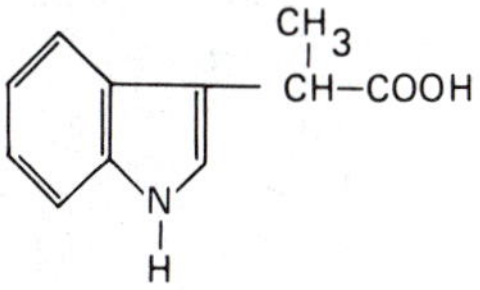

IPA (Indole-3-propionic acid)
(A synthetic auxin)

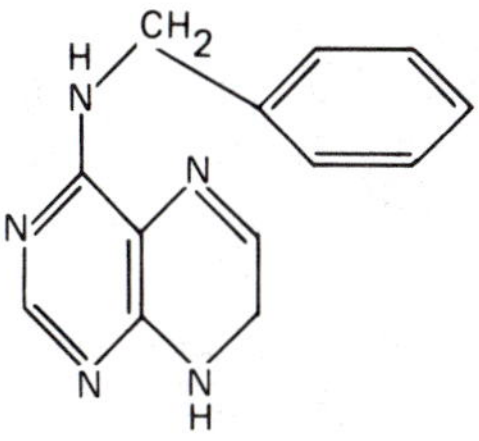

BA (Benzyl adenine or 6-benzyl aminopurine)
(A synthetic cytokinin)

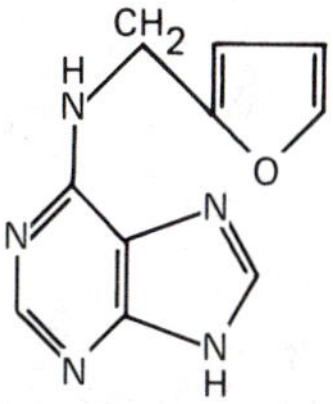

K (Kinetin or 6-furfurylaminopurine)
(A synthetic cytokinin)

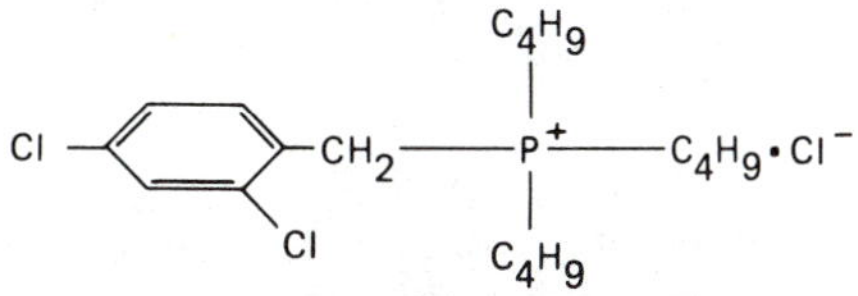

Phosphon D (An anti-gibberellin)
(2, 4 - dichlorobenzyltributyl phosphonium chloride)

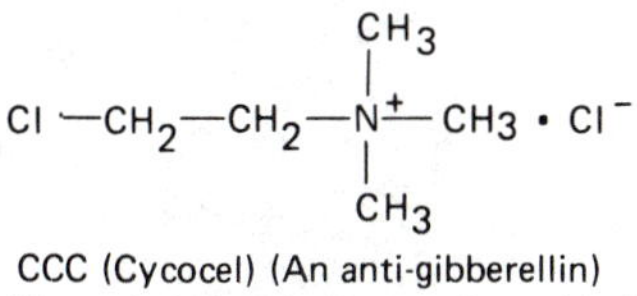

CCC (Cycocel) (An anti-gibberellin)
(2 - chloroethyltrimethylammonium chloride)

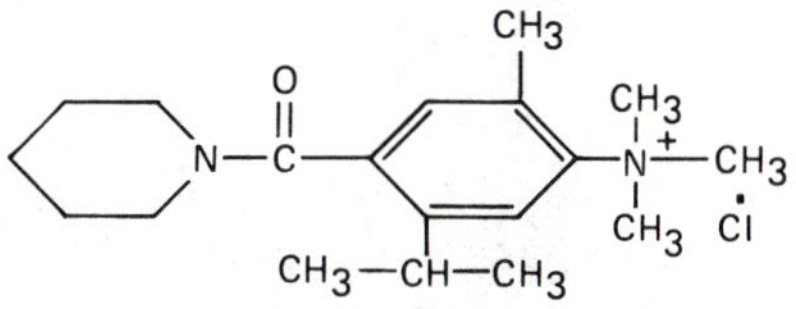

Amo-1618 (An anti-gibberellin)
Ammonium (5-hydroxycarvacryl) trimethylchloride
piperidine carboxylate

*FIGURE 3–2*
Chemical structures of representatives of some of the synthetic plant hormones used to regulate plant growth and development.

## HORMONES USED AS HERBICIDES

Synthetic auxin hormones, such as 2,4-D and 2,4,5-T, had their origin about 1945 in England and the U.S.A. Since then, 2,4-D has been widely used to control weedy broadleaf dicots, including dandelions *(Taraxacum officinale)*, plantains *(Plantago lanceolata, P. major)* in lawns, and mustard *(Brassica nigra, B. campestris)* and wild radish *(Raphanus raphinistrum)* in grain fields, while 2,4,5-T has been widely used to control woody brush along roadsides, railways, and power lines. This latter herbicide is now banned by the Environmental Protection Agency (EPA) because it contains dioxin, a contaminant that is very toxic to humans, suspected of causing birth defects. The widespread use of 2,4-D has caused problems: it can cause damage to monocots, especially grasses, in the seedling and flowering stages of development. When 2,4-D sprays are applied by airplane on windy days to control weeds in grain fields, they can also cause serious injury to nearby dicots such as tomatoes, grapes, cotton, roses, spinach, willows, and many other plants in the garden or on farms. The drift of the herbicide into areas where these plants are growing can be up to 21 miles (35 kilometers), as has been recorded in California. The same result can happen when one is spraying the home lawn with 2,4-D to control broadleaf weeds, and the spray is carried by wind into the vegetable and flower garden. This kind of damage from 2,4-D is minimized when one applies it carefully and directly to the lawn in pelleted form with fertilizer such as "weed and feed" preparations. Such preparations work best during warm weather when the weeds are actively growing. However, one should *never* harvest wild edible weeds such as dandelions for salads or wine making in areas that have been sprayed or treated with herbicides such as 2,4-D or other synthetic auxins.

## HORMONES USED TO INDUCE ROOTING IN CUTTINGS

Synthetic auxin-type hormones are also important to induce initiation of roots in cuttings that otherwise are difficult to root. This effect of auxin, discovered in the early 1930s, was one of the earliest practical applications of plant hormones. The natural auxin (IAA) hormone works fairly well, but the synthetic auxins, IBA (indolebutyric acid) and IPA (indolepropionic acid) work better. This auxin effect is manifested in two ways: it increases the numbers of roots, and roots are initiated faster than would normally occur. Hormone treatment of a cutting is relatively simple. The base of the cutting is first immersed in water, then dipped in hormone powder. After this, the cutting is planted in an appropriate rooting medium such as vermiculite/perlite (3 parts to 1 part, respectively). Fungicide is often included in commercial hormone preparations to prevent rotting of the cuttings' bases by fungi. Hormone treatment of cuttings is especially effective with hard-to-root woody plants, both coniferous (Figure 3–3) and deciduous.

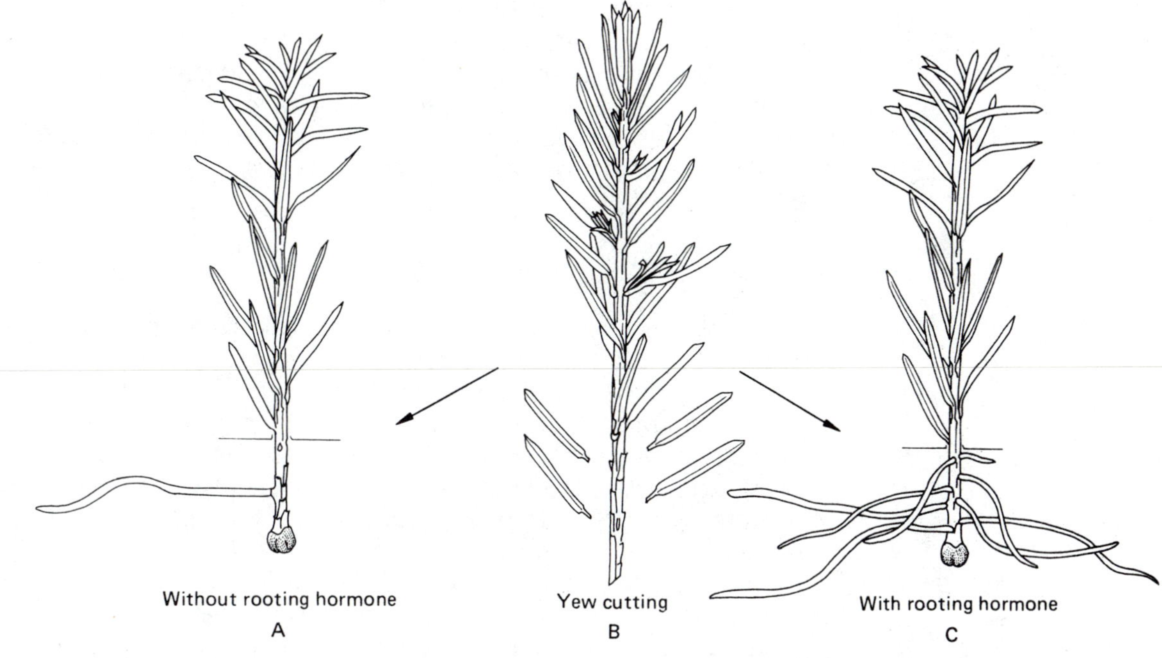


*FIGURE 3–3*
Cuttings of Japanese yew *(Taxus cuspidata)*. The effect of auxin hormone treatment of a cutting to increase the number of roots is shown by (A) cutting without hormone treatment, (B) initial cutting, (C) cutting with hormone treatment.

## HORMONES USED TO INDUCE CONE FORMATION

Another group of hormones, the gibberellins, are used to stimulate male and female cone development in young conifers. (Conifers produce separate male and female cones on the same tree.) Richard Pharis at the University of Calgary in Alberta, Canada has shown that cone formation in members of the pine family (Pinaceae) can be induced by repeated applications of a gibberellin mixture ($GA_4$ and $GA_7$). Likewise, in members of the cypress family (Cupressaceae), such as Arizona cypress *(Cupressus arizonica)*, and the taxodium family (Taxodiaceae), such as redwood *(Sequoiadendron)* and the giant sequoia *(Sequoia gigantea)*, repeated applications (weekly over one to two months) of gibberellic acid ($GA_3$) will induce cone development in the seedlings of these trees. What is interesting about these gibberellin treatments is that they can induce conifer seedlings to produce cones years before they ordinarily do so in nature. A classic example is found with Arizona cypress. Here, seedlings 6 to 8 weeks old can be induced to produce cones when grown under long days (16 to 18 hours) and treated with $GA_1$ or $GA_3$. The gibberellin applications must be repeated several times, using relatively high concentrations of the hormone, around 10,000 parts per million (ppm). The practical inference is that the forest tree plant breeder is able to cross-pollinate species or cultivars of young conifer trees grown in the greenhouse, a growth chamber, or in a nursery rather than high up in trees that are 15 to 20 years old at the time they are normally first producing male and female cones. Thus, GA treatment makes the work of the tree breeder more convenient and productive because it shortens the time required for cone formation.

## HORMONES USED TO INDUCE FLOWERING

Plant physiologists have shown that the flowering response in long-day and short-day plants is under hormonal control. The evidence for this is as follows. First, night length, the environmental signal that brings about flowering, can be perceived by a single mature leaf or even a part of a mature leaf. Second, the stimulus is transmissible through a graft. When a branch of a short-day plant is placed under short days, it will induce flowering in a graft onto a long-day plant that has received short days and long nights. Third, the flower induction response occurs at the shoot apex, a site which is considerably removed from the mature leaf or leaves that perceive the short-day or long-day signal, thus implicating a hormonal messenger at work.

No flowering hormone has yet been isolated from plants. Several hormones are used to induce flowering in restricted groups of plants. For example, it has been shown in bromeliads that ethylene or auxin (a hormone that induces ethylene formation in plants) applied in an enclosed environment may bring about flowering in these plants. Also, gibberellin hormones may bring about flowering in long-day biennials[1] the first year. The current idea espoused by the plant physiologist Anton Lang, at Michigan State University, is that ethylene or gibberellins may be the plant hormones that lead to the production of the special flowering hormone, tentatively called *florigen*, and further that antiflorigen-type hormones may be acting to inhibit the formation of florigen under noninductive day-length conditions. Thus, we can visualize the flowering process in day-length sensitive plants as being under the control of both promoter- and inhibitor-type hormones. Isolation and characterization of either a florigen or antiflorigen hormone would be a major step forward in helping us to understand the flowering process and how it is controlled in plants. It would also be of immense practical benefit to all growers of plants who wish to induce flowering in their plants!

Some applications have accrued from our knowledge that both ethylene (or auxin) and gibberellin hormones can stimulate flowering in certain groups of plants. One notable case is that the synthetic auxin NAA (naphthaleneacetic acid) will induce flowering in pineapple. The NAA causes the pineapple plants to produce ethylene, which in turn induces vegetative pineapple plants to flower. In commercial production, NAA and other synthetic auxins are applied to young vegetative pineapple plants to induce

---

[1] *Biennials* are plants which require more than one year but less than two to flower and set seed.

flowering earlier than would normally occur, and the resultant flowering and fruiting are more uniform. Since pineapple *(Ananas comosus)* is a bromeliad, the application of auxin was tried on the shoots of other bromeliads that are grown as houseplants (see Chapter 5). As with pineapple, such treatments would also bring about flowering in these plants. Thus, today, such common bromeliads as *Billbergia, Tillandsia, Aechmea,* and *Vriesia* can be induced to flower earlier than normal when treated with auxin or ethylene-producing products such as Ethrel or Ethephon. When a product such as Ethephon [(2-chloroethyl) phosphonic acid] is applied as a spray, it releases ethylene gradually into the plant tissues. However, to be effective the plants must be sprayed several times. A concentration of Ethephon that is effective in inducing flowering in bromeliads is approximately 2,500 parts per million (ppm).

In flowering plants, the development of female (pistillate) parts (pistils) of the flower is promoted by auxin, gibberellins, cytokinins, and sometimes ethylene, depending on the species of plant. Development of male (staminate) parts (stamens) of the flower is controlled by gibberellins or cytokinins. These observations have led to the use of hormones by plant breeders and horticulturists to regulate sex expression in plants which have separate staminate and pistillate flowers on the same plant (monoecious condition) or on different plants (dioecious condition).

Let us cite a few examples. Cytokinins can be used to cause hemp *(Cannabis sativa)* plants to produce 80 to 90 percent female flowers. Ethephon as well as auxin treatments will cause only pistillate flowers to develop on cucumber vines, whereas gibberellin treatment will induce only staminate flowers to develop. In begonia, both gibberellic acid ($GA_3$) and auxin (IAA) induce pistillate flower formation. Normally, begonia produces staminate and pistillate flowers on the same plant. With hops *(Humulus lupulus)* gibberellin treatment increases the number of pistillate flowers which develop, resulting in significant increases (as much as 25 percent) in cone development. Hop cones represent the overlapping bracts produced by the pistillate flowers. They are used to provide an extract that flavors beer. The primary benefit to the plant breeder or grower is that such hormone treatments can be used to produce only staminate or pistillate flowers on plants which normally produce both.

To produce larger "show" flowers, hobby growers of camellias use gibberellin. "Gibbing" is accomplished by removing one flower bud of a close pair, then applying gibberellin to the wound. The hormone is absorbed and causes the other bud to open into a much larger flower.

## HORMONES USED TO IMPROVE FRUIT SET, INCREASE FRUIT SIZE, AND PROMOTE FRUIT RIPENING

In the flower, once pollination has occurred, and if fertilization of eggs in the ovules has taken place, the ovary starts to enlarge. This marks the beginning of fruit development. The function of the fruit is to contain, protect, and sometimes even help disperse the seeds of the plant. The initial swelling of the ovary results from natural auxin (IAA-produced by the pollen) stimulating cell division and enlargement of the ovary wall tissue. Later, the ovary enlarges due to auxin and/or gibberellin produced in the developing seeds of the fruit.

From this, you can see that there might be some practical applications based on the fact that certain hormones may regulate ovary enlargement in flowering plants. Synthetic auxins are used to cause fruit set, that is, to stimulate the initial swelling of the ovary, and hence to prevent fruit drop which would occur if effective pollination and fertilization had not occurred.

Fruit size may be increased by either auxin or gibberellin hormone treatment, depending on the plant species. The classical example is seen with auxin-treated strawberry fruits (Figure 3–4). Normally, the fruits (achenes) produce auxin that causes the fleshy portion of the fruit (the flower receptacle) to enlarge, producing the strawberry fruit that we all know. If the seeds only develop on one side, one obtains a misshapen fruit, and if no achenes develop, the fruit aborts. The proof is seen if one removes all the developing ovules (future achenes) on an undeveloped fruit, then treats the young fruit with auxin. Results show the ovary will swell normally and produce a seedless fruit. This is called *parthenocarpic fruit development*—literally, seedless fruit development. You may already be aware that such fruits exist in nature, as evidenced by genetic variants of seedless citrus, seedless grapes, and seedless watermelons.

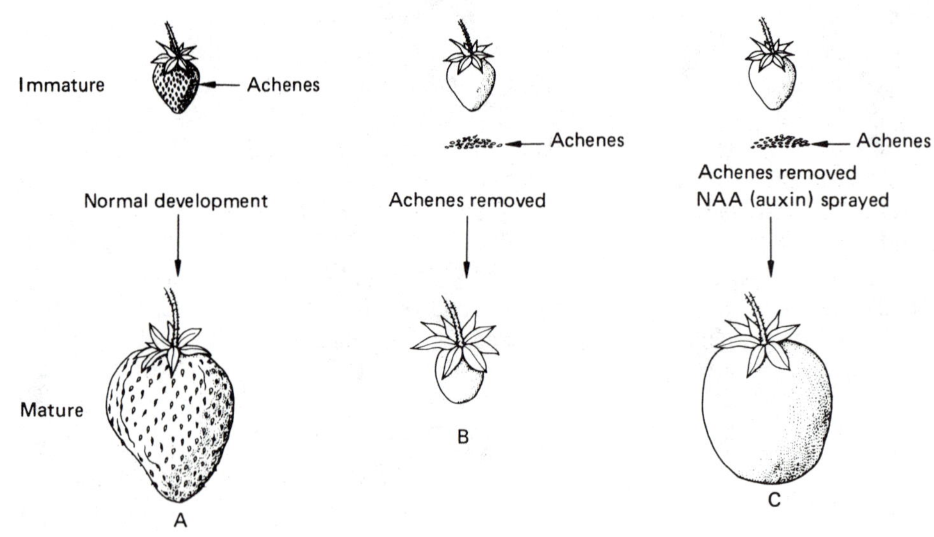

Auxin treatments can also cause swelling of ovary tissue in other fruits that will be seedless at maturity. Examples include tomato, fig, and squash. For fig, the auxin treatment has been used to substitute for pollination by the fig wasp *(Blastophaga)* in California (where the fig wasp is not native). For tomato, auxin treatment not only induces parthenocarpic fruits to develop, but also increases "fruit set" by preventing normal drop of developing flowers in the flower cluster. Ovary size may also be increased by externally applied gibberellins on developing apples, pears, grapes, and stone fruits (cherry, plum, peach).

The maturation of fruit, like that of fruit enlargement, is under hormonal control. The primary fruit-ripening hormone is ethylene. In maturing fruits, it markedly stimulates the rate of respiration in many species, resulting in significant production of carbon dioxide gas, release of sugars and organic acids, and the beginning of softening of the fruit pericarp (fleshy layers outside the central seed-bearing region). The production of ethylene in ripening fruits is triggered by increases of auxin-type hormones such as IAA. Applications of synthetic auxins, such as NAA, will also stimulate ethylene production in ripening fruits. These hormones are used commercially to ripen tomatoes and bananas which are shipped green for the market. The response in bananas is very rapid—less than 24 hours to turn from green to yellow when IAA or NAA is applied. Another way to achieve this result is to enclose green-bananas in a plastic bag with ripening apples which give off large amounts of ethylene.

*FIGURE 3–4*
The effect of achene removal and auxin treatment on strawberry *(Fragaria):*
(A) normal fruit development;
(B) with achenes removed, the fleshy receptacle does not enlarge;
(C) with achenes removed and sprayed with auxin (NAA), the receptacle does enlarge resulting in a seedless (parthenocarpic) fruit. This experiment shows that achenes produce auxin which stimulates receptacle enlargement.

### HORMONES USED TO DWARF PLANTS

Several compounds which block gibberellin biosynthesis in the plant, and hence result in suppressed stem elongation, include antigibberellins (CCC, Phosphon D, and Amo-1618) (Figure 3–2). They are used to reduce shoot height, and thus to obtain more compact plants such as in chrysanthemums, azaleas, and other commercial florist "crops." One of the widest uses of antigibberellins is for cereal crops, such as wheat. One commercial product, Cycocel (CCC), is sprayed on wheat plants to dwarf the shoots (culms), preventing the malady called *lodging*. Lodging refers to the falling over of the shoots due to the action of wind and rain, resulting in significant losses in grain yield. The dwarfed plants, in contrast, do not fall over in wind and rain storms because the shoots can support themselves better. This is largely due to the fact that the shoot is stiffer because of the presence of more silica. Grain yields are greatly increased by these treatments. The compounds used, relatively innocuous to humans, have low mammalian toxicity.

### HORMONES USED TO BREAK SEED DORMANCY

Hormones are frequently used to treat seeds which fail to germinate because of the presence of chemical inhibitors in their seed coats (see Chapter 2). Both cytokinins (benzyl adenine and kinetin) and gibberellins ($GA_3$) are used for this purpose. Such treatments (soaking the seeds for 24 to 48 hours in aqueous solution of $GA_3$ or cytokinin at 0.01 percent = 100 parts per million) will often substitute for the low temperature stratification treatments used to break dormancy.

## How plants harvest energy from the sun through photosynthesis

*Photosynthesis* is the basic process by which plants harvest solar energy from the sun, using carbon dioxide and water to produce oxygen and carbohydrates (mainly sugars) that can be used for plant growth. The overall equation for photosynthesis follows:

$$\underset{\text{carbon dioxide}}{CO_2} + \underset{\text{water}}{H_2O} \xrightarrow[\text{chlorophyll}]{\text{light}} \underset{\text{oxygen}}{O_2\uparrow} + \underset{\text{carbohydrate}}{CH_2O} + \text{heat}$$

In the process of photosynthesis, light is used to make the high energy compounds, ATP (adenosine triphosphate) and $NADPH_2$ (nicotinamide adenine dinucleotide phosphate); these compounds then provide energy for the synthesis of carbohydrates and for other energy-requiring processes in the plant.[2]

---

[2]The most effective portions of the light spectrum in photosynthesis are in the blue and red regions.

In growing plants, we can capitalize on our knowledge of the basic photosynthetic process by providing growing plants with some regulatory inputs which maximize rates and total amounts of photosynthesis. In the greenhouse we can use carbon dioxide ($CO_2$) generators to step up the atmospheric $CO_2$ partial pressure from its average value of 0.027 percent—an amount which typically limits photosynthesis. Fermenting manure will also increase $CO_2$ partial pressure in a greenhouse (as under a greenhouse bench). The next obvious way to increase photosynthesis is to add overhead lights indoors or in a greenhouse. This is effective during periods of the year when light (intensity and duration) is limited, especially as occurs in temperate regions.

## Light in the regulation of plant growth and development

The blue portion of the white light spectrum is directly involved in bringing about *phototropic responses* in plants—that is, the bending of a plant toward light of higher intensity. An example of this response is seen in plants on windowsills. It is a way by which the plant can orient its primary photosynthetic organs (leaves and stems) toward the light and thus harvest more light energy from the sun in photosynthesis. Hormones are also involved in this response, with higher amounts of auxin moving to the shaded side of the plant shoots, causing cell elongation on that side and consequent bending of the plant toward the lighted side or side with higher light intensity.

For most plants, the initiation of flowers is not under *day-length (photoperiod) control.* Such plants are called day-neutral in their flowering responses. Practical benefits of this discovery are (1) that tomatoes and cucumbers can be grown even in the short days of winter in northern greenhouses, and (2) most plants native to the tropics are day-neutral because there is little change in day-length throughout the year; so they are good houseplants.

However, there are some plants, such as chrysanthemum, Christmas cactus, and poinsettia, which, after a period of vegetative growth, will only flower under conditions of short days (long nights). Plant biologists have learned that it is the length of the dark period that is critical to flower initiation in these plants. There must be an uninterrupted long nighttime, of greater than some critical length, that is specific for each plant species. The control is a bit more complex, however. If the long night, for example longer than 12 hours, is interrupted by a short period of white light, flowering will be inhibited by the red wavelengths in the white light. On the other hand, interruption by far-red light does not affect the flowering response, and the short-day plant will bloom. The opposite is true for long-day plants such as spinach, barley, and henbane. What all this means is that the flowering response in long-day and short-day plants is controlled by the red/far-red phytochrome pigment system. Phytochrome has two chemical forms, a red, light-absorbing form ($P_r$) and a far-red, light-absorbing form ($P_{fr}$). The slow, dark conversion of the far-red form to the red form can measure the length of the dark period. Thus, it works as a biological clock (a time-keeping mechanism)—a way by which plants respond to time periods of given lengths and in this case to the lengths of the dark periods.

In practical botany we are not so much concerned with the phytochrome reaction, but we are very much concerned with its consequences. Growers of chrysanthemums can obtain flowers at any time of the year by the use of overhead lamps in winter to provide artificially long days to maintain vegetative growth, then use black cloth to cover the plants in order to shorten the day-length to about eight hours which induces flowering (Figure 3–5).

Some plants require only one short day to cause flowering, while others need as many as three to five days as is the case for soybeans, tobacco, and chrysanthemum. It is important here to stress one point in connection with photoperiodic requirements. Some plants, such as chrysanthemum, have an absolute short-day requirement for the induction of flowering; that is, after the minimum requirements are satisfied, all of the flowers are produced—an all-or-none type response. These plants are called *qualitative* short-day plants. In contrast, other plants do not have an absolute short-day requirement for flower induction. Instead, short days, in increasing numbers, correspondingly increase the number of flowers which develop, so that the more short days past the critical stage, the more flowers will be produced. Such plants are called *quantitative* short-day plants. The same types of responses apply to long-day plants (Table 3–1).

*FIGURE 3–5*
Greenhouse interior with short day and long day provisions. To provide long days, overhead lights illuminate plants. To provide short days, black cloth is draped over benches of plants.

TABLE 3–1
*Examples of important crop and ornamental plants which are short-day, long-day, or day-neutral in their flowering response*

| Short-day Plants (require long nights) | |
|---|---|
| *Qualitative (Absolute)* | *Quantitative* |
| Coffee *(Coffea arabica)* | Cotton *(Gossypium hirsutum)* |
| Tobacco *(Nicotiana tabacum)* | Rice *(Oryza sativa)* |
| Chrysanthemum *(Chrysanthemum morifolium)* | Sugarcane *(Saccharum officinale)* |
| Thanksgiving cactus *(Schlumbergera truncatus)* | Scarlet sage *(Salvia splendens)* |
| Christmas cactus *(Schlumbergera bridgesii)* | Cosmos *(Cosmos bipinnatus)* |
| Poinsettia *(Euphorbia pulcherrima)* | Hemp *(Cannabis sativa)* |
| Soybean *(Glycine max)* | |
| Strawberry *(Fragaria ananassa)* | |
| Jerusalem artichoke *(Helianthus tuberosus)* | |
| Kalanchoë *(Kalanchoe blossfeldiana)* | |

| Long-day Plants (require short nights) | |
|---|---|
| *Qualitative (Absolute)* | *Quantitative* |
| Spinach *(Spinacea oleracea)* | Spring barley *(Hordeum vulgare)* |
| Sedum *(Sedum spectabile)* | Petunia *(Petunia hybrida)* |
| Rye grass *(Lolium temulentum)* | Garden pea *(Pisum sativum)* |
| Carnation *(Dianthus caryophyllus)* | Kentucky bluegrass *(Poa pratensis)* |
| Oat *(Avena sativa)* | Spring rye *(Secale cereale)* |
| Dill *(Anethum graveolens)* | Spring wheat *(Triticum aestivum)* |
| Sweet clover *(Melilotus alba)* | Garden beet *(Beta vulgaris)* |
| Peppermint *(Mentha piperita)* | Snapdragon *(Antirrhinum majus)* |
| Timothy *(Phleum pratense)* | Turnip *(Brassica rapa)* |
| Radish *(Raphanus sativus)* | Lettuce *(Lactuca sativa)* |
| Cone flower *(Rudbeckia bicolor)* | Evening primrose *(Oenothera* spp.*)* |
| Clover *(Trifolium* spp.*)* | |

| Day-neutral Plants | |
|---|---|
| Tomato *(Lycopersicon lycopersicum)* | African violet *(Saintpaulia ionantha)* |
| Garden bean *(Phaseolus vulgaris)* | Cucumber *(Cucumis sativus)* |
| Buckwheat *(Fagopyrum esculentum)* | Most tropical flowering plants |
| Corn or maize *(Zea mays)* | |
| Most woody plants | |

## HOW TO SET UP A BALANCED INDOOR LIGHTING SYSTEM FOR GROWING PLANTS

The question is how to obtain an artificial light system that will provide light over the range from violet to far-red (400 to 800 nanometers). Cool white fluorescent lamps alone are not the best because they emit only up to and including red light with no appreciable far-red light component that is necessary for normal plant growth. Incandescent lamps emit over the entire white light spectrum, including far-red; however, they also give off a lot of heat as infrared. Early-make indoor lighting systems thus made use of a combination of fluorescent and incandescent lamps to provide a full-range white light spectrum. Now, a single type of wide-spectrum lamp bulb is available which emits over the full white light spectrum. These lamps are tubular, like fluorescent lamps, and some companies have even made them fluted to provide a greater illumination surface. Their big advantage is that they do not give off as much heat as tungsten or incandescent lamps.

After you have set up your lamps, it is a good idea to include a timer and set it so that plants get 16 to 18 hours of light and 6 to 8 hours of darkness. Alternatively, illuminated shelves mounted on carts are available from commercial sources which are especially useful for growing houseplants, herb seedlings, flowers, and vegetables.

## PRACTICAL APPLICATIONS OF LIGHT REGULATION

One of the first applications for growing plants under lights in greenhouses was to extend the day-length in order to delay flowering in short-day plants. Ordinarily, poinsettia, chrysanthemum and kalanchoe would begin to flower in response to the short days of northern autumns and southern springs in the respective temperate regions. Extending the day-length delays flower initiation. After long-day light treatment, these plants can be induced to flower with a short-day lighting regime (typically about eight hours).

A second application for the use of overhead lamps is also found in greenhouses. A number of crop plants such as tomatoes, cucumbers, and lettuce are raised in greenhouses during cold winter months in northern temperate and southern temperate countries. Since the days are so short during these months, it is advantageous to lengthen the days with overhead lights so that the plants make more food via photosynthesis and hence give higher yields. Usually, incandescent lamps or high energy halide lamps are used for this purpose, as in the case of controlling flowering in short- and long-day plants. It would be far too expensive to use combinations of fluorescent and incandescent lamps or the new full-spectrum lamps that are used for indoor lighting systems for growing these vegetables.

Day-length is also important in the control of vegetative growth in deciduous trees and conifers in temperate regions. In such plants, the advent of short days brings about the onset of bud dormancy due to the production of high levels of abscisic acid (ABA), and a decrease in the levels of gibberellins. Dormancy in such plants is usually broken by a period of low winter temperature followed by long days. Coincident with long days is a rise in the levels of natural gibberellins and a decrease in the levels of abscisic acid. Growers of nursery stock often capitalize on long-day treatments. They provide woody plant cuttings with supplementary overhead light in the greenhouse. This keeps the plants growing, usually in flushes of growth, and prevents them from becoming dormant and setting terminal and lateral buds.

A third application for growing plants under lights is in the home. The devices vary from single lamp fixtures to rather elaborate tiers of illuminated shelves. The uses of such lamps are many, including starting seedlings in the spring for the garden; forcing bulbs in the winter; growing rooted cuttings; illuminating plants in hanging pots; starting begonias and dahlias in the spring before setting them out in the garden; illuminating aquarium plants, and growing cacti and succulents, annuals, and such gesneriads as African violets and gloxinias. Some plants such as orchids, bromeliads, and succulents cannot be grown under continuous light as it interrupts certain plant metabolism processes. Excess light in the home must be controlled if one is trying to flower short-day plants such as poinsettia and Christmas cactus.

If you want a fancy lighting device, you can purchase special cabinets (see Figure 9–14) which have timers for regulating day-length and temperature, and which have controls for regulating relative humidity of the air in the chamber. Most people do not need such devices, but they have proved to be invaluable for plant research in agricultural experiment stations, commercial companies, colleges, universities, and in high schools. But for the home, all you need are hooks, suitable full-spectrum lamps, and an electric timer.

Finally, we should mention that overhead lighting devices can be used to great advantage for illuminating plants in restaurants, business offices, and hospitals; for holding plants in special illuminated rooms in schools and nature centers; and for illuminating aquarium plants.

# Regulation of plant growth by temperature

## BASIC ROLES OF TEMPERATURE IN THE GROWTH PROCESS

Temperature, next to light, is one of the most important environmental factors affecting plant growth. It is one of the key regulators of the rate of enzyme reactions and a primary determinant of plant hardiness and the length of the growing season. Temperature also regulates germination of seeds and the onset and breaking of dormancy in buds of woody plants in temperate regions. In conjunction with day-length (or night-length), temperature influences the induction of flowering in most biennials, hardy bulbs, and cereal grains such as winter rye and winter wheat. The temperature range over which most plants grow is 5° to 40° C (40° to 104° F). Of course, exceptions occur beyond the limits of this range, as for example in thermophilic (heat-loving) algae and bacteria and a number of cold-tolerant mosses and liverworts (bryophytes) and marine algae.

## PLANTING BY LATITUDE

In selecting plants for the outdoor landscape, it is essential that you use trees or shrubs that are native to or adapted to your particular latitude. This is necessary because many southern latitude woody species or cultivars of a species will not survive in northern latitudes. They get frost-killed or winter-killed while they are still actively growing during the short days of autumn or because they are not winter-hardy. Likewise, many northern species, or cultivars of a species, of woody plants will not survive in southern latitudes because they must experience a much colder winter to break dormancy in the spring. It is a mistake to try to grow plants in northern latitudes that were collected in southern latitudes or obtained from southern latitude nurseries (such as peaches or apples). And the converse is true for northern varieties grown in southern latitudes.

Thus, in landscaping (see Chapter 7), one is most successful, in terms of plant survival, when one plants woody species that are native to the particular region or are well-adapted to the winter temperatures that prevail there (as when one grows woody plants obtained from other regions of the world but that occur at similar latitudes).

## PRACTICAL APPLICATIONS OF TEMPERATURE REGULATION

In practical terms, one can prevent serious frost damage to plants caused by late spring or early autumn frosts by spraying the plants with water early in the morning when the sun comes up. Another expedient with seedlings or new transplants is to cover the plants with paper "hot caps" or one-gallon plastic milk jugs whose caps are removed and bottoms cut off. Covers are removed when danger of frost is over and during warmer periods of the day. A much more expensive treatment to prevent frost damage to flowers or fruits of orchard trees is to use smudge pots (round pots containing oil, which is burned to produce a dense smoke for the purpose of protecting plants against frost) spaced at appropriate intervals in the orchard. A less expensive treatment, useful where trees are relatively small and few in number, is to cover them with a blanket or plastic film during frosty nights.

Bulbs and corm-producing plants native to temperate regions require a chilling period in order for flowering to occur. Low temperatures of winter accomplish this. One can take a cue from this observation in order to force these types of plants into flower. In essence, the forcing treatment is first to chill for a given period of time (usually six to eight weeks), then to subject the plants to warm temperatures which stimulates stem elongation and further flower development. As a result of forcing treatments, one can have bulbs in flower indoors during the cold, dark days of winter in temperate latitudes from December through March (see Chapter 6).

Twigs of many flowering shrubs or trees can also be forced into bloom during the winter months. They will only force if they have experienced chilling temperatures below freezing outdoors for at least a couple of months before the twigs are brought indoors. Plants that work well for forcing include forsythia, apple, and quince. You can also force twigs of catkin-producing plants such as willow, birch, aspen, and oak.

One can take advantage of the fact that winter wheat or winter rye requires a low temperature period (called a *vernalization* or low temperature exposure) to flower the following spring. This means that they are winter-hardy in temperate regions. Thus, we can plant winter rye or wheat in the fall and plow or rototil it under in the spring as a "green manure" or cover crop to build up soil humus (see Chapter 4).

## REFERENCES

Bickford, E. B., and Dunn, S. 1972. *Lighting for plant growth.* Kent, Ohio: Kent State University Press. (Covers major topics on characteristics of light, the fundamentals of photochemistry, electrical terminology, light measurement and control, effects of light on plant processes such as photosynthesis, phototropism, and photomorphogenesis, spectral design of plant growth lamps, growth room lighting, and horticultural lighting.)

Brooklyn Botanic Garden. 1970. *Gardening under artificial light.* Plants and Gardens 26:1–65. (A good primer on how to grow plants indoors under artificial lights. Good on specific responses of seedlings and on flowering in houseplants.)

Gaines, R. L. 1977. *Indoor plantscaping.* New York: Architectural Record Books. (Covers houseplants and their use in indoor landscaping and the use of lights over plants in the indoor landscape.)

Kaufman, P. B. 1953. Gross morphological responses of the rice plant to 2,4-D. *Weeds* 2 (4):223–53. (Article which covers over 20 different symptoms of damage caused to rice plants in the field when sprayed with the herbicide, 2,4-D, to control aquatic dicot weeds in rice paddies.)

Kaufman, P. B., and LaCroix, J. D. 1979. *Plants, people, and environment.* New York: Macmillan Co.

Kranz, F. H., and Kranz, J. 1971. *Gardening indoors under lights.* Rev. ed. New York: Viking Press. (A basic text on how light controls plant growth and is especially good on applications for growing houseplants under artificial lights.)

Kramer, J. 1974. *Plants under lights.* New York: Simon and Schuster. (Similar to previous text.)

Leopold, A. C., and Kriedemann, P. E. 1975. *Plant growth and development.* New York: McGraw-Hill Book Co. (A basic textbook which covers assimilation and growth in plants, growth regulation by different types of plant hormones, major aspects of plant development, and chemical modification of plant growth and development.)

Manaker, G. H. 1981. *Interior plantscapes.* Englewood Cliffs, N. J.: Prentice-Hall. (Similar to Gaines book above.)

Mellichamp, T. L. 1976. Forcing northern woody plants into flower. *The Michigan Botanist* 15 (4):205–14. (Excellent reference on how to force branches of woody plants into flower in the greenhouse to study their flowers and to use for interior arrangements in wintertime.)

Serle, S. A. 1973. *Environment and plant life.* London: Faber and Faber. (Covers basic material on soil temperature, moisture; solar radiation and day-length; humidity, air temperature, wind, precipitation, plant transpiration, frost, winter dormancy in plants, protected cultivation, juvenility in plants, and the soil.)

Thimann, K. V. 1977. *Hormone action in the whole life of plants.* Amherst, Mass.: University of Massachusetts Press. (Covers in an excellent historical fashion the major effects of hormones on developmental processes in plants and on the mechanisms of action of plant hormones so far as they are known.)

*Time-Life Encyclopedia of Gardening.* 1972–1980. Gardening under lights. New York: Time-Life Books. (Similar to Kranz and Kranz text.)

U.S. Forest Service. 1974. *Seeds of woody plants of the United States.* U.S. Department of Agriculture Handbook No. 450. Washington, D.C.: U.S. Govt. Printing Office. (Good discussion of the primary temperature requirements for breaking dormancy in seeds and for bringing about their germination for woody plant species.)

van der Veen, R., and Meijer, G. 1959. *Light and plant growth.* Eindhoven, The Netherlands: Philips' Technical Library. (Covers basic studies on the effects of light on plant growth and development and presents many applications for the use of artificial light in growing horticultural plants.)

Wareing, P. F., and Phillips, I. D. J. 1981. *Growth and differentiation in plants.* New York: Pergamon Press.

Weaver, R. J. 1972. *Plant growth substances in agriculture.* San Francisco: W. H. Freeman and Co. (Covers the major uses of herbicides and growth promotors and retardants in agriculture.)

# Soils and their proper management 4

Soils are the basic materials of practical botany. They are extremely fragile and must be handled with care. Soil is the medium in which the roots grow and obtain nutrients and water. Consequently, soil condition can be one of the most important factors in successful plant growth. We need to learn about soil properties, how to modify soils to improve their fertility and structure. We also need to preserve them, to prevent their loss or deterioration through wind and water erosion, pollution, or mismanagement. So, let us "soil our hands" and examine the basic "stuff" in which plants grow.

## Soil contents

Soil texture depends on the proportion of sand, silt, and clay particles present. These particles are of various sizes. They are derived from weathered rock (Figure 4–1). Clay particles are the smallest, less than 0.002 millimeter in diameter. Silt particles are larger, 0.02 to 0.002 millimeter in diameter. Sand particles are the largest, 0.2 to 0.02 millimeter in diameter. Clay gives a soil the property of high water-holding capacity and high mineral nutrient-binding capacity. Sand makes the soil more porous, allowing for air penetration between the particles (aeration). Plants do not grow well when clay holds too much water and excludes oxygen. Then, too, clay can be as hard as brick and almost impossible to break up with a spade when it is dry. At the other extreme, sandy soils can dry out too quickly and must be watered more frequently to prevent plants from wilting. Stunted plants may result from essential plant nutrients being leached out of sand too quickly.

*FIGURE 4–1*
A vertical soil profile. At the top is the root zone soil; successive layers lead downward to parent rock. Weathering of the parent rock forms soil over a long period of time (an average of 1,000 years for 2.54 cm [1 inch] of soil).

The best way to avoid these problems is to enrich the soil with humus (decayed organic matter) from a compost source or animal manure. The addition of humus makes the soil much easier to cultivate, saves water, improves aeration, and adds nutrients for better plant growth. Soils, such as muck, contain 20 to 65 percent organic matter. Muck is derived from decayed twigs, roots, leaves, sedges, sphagnum moss and the like from old lake beds or river bottoms. Many vegetables are grown on such soils. Peat is derived from sphagnum moss and other plants which grow in acid bogs. During plant succession from wet bog to soil-plant fill-in, peat is formed primarily from the dead plants (see Chapter 10). Peat moss is frequently used as a soil amendment to increase water-holding capacity, improve soil texture, and to increase acidity (lower soil pH). Plants such as rhododendrons, azaleas, heather, blueberries, and cranberries grow well in acid peaty soils.

A third and important component for soils is represented by a large assemblage of microscopic organisms, bacteria, and fungi. Many of these creatures are essential because they are involved as decomposers in bringing about the breakdown of plant and animal organic matter to form humus. Leaf mold is humus formed from leaves in the forest floor litter that are decomposed by fungi. It is commercially packed for sale. Some bacteria and blue-green algae are involved in the very important nitrogen cycle, where nitrogen from the air is converted into a form (ammonia and other nitrogen compounds) usable to plants. Mycorrhizal fungi live symbiotically (mutualistically, in combination) with some plant roots and help provide greater nutrient and water uptake.

Some soil creatures, on the other hand, can be very obnoxious and destructive—such as nematodes, larvae, disease-causing bacteria and fungi, moles, and rodents. For example, when fresh manure, straw, or a green cover crop (e.g., rye) is added to the soil, growth of the decomposers is stimulated. These decomposers use a large percentage of the soil nitrogen and make it unavailable for growth of garden flowers and vegetables.

Besides weathered rock particles, organic matter, and soil creatures of varying sizes, the fourth component of soil is air space. Porous, sandy soils possess far more air space than heavy clay soils. Air space is important in providing oxygen for plant root growth. Of course, some plants such as rice, cattails, and other water-loving plants grow best in water-logged or flooded soils. These plants obtain oxygen via air spaces in the leaves which extend through the stems to the roots.

## Soil nutrients

### ELEMENTS ESSENTIAL FOR PLANT GROWTH

Elements essential for plant growth that are required in relatively large amounts are called *macronutrients*. They include hydrogen (H), carbon (C), oxygen (O), nitrogen (N), phosphorus (P), potassium (K), sulfur (S), magnesium (Mg), and calcium (Ca). In contrast, those elements that are required in relatively small amounts are called *micronutrients*. They include manganese (Mn), boron (B), molybdenum (Mo), copper (Cu), chlorine (Cl), zinc (Zn), iron (Fe), and for some plants (grasses, sedges, scouring rushes) silicon (Si). These elements play many different roles in the normal functioning of the plant. The following are a few examples. Magnesium is a constituent of chlorophyll, the light-absorbing pigment. Potassium and chlorine help to regulate the opening and closing of plant pores (stomata). Sulfur is a constituent of some of the amino acids which build proteins. Phosphorus is a constituent of essential cellular compounds involved in heredity (the nucleic acids—DNA and RNA) and energy (ATP) for plant processes. Molybdenum, potassium, and manganese are important regulators of enzyme activity. Copper and iron are constituents of enzymes. Nitrogen is a part of many important compounds in the plant, such as proteins, nucleic acids, several plant hormones, and plant alkaloids. Carbon, hydrogen, and oxygen occur abundantly in plant carbohydrates and proteins. Carbon and hydrogen are present in plant fats and oils (lipids). Each of these elements is very important in the normal growth and functioning of plants.

## SOIL pH AND NUTRIENT AVAILABILITY

The pH range extends from 1 to 14. It is a measure of the acidity or alkalinity of the soil. For most soils the pH range of importance for plant growth is 4 to 8. Seven is neutral, indicating neither an acidic nor a basic condition. Above 7, pH's are basic (alkaline), while pH's below 7 are acidic.

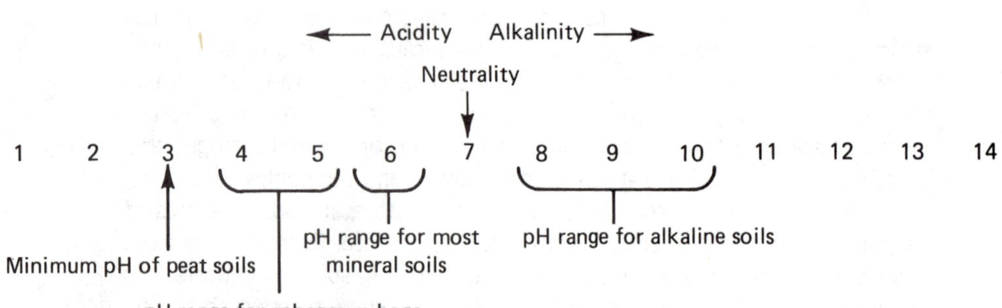


On the scale, a pH of 3 is ten times more acid than a pH of 4 and one hundred times more acid than a pH of 5; hence, this is a logarithmic scale.

The elements carbon, hydrogen, and oxygen are abundant in gaseous form in air absorbed by plants. The other macronutrients, in general, are available for uptake from the soil by plant roots over a wide pH range from pH 4 to 9. In contrast, six of the micronutrients have much narrower pH ranges of availability. Availability means the elements are present in the soil in soluble form and are not tied up as insoluble compounds. Iron, manganese, copper, and zinc are most readily available at pH's below 7, whereas molybdenum is more available at low and high ends of the pH scales: 3 to 7 and 9 to 14. When an element is unavailable, it is usually in an insoluble, precipitated form and only goes into solution at favorable pH's. For an element in the soil to be absorbed by a plant's roots, the element must be dissolved in soluble form in the soil solution.

It is very important to have a soil test done before beginning gardening. A soil sample may be sent to the state department of agriculture, or a simple home soil test kit can be purchased to use. Knowing the pH and the relative amounts of the major elements will help you determine which nutrient additions are needed for an optimally fertile soil for a particular crop. Table 4–1 indicates pH preferences for various crop and garden plants.

Altering soil pH is a relatively simple job. If your tests indicate the soil to be too alkaline (a high pH), you will need to make it more acid. This can be achieved using inorganic salts such as aluminum sulfate and iron sulfate

*TABLE 4–1*
*pH Preferences for various crop and garden plants*

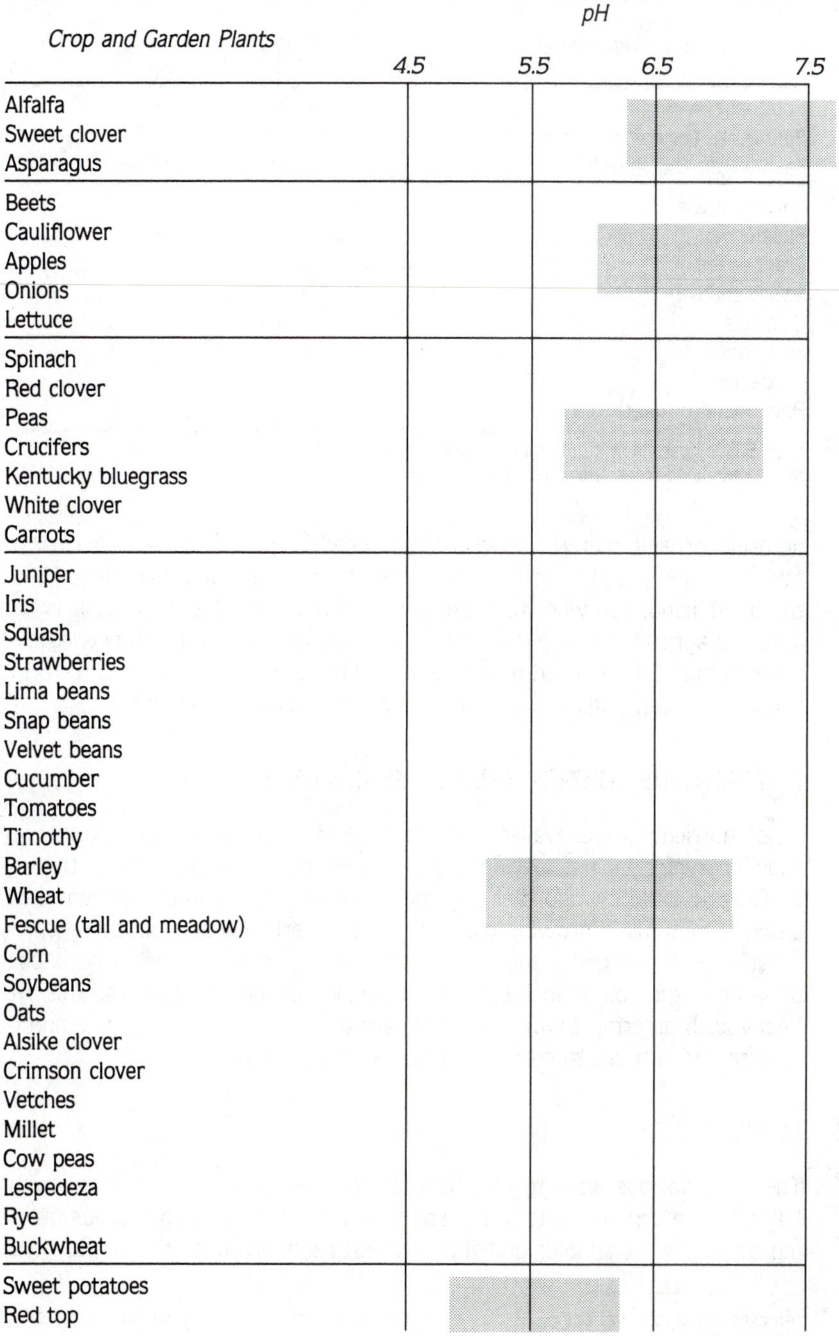

| *Crop and Garden Plants* | *pH* 4.5 | 5.5 | 6.5 | 7.5 |
|---|---|---|---|---|
| Alfalfa<br>Sweet clover<br>Asparagus | | | | |
| Beets<br>Cauliflower<br>Apples<br>Onions<br>Lettuce | | | | |
| Spinach<br>Red clover<br>Peas<br>Crucifers<br>Kentucky bluegrass<br>White clover<br>Carrots | | | | |
| Juniper<br>Iris<br>Squash<br>Strawberries<br>Lima beans<br>Snap beans<br>Velvet beans<br>Cucumber<br>Tomatoes<br>Timothy<br>Barley<br>Wheat<br>Fescue (tall and meadow)<br>Corn<br>Soybeans<br>Oats<br>Alsike clover<br>Crimson clover<br>Vetches<br>Millet<br>Cow peas<br>Lespedeza<br>Rye<br>Buckwheat | | | | |
| Sweet potatoes<br>Red top | | | | |

(Continued)

*TABLE 4–1 (Continued)*
*pH Preferences for various crop and garden plants*

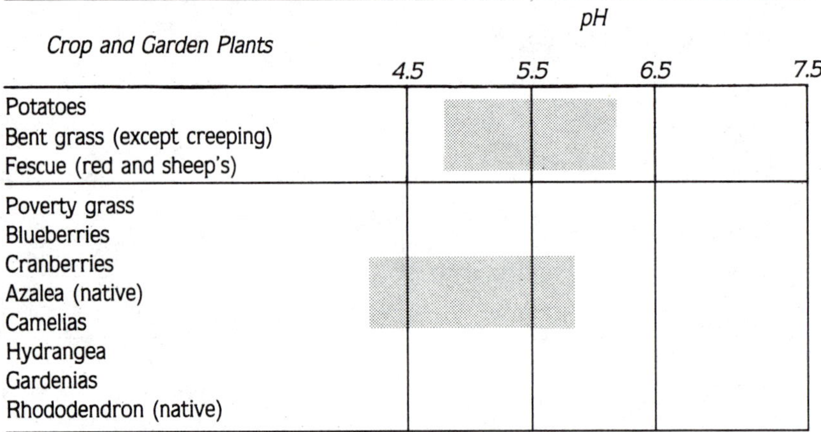

| *Crop and Garden Plants* | *pH* 4.5 | 5.5 | 6.5 | 7.5 |
|---|---|---|---|---|
| Potatoes<br>Bent grass (except creeping)<br>Fescue (red and sheep's) | | | | |
| Poverty grass<br>Blueberries<br>Cranberries<br>Azalea (native)<br>Camelias<br>Hydrangea<br>Gardenias<br>Rhododendron (native) | | | | |

*Note:* Shaded bars indicate pH ranges most suitable for growth.
*Source:* Modified from Janick 1979.

or using organic materials such as pine needles, peat moss, oak leaves, or sawdust. The inorganic salts work faster than the organic materials. If, on the other hand, you want to make your soil more alkaline, because it is too acid, use agricultural lime (calcium oxide). This also adds a much-needed element, calcium. If you need to add calcium without raising the pH, as for acid-loving rhododendrons and azaleas, add calcium sulfate (gypsum).

### NUTRIENT DEFICIENCY SYMPTOMS IN PLANTS

When nutrients are unavailable or are in short supply in the soil, plants may show characteristic nutrient deficiency symptoms. These are listed in the key in Table 4–2. It is important for you to be able to diagnose nutrient deficiency symptoms in growing plants so that corrective action can be taken by modifying soil pH or by adding the appropriate fertilizers. Fertilizing is best done after the soil is analyzed for suspected nutrient deficiencies. This will help you to determine which nutrients are deficient and how much to add of a particular nutrient to correct the deficiency symptoms.

## Adding soil nutrients

### FERTILIZERS

There are various ways to maintain soil for best plant growth. It is always helpful to return nutrients to the soil that are removed by continuous growing of annual[1] vegetables and flowers. Fertilizers are used to add nutrients

[1]Annuals are plants that complete their life cycle (seed germination to flowering and seed formation) in a year or less.

*TABLE 4–2*
*A key to plant nutrient deficiency symptoms*

| *Symptoms* | *Element Deficient* |
|---|---|
| 1. Older or lower leaves of plant mostly affected; effects localized or generalized | |
| 2. Effects mostly generalized over whole plant; more or less drying or firing of lower leaves; plant light or dark green | |
| 3. Plant light green; lower leaves yellow, drying to light brown color; stalks short and slender if element is deficient in later stages of growth .......... | Nitrogen |
| 3. Plant dark green, often developing red and purple colors; lower leaves sometimes yellow, drying to greenish brown or black color; stalks short and slender if element is deficient in later stages of growth .......... | Phosphorus |
| 2. Effects mostly localized; mottling or chlorosis with or without spots of dead tissue on lower leaves; little or no drying up of lower leaves | |
| 4. Mottled or chlorotic leaves, typically may redden, as with cotton; sometimes with dead spots; tips and margins turned or cupped upward; stalks slender .......... | Magnesium |
| 4. Mottled or chlorotic leaves with large or small spots of dead tissue | |
| 5. Spots of dead tissue small, usually at tips and between veins, more marked at margins of leaves; stalks slender .......... | Potassium |
| 5. Spots generalized, rapidly enlarging, generally involving areas between veins and eventually involving secondary and even primary veins; leaves thick; stalks with shortened internodes .......... | Zinc |
| 1.′ Newer or bud leaves affected; symptoms localized | |
| 2.′ Terminal bud dies, following appearance of distortions at tips or bases of young leaves | |
| 3.′ Young leaves of terminal bud at first typically hooked, finally dying back at tips and margins, so that later growth is characterized by a cut-out appearance at these points; stalk finally dies at terminal bud .......... | Calcium |
| 3.′ Young leaves of terminal bud becoming light green at bases, with final breakdown here; in later growth, leaves become twisted; stalk finally dies back at terminal bud ..... | Boron |
| 2.′ Terminal bud commonly remains alive; wilting or chlorosis of younger or bud leaves with or without spots of dead tissue; veins light or dark green | |
| 4.′ Young leaves permanently wilted (wither-tip effect) without spotting or marked chlorosis; twig or stalk just below tip and seedhead often unable to stand erect in later stages when shortage is acute .......... | Copper |

(Continued)

## TABLE 4–2 (Continued)
## A key to plant nutrient deficiency symptoms

| Symptoms | Element Deficient |
|---|---|
| 4.′ Young leaves not wilted; chlorosis present with or without spots of dead tissue scattered over the leaf | |
| 5.′ Spots of dead tissue scattered over the leaf; smallest veins tend to remain green, producing a checkered or reticulated effect ........ | Manganese |
| 5.′ Dead spots not commonly present; chlorosis may or may not involve veins, making them light or dark green in color | |
| 6.′ Young leaves with veins and tissue between veins light green in color ........ | Sulfur |
| 6.′ Young leaves chlorotic, principal veins typically green; stalks short and slender ........ | Iron |

*Source:* McMurtrey. 1950. *Diagnostic techniques for soils and crops.* American Potash Institute. Modified from Kaufman et al. 1975.

to the soil or to correct nutrient deficiencies in plants. They can be applied to the plant directly as a foliar spray in highly soluble form, or they can be added to the soil in granular or pelleted form. Fertilizers can also be added to irrigation water. When trickle irrigation is employed, for example, the water-fertilizer solution is directed to the bases of plants through the use of perforated plastic hoses.

Organic gardeners and farmers prefer to add nutrients in organic form to ensure that the nutrients are released slowly. This also serves to conserve energy because the manufacture of inorganic fertilizers requires high energy inputs. Inorganic fertilizers, in general, are more soluble and quicker acting. They are now available in pellets, which release the nutrients more slowly. A typical N-P-K (nitrogen-phosphorus-potassium) fertilizer, listed as 10-10-10, means that it contains 10 percent total nitrogen, 10 percent available phosphoric acid, and 10 percent water-soluble potash. The other 70 percent is made up of inert matter or filler.

Table 4–3 gives a comparison of typical organic and inorganic forms of N, P, and K fertilizers commonly applied to soil while plants are actively growing. Nitrogen is generally the easiest element lost from the soil, but all three, N, P, and K, are fairly easily leached with repeated waterings. Because they are so essential, they must be provided to the actively growing plant at regular intervals.

## NATURAL NITROGEN SOURCES

Nitrogen fixation is the conversion of gaseous nitrogen ($N_2$) in the air into ammonia ($NH_3$) and eventually organic compounds containing nitrogen. While nitrogen gas cannot be used by plants, ammonia and its products can be ab-

*TABLE 4–3*
*Fertilizer sources of nitrogen, phosphorus, and potassium*

| *Type of Fertilizer* | *N (Nitrogen)* | *P (Phosphorus)* | *K (Potassium)* |
|---|---|---|---|
| Inorganic fertilizers | Liquid ammonia $NH_3$<br>Ammonium nitrate $(NH_4) NO_3$<br>Potassium nitrate $KNO_3$ | Superphosphate<br>Rock phosphate | Muriate of potash<br>Greensand (processed algae) |
| Organic fertilizers | Urea<br>Manure<br>Processed sewage<br>Fish emulsion<br>Blood meal | Bone meal | Wood ashes |

sorbed. The pathway from $N_2$ to $NH_3$ involves many steps and is achieved by nitrogen-fixing microorganisms such as certain blue-green algae (*Nostoc* and *Anabaena*) and bacteria (*Rhizobium* and *Azospirillum lipoferum*).

Legumes, both woody and herbaceous, are capable of carrying out nitrogen fixation via bacterial partners (symbionts) that live in nodules on their roots. The herbaceous legumes include soybeans, lupine, the clovers, lespedeza, birds-foot trefoil, alfalfa, and vetch. Woody leguminous plants include black locust, Kentucky coffee tree, golden rain tree, redbud, and many others. Clovers are useful in the lawn to provide nitrogen for the grass. Lupine is used to provide fixed nitrogen and shade to young pine and spruce tree seedlings planted on reclaimed land after strip mining. The clovers, alfalfa, and vetch are used as soil cover crops in rotation with food crops such as corn or wheat. The cover crops are plowed under at the end of their rotation cycle and thus provide humus (green manure) and fixed nitrogen to the soil for the next food crop in rotation.

As a practical matter, when sowing seeds of herbaceous legumes such as vetch, clovers, birds-foot trefoil, alfalfa, and lespedeza, it is important to add to the seeds the correct strain of *Rhizobium* bacterium as an inoculant. Each of these crops has a specific strain of *Rhizobium* which is able to elicit a successful symbiotic association in the roots of its host plant. Inoculants can be purchased at most seed stores and from seed companies.

Leguminous trees, such as black locust, are used in conservation plantings and in forests where the soil is poor or has been depleted by erosion. Three other leguminous trees, mesquite *(Prosopis)*, acacia, and *Pithecelobium* are grown as food for livestock or humans because of the high nitrogen content in seeds and pods (up to 69 percent protein in mesquite pods). These plants are being proposed for culture in "orchards" as an agricultural system requiring minimal input of fossil fuel, capital, and machinery but producing a relatively high source of seed protein (16 to 69 percent). (See Felker and Bandurski, 1979.)

As a final example of nitrogen fixation, we cite the instance where rice farmers in Viet Nam, China, the Philippines, and California are exploiting the use of the floating aquatic fern, *Azolla*. This fern fixes nitrogen via a blue green alga symbiont, *Anabaena. Azolla* is grown in rice paddies for several weeks where it doubles its volume (biomass) every two to three days during the growing season. It is then plowed into the soil as a "green cover crop." The nitrogen fixed by the *Anabaena* in the *Azolla* plants is then released in the paddy, available for rice plants which are grown next. This is a far cheaper way of providing nitrogen for the rice than through use of the much more expensive commercially produced urea or liquid ammonia. Also, the addition of more humus to the soil from the *Azolla* plants improves the soil's physical structure (makes it more friable or easier to cultivate because of the coarser texture).

### WHEN TO FERTILIZE

Applications of fertilizers should be timed to coincide with periods of active plant growth, both for houseplants and plants grown outdoors. One rule of thumb is not to apply fertilizers to perennials, shrubs, and trees when they are dormant or about to enter dormancy in the fall, as this may stimulate growth, and plants can be injured by early frost. If plants are fertilized optimally during the active growing season, they will enter winter dormancy in a much healthier condition. Another rule of thumb is that when indoor plants, such as cacti and succulents, are not growing actively, they should not be fertilized; otherwise, they may be killed due to the accumulation of the fertilizer salts in the roots and shoots. Potted plants can be flushed periodically with pure (distilled) water to wash out accumulated fertilizer and salts.

As a general rule, nitrogen compounds stimulate vegetative growth of stems and leaves of plants, while phosphorus stimulates flowering and potassium root growth. Use of varying ratios of N, P and K, such as 10-10-10 or 5-10-5, may make some differences depending on the stage of growth the plant is in.

## Improving soil condition

### COMPOSTING

One of the primary ways to increase the organic matter content of garden soils is to add compost. Compost improves the water-holding capacity of the soil, increases the population of beneficial microorganisms such as bacteria and fungus decomposers, helps to prevent excessive soil drying and caking, and improves soil aeration. It is thus one of the mainstays of gardeners and farmers who want to improve their soils.

Composting involves the breakdown of organic materials such as leaves, grass clippings, and straw by bacteria and fungi, the decomposers in the food chain. They work best when the compost pile is moist and well aerated. Periodic turning over of the different layers in a compost pile adds oxygen. The decomposers, in breaking down organic matter, cause the generation of heat in a compost pile. At these warm temperatures, decomposition proceeds rapidly. It is accelerated by the addition of moisture, layers of soil (containing microorganisms), and fertilizer to the compost pile. The fertilizer and moisture help the decomposers to multiply more rapidly. You can obtain humus from a compost pile in several months' time, in the same growing season that you make it.

*MAKING A COMPOST PILE.* When constructing a compost pile, use alternate layers of vegetation and soil; add some rotted animal manure to the vegetation layers to help accelerate decomposition (Figure 4–2). In addition,

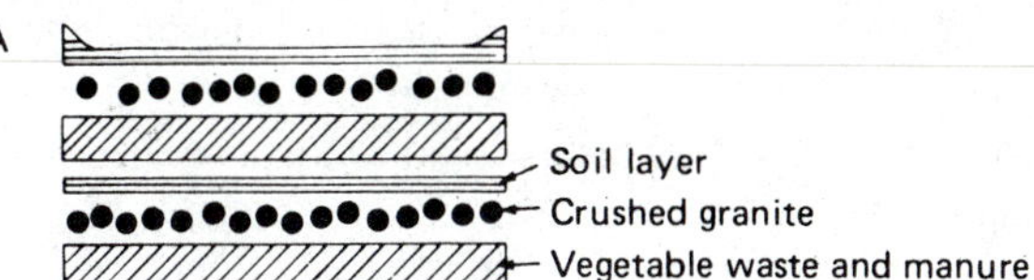


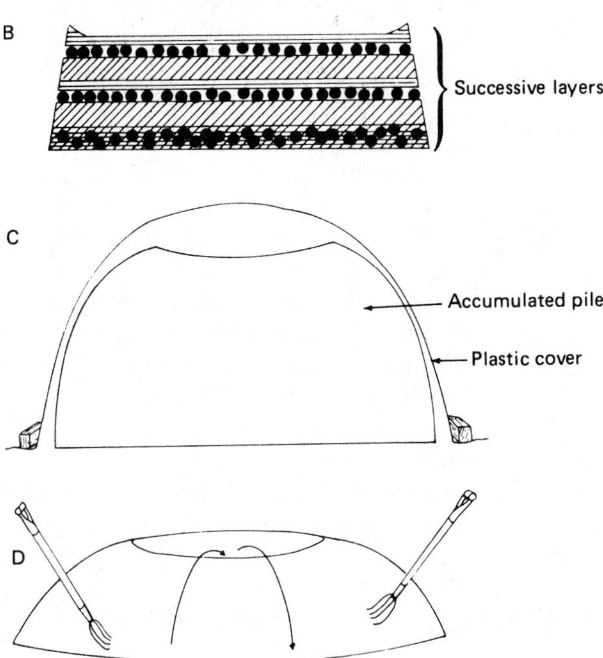


*FIGURE 4–2*
Making a compost pile. (A) An area is enclosed, and layer by layer of compost is built up. (B) An unenclosed compost pile with sloping sides and a concave depression at the top to trap water. (C) A rounded compost pile, covered with plastic to retain heat generated by microorganisms. (D) To provide aeration to microorganisms, the compost pile is periodically turned over.

crushed granite can be sprinkled between the soil and vegetation layers to enhance the phosphate content of the final humus product. Large plant stems such as sunflowers, goldenrod, or Jerusalem artichokes can be added above the soil layer to improve aeration. The oxygen will accelerate the breakdown process by the fungal and bacterial decomposers.

After you build up the successive layers to a manageable height, usually about one to one and a half meters, make the top concave to trap rainwater. You can hasten the rate of decomposition by covering the compost pile with black plastic which is heat absorbant and helps to retain heat generated by the decomposing vegetation. (Black surfaces absorb more heat than light surfaces.) Decomposition rate is also speeded up by using a shredder to fragment the plant debris you add to the compost pile. This increases the surface area of the material on which the organisms will work. One can also add garbage such as egg shells, coffee grounds, and vegetable matter (avoid meat scraps) from the kitchen as well as leaves and grass clippings to the compost pile.

The humus that one obtains from a compost pile is usually low in nutrient value, especially of N, P, K, and Ca. To overcome this, add to a compost pile certain supplements to provide these macronutrients: manure for nitrogen, rock phosphate or bone meal for phosphorus, ashes or greensand for potassium, and lime for calcium. Leaves and other vegetable matter provide adequate micronutrients to the compost pile.

*USING COMPOST.* Finished compost, as humus, has many uses in the garden. One of the chief uses is to improve the soil structure, to make it easier to cultivate if very heavy clay, or to hold more mositure if very sandy. Another use of compost is for mulch, which provides winter protection or helps to conserve moisture near the soil surface. As a mulch, compost is also used to curtail growth of weeds, provided it has been properly pasteurized and is not full of weed seed. Compost is used in preparing seedbeds for lawns and in making up potting soil mixes for plants in baskets, pots, tubs, or planter boxes. Compost is also employed to advantage when mixed in the soil used to fill holes made for planting new trees and shrubs. It provides nutrients and improves moisture-holding capacity.

Finally, compost can serve as a "top-dressing" around newly planted trees and shrubs to trap surface moisture, curtail weed growth, provide nutrition, and improve appearance. This natural surface cover around these plants is less expensive and much more attractive than gravel. One alternative to compost, if you prefer, is using longer lasting wood chips as a top-dressing.

## MULCHES

Mulches are also beneficial to soil. They prevent excessive growth of competing weeds, help prevent fruit rot when used around tomatoes, squash, melons, and cucumbers, retain soil moisture, and help prevent soil erosion. Besides compost, there are many other types of mulches, both synthetic and natural. The synthetic ones include plastic film, gravel and stones used separately or in combination. Plastic film is a temporary mulch, especially useful in strawberry beds; but being synthetic, it does not break down readily. It is very effective in preventing weed growth, however. Gravel and stones have the advantage of withstanding a great deal of wear and persisting for a long time. They are especially useful in urban areas around street trees, in medians, and along curb strips and parking lots.

However, for the home gardener, natural mulches are generally cheaper and aesthetically more pleasing. They also can be recycled into the soil. The types of natural mulches are manifold: sawdust, pine needles, leaves, compost, grass clippings (without weed killer), buckwheat hulls, cocoa bean hulls, rice hulls, peanut hulls, newspaper, straw or hay, excelsior, peat moss, corn stalks, corncobs, wood chips, and fir boughs. Pine needles are especially useful around acid-loving plants such as blueberries, azaleas, rhododendrons, huckleberries, and heather. Probably one of the cheapest and most readily available mulches is derived from deciduous tree leaves. They are especially useful for winter protection and readily break down to form humus the following growing season. Some leaves, such as those of oak, tend to provide a more acid mulch than others such as maple.

Newspaper, anchored down with stones, sawdust, or large clumps of soil, can serve as an excellent seasonal mulch under cucumbers, melons, and squash (Figure 4–3). It does not provide much humus, but it very effectively helps to conserve soil moisture and curtail weed growth. Straw is one of the

*FIGURE 4–3*
Mulching a home vegetable garden. Between the rows have been placed layers of newspaper which have been covered over with sawdust, both of which will eventually decompose and add to the organic matter content of the soil. Alternatively, one can anchor the newspaper with stones or large clumps of soil instead of with sawdust.

other relatively cheap mulches. It has high insulating capacity and is thus good for winter protection around roses, nut and fruit trees, and other plants. However, it can be a fire hazard, and it can tie up a lot of nitrogen in the soil due to use by microorganisms that multiply rapidly while breaking down straw to humus.

## Soil mixes for houseplants

Many natural and commercially made materials are used to improve potting soil or are used alone or in various combinations in soilless mixes. Peat moss and leaf mold are used to add organic matter to soil. Perlite is a pumicelike material used to lighten the soil and to provide aeration. It has essentially no nutrients. Vermiculite (heat-expanded mica) improves soil water-holding capacity. It, too, has no nutrients. Vermiculite plus perlite is often used as a medium to root cuttings, to germinate seeds, and as a soil conditioner. These substances are naturally sterile and essentially inert. When using soilless mixes, one has to be careful to fertilize regularly. In soilless mixes, peat is used to simulate clay; perlite is used to simulate sand; and vermiculite is used to simulate humus. Loam is composed of a proportional mixture of clay, silt, sand, and some organic matter.

Some plants such as epiphytes do not grow well in soil. These are "air plants" that grow on the trunks of trees and shrubs in their native habitats. Epiphytic bromeliads and orchids are thus grown in well-aerated media such as osmunda fiber (roots of osmunda fern) or fir bark plus perlite.

Table 4–4 indicates some basic soil mixes that are important for growing different groups of houseplants. Table 4–5 is a listing of components that are used to prepare sterile, soilless mixes for growing garden plants, vegetable seedlings, and flowers (mix A), foliage houseplants (mix B), and epiphytes (mix C).

## Soil water

### TYPES OF SOIL WATER

Soil water supply can be one of the major limiting factors for plant growth, particularly during the growing season under conditions of high air and soil temperature. In the soil, water that percolates freely through the soil is called *gravitational water*. It is usually not available for plant growth except in heavy soils. Water which is held by the force of adhesion between soil particles is called *capillary water;* this water is primarily available for plant growth. Water which is tightly bound to soil particles and unavailable for plant growth is called *hygroscopic water*.

When drought occurs, the capillary water can be exhausted from the soil through uptake by and evaporative loss (transpiration) from the plants. When the plants permanently wilt under these conditions and will not recover, even when watered, then one reaches what is called the PWP or *permanent wilting percentage* for a soil. It varies between 3 and 15 percent water (dry weight basis), the PWP being much higher for clay soils than for

*TABLE 4–4*
*Basic soil mixes for different types of houseplants*

*General potting soil:*
1 part peat
1 part sand (can use perlite)
1 part loam
To make good loam from the clay soil, use the following mixture:
2 parts soil
1 part leaf mold
1 part peat
This mixture should be sterilized before using in potting media as loam. You can do this at home by baking the mix in a flat pan in a slow oven for several hours (93° C or 200° F)

*Cacti and succulents* (except Christmas cactus which needs the general potting soil mix):
2 parts loam
1 part sand

*Most tropical plants:*
1 part peat
1 part leaf mold
1 part sand
1 part loam

*Ferns:*
2 parts leaf mold
2 parts loam
1 part sand and perlite
1 part peat

*Propagating pot for cuttings:*
1 part peat or vermiculite
1 part perlite

*Orchids and bromeliads:*
Fir bark or osmunda fiber for epiphytes; finely chopped fir bark, soft brown osmunda, or any mixture which drains well but retains moisture for terrestrial orchids; and normal potting soil for terrestrial bromeliads (see general potting soil above)

sandy soils because clay particles are much smaller than those of sand, and thus, have a higher percentage of water that is held hygroscopically.

## USING WATER

Clearly, one must understand the water requirements of each type of plant that is to be grown indoors or outdoors before trying to manipulate the water supply to improve vegetative growth. Cacti, for example, require very little water in general, and during their winter dormant season, they have very little need for water, maybe once a month. But when they are actively growing in the spring and summer, they need to be thoroughly watered once a week. Other plants respond favorably to more frequent watering such as impatiens, sunflowers, rhubarb, raspberries, squash and cucumbers, and newly planted trees and shrubs.

Watering should be done in such a way that it is *directed to the roots,* as for example with trickle irrigation for row crops. Here, the water is held in the soil in proper amounts over as long a period as possible (as in soil well supplied with humus). It is *applied in sufficient amounts* to penetrate deep to the roots and not just run off or shallowly penetrate the surface, as occurs with short-term sprinkling. Even in late fall or early winter, watering may be necessary in temperate regions when there is insufficient water in the soil to

*TABLE 4–5*
*Sterile, soilless mixes (Cornell University peatlike mixes for 35 liters or one bushel)*

| *Mix Component* | *Mix A*[1] | *Foliage Plant*[2] *(Mix B)* | *Epiphytic*[3] *(Mix C)* |
|---|---|---|---|
| Sphagnum peat moss | 17.5 liters (½ bushel) | 17.5 liters (½ bushel) | 11.7 liters (⅓ bushel) |
| Vermiculite | 17.5 liters (½ bushel) | 8.7 liters (¼ bushel) | — |
| Perlite (medium grade) | — | 8.7 liters (¼ bushel) | 11.7 liters (⅓ bushel) |
| Douglas fir bark (⅛″ to ¼″) | — | — | 11.7 liters (⅓ bushel) |
| Ground dolomitic limestone | 74 ml (5 tbsp) | 118.4 ml (8 tbsp) | 118.4 ml (8 tbsp) |
| 20 percent superphosphate | 29.6 ml (2 tbsp) | 29.6 ml (2 tbsp) | 88.8 ml (6 tbsp) |
| 10-10-10 fertilizer | 44.4 ml (3 tbsp) | 44.4 ml (3 tbsp) | 44.4 ml (3 tbsp) |
| Potassium or calcium nitrate | — | 14.8 ml (1 tbsp) | 14.8 ml (1 tbsp) |
| Granular wetting agent | 44.4 ml (3 tbsp) | 44.4 ml (3 tbsp) | 44.4 ml (3 tbsp) |
| Liquid wetting agent | 4.8 ml (1 tsp) | 4.8 ml (1 tsp) | 4.8 ml (1 tsp) |
| Soluble trace element mix[4] | | | |

[1]Mix A is recommended for general garden plants, for vegetable seedlings, and for flowers.

[2]Foliage Plant Mix B is recommended for those plants which need growing media with high moisture-retention characteristics. Plants having a fine root system or possessing many fine root hairs are included in this group. Some examples are: *Amaryllis, Aphelandra, Begonia, Beloperone, Buxus, Caladium, Cissus, Citrus, Coleus,* ferns, *Ficus, Hedera, Helxine, Maranta, Oxalis,* palms, *Pilea, Sansevieria,* and *Tolmiea.*

[3]Epiphytic Mix C is recommended for plants that require good drainage and aeration and have the ability to withstand drying between waterings. Plants having coarse, tuberous, or rhizomatous roots are included in this category. For example: African violets, *Aglaeonema, Aloe,* bromeliads, cacti, *Crassula, Dieffenbachia, Episcia,* geraniums, gloxinias, *Hoya, Monstera, Nephthytis, Peperomia, Philodendron, Pothos, Syngonium.*

[4]Trace elements are needed in extremely small quantities. Overapplication can result in severe injury to the plant. Prepare a stock solution by dissolving one ounce of Soluble Trace Element Mix in one quart of water. Use one-third cup of stock solution added to sufficient volume of water (one gallon) to get good distribution in one bushel of mix. Do not add more than this amount. Do not repeat the application.

compensate for that lost by evaporation on bright sunny days. This is especially critical for established trees and shrubs in foundation plantings around a house or building, where the soil tends to dry out.

## CONSERVING SOIL WATER

There are several ways to conserve water in the soil. The first involves making "saucers" (shallow depressions) in the soil to hold water around trees and shrubs following rainfall or irrigation. The second involves laying down straw, plastic, or newspaper mulch between rows or hills of plants such as strawberries, squash, pumpkins, and cucumbers. Third, it is far better to use "deep watering" less often than "one-shot" sprinkling. Plastic water soakers and use of trickle irrigation have been shown to be much more effective for plant growth, water conservation, and as a means of achieving deep watering. *Trickle irrigation* is water provided by perforated pipes, mostly plastic, that permit the water to trickle out slowly only where the plants are growing. As mentioned previously, with trickle irrigation one can add fertilizers directly in the irrigation system. Filters are necessary to prevent plugging of the orifices when water leaves the pipes. Trickle irrigation is most useful for beds enclosed by concrete borders or railroad ties, row crops, orchards, and nurseries. Water soakers (perforated plastic flat strips) work best for most home gardens.

Trickle irrigation by perforated pipes or soakers conserves water by getting it directly into the soil near the bases of the plant instead of allowing it to be dissipated into the atmosphere, as occurs with overhead sprinkler irrigation or irrigation ditches. Trickle irrigation also cuts down on the spread of airborne spores of disease-causing fungal and bacterial pathogens, thus reducing the incidence of disease.

A fourth way to conserve water is to grow flowers and vegetables in raised beds of soil (usually 20 to 30 centimeters high and 1.0 to 1.5 meters wide) that are filled with humus from compost. Such beds are well aerated and advantageous over clay soils where one can have too much water present, or over sandy soils, where too little water may be present. They also conserve growing space normally used for paths or to divide row crops.

## WATER-LOGGED SOIL

Clay soils are notorious for their poor drainage. Many plants growing in such soils perform poorly, whereas others thrive. Cacti and succulents often perish in water-logged clay soils. On the other hand, plants such as pears, apples, roses, elderberries, and raspberries thrive in clay soils. Plants such as peaches, melons, cucumbers, grapes, asparagus, and strawberries tend to do better on well-drained sandy-loam soils.

For those plants that do poorly on clay soils, one can improve both the drainage and aeration. There are two methods that can be used to accomplish this. One is to add copious amounts of sand and compost (humus) to the clay soil. The other is to fill raised beds over the clay with four to six inches of a mixture of humus from compost and sand, or just humus alone, and to dig this in, using the double digging technique (digging to a depth of about 0.5 meter, first down one spade length, then another). This method is widely used in home vegetable gardens, in home landscapes, and on farms in modified form where row crops such as peppers, potatoes, lettuce, and onions are irrigated.

Some plants prefer "wet feet." In terrain with heavy clay soil, peat, or muck soil in depressions, it is possible to grow such plants. A boglike depression that has acid peat or muck (pH 4 to 5) is ideal for growing blueberries, cranberries, and huckleberries. Some soils can be made into modified miniature acid bogs with peat moss, sphagnum moss, pine needles and aluminum sulfate or sulfur to increase soil acidity. Heavy clay soils that tend to stay wet are good for growing edible wild food plants such as arrowhead *(Sagittaria)*, cattails *(Typha)*, and wild rice *(Zizania aquatica)*. A stream bank is a good place to grow watercress *(Nasturtium officinale)* and mints *(Mentha* spp.). A pond or lakeshore provides a suitable habitat for wild rice in the Great Lakes region and in almost any region for cattail.

## Preventing soil erosion

Soil must be protected from erosion, both wind and water. Some farmers who grow vegetables on peat soils use snow fences to prevent wind erosion. Others use windbreaks of tree plantings. Growing plants on the contour in hilly areas helps to prevent gully and sheet (surface) erosion caused by water. On hills, one must rely on heavy mulches, ground covers, and even dikes to prevent water erosion. Good soil binders include American beach grass *(Ammophila breviligulata)* for sandy dunelike soils and crown vetch for embankments of heavier soils. Kudzu vine was imported from Japan for retaining soil on slopes in the southeastern United States, but quickly got out of hand by "taking to the woods" and becoming a pernicious, highly competitive weed. The same can be said for Japanese honeysuckle, but it is not quite as aggressive as kudzu vine. Finally, grass sod can be used to retain soil on embankments.

In the garden, one can use different kinds of ground covers to prevent erosion and to improve the landscape. These include periwinkle *(Vinca minor, V. major)*, English ivy *(Hedera helix)*, Japanese spurge *(Pachysandra terminalis)*, lily of the valley *(Convallaria majalis)*, dichondra *(Dichondra)*, sandwort *(Arenaria verna)*, ajuga *(Ajuga reptans)*, and ice plant *(Mesembryanthemum)*. Dichondra, sandwort, and ice plant can only be planted in tropical and subtropical areas.

# REFERENCES



Brill, W. J. 1979. Nitrogen fixation: basic to applied. *Amer. Sci.* 67:458–66. (Article which covers the basic chemistry of nitrogen fixation, the microorganisms involved in capturing atmospheric nitrogen, and the utilization of nitrogen-fixing legumes, both woody and herbaceous, in agriculture.)

Brooklyn Botanic Garden. 1957. *Handbook on mulches.* Plants and Gardens 13(1): 1–79. (This practical handbook discusses why we use mulches, what kinds of mulches are available, and the best ways and times of year to apply them in the home flower and vegetable garden.)

Brooklyn Botanic Garden. 1956. *Handbook on soils.* Plants and Gardens 12(1):1–81. (Discusses the nature and properties of soils, how to keep them fertile, how to improve their structure, how to maintain soil moisture and test soil pH, and how to prevent soil erosion.)

Brooklyn Botanic Garden. 1975. *Natural gardening handbook.* Plants and Gardens 31(1):1–73. (A basic primer on how to use the basic methods of the organic gardener in the home garden—biological means of pest control, use of organic fertilizers, use of mulches, intercropping and crop rotation, composting, and soil improvement through use of compost, nitrogen-fixing legumes, and nontillage.)

Buckman, H. O., and Brady, N. C. 1969. *The nature and properties of soils.* 7th ed. New York: Macmillan Co. (One of the finest books available on soils and fertilizers. It covers the physical properties of soils, soil chemistry and pH, soil microorganisms, soil nutrients and nitrogen fixation, and soil improvement.)

Delwiche, C. C. 1978. Legumes—past, present, and future. *Bio Science* 28(9):565–70. (Covers the use of legumes to improve soil nitrogen supply through nitrogen fixation and the basic chemistry of the nitrogen-fixation process.)

Esau, K. 1977. *Anatomy of seed plants.* 2d ed. New York: John Wiley and Sons.

Evans, H. J., and Barber, L. E. 1977. Biological nitrogen fixation for food and fiber production. *Science* 197:332–39. (Covers same type of material as in Brill article cited above.)

Fahn, A. 1974. *Plant anatomy.* 2d ed. New York: Pergamon Press.

Felker, P., and Bandurski, R. S. 1979. Uses and potential uses of leguminous trees for minimal energy input agriculture. *Economic Botany* 33 (2):172–84. (An excellent article on the kinds and amounts of amino acids found in the seeds of leguminous trees; discusses also the characteristics of leguminous trees, their nitrogen-fixation potential in terms of percent protein in the seeds and the uses of such trees in "orchards" to produce high protein feed via the seeds.)

Janick, J. 1979. *Horticultural science.* San Francisco: W. H. Freeman and Co.

Jenny, H. 1980. *The soil resource: origin and behavior.* New York: Springer-Verlag. (An excellent recent text on how soils have their genesis from parent rock and from various types of erosion processes; also covers the basic properties of soils, both physical and chemical.)

Kaufman, P. B.; Labavitch, J.; Anderson-Prouty, A.; and Ghosheh, N. S. 1975. *Labortory experiments in plant physiology.* New York: Macmillan Co. (Includes a series of experiments in the major areas of plant physiology—structure and function of cell organelles; photosynthesis; fermentation and respiration; metabolites and enzymes; mineral nutrition; water, nutrient, and organic solute transport; plant growth and differentiation and their regulation by hormones, light, and temperature.)

Langer, R. W. 1972. *Grow it.* New York: Galahad Books. (Practical book on growing all kinds of plants in the home garden and on the farm and includes how to manage your soil properly, composting, legume crops, soil fertility and water conservation, and different types of cropping systems.)

Nisbet, F. J. 1979. Sources of compost. *Country Journal,* July 1979, 53–55. (One of the most complete articles available on the process of composting and the different kinds of materials one can add to the compost pile to make good humus supplemented with different sources of nitrogen, phosphorus, and potassium as well as other macronutrients and the micronutrients essential for plant growth.)

Peters, G. A. 1978. Blue green algae and algal associations. *Bio Science* 28 (9):580–85. (Discusses the role of the blue green algae, such as *Nostoc* and *Anabaena,* in nitrogen fixation in other plants, especially in the water fern, *Azolla.*)

Phillips, R. E.; Blevins, R. L.; Thomas, G. W.; Frye, W. W.; and Phillips, S. H.; 1980. No-tillage agriculture. *Science* 208:1108–13. (This article discusses the merits and disadvantages of no-tillage farming and gardening in terms of effects on soil structure, weed control, and crop yield. No-tillage agriculture is one of the methods used by the organic farmer to decrease energy inputs and to improve soil structure.)

Rodale, J. I. 1960. *The complete book of composting.* Emmaus, Pa.: Rodale Books. (One of the early books published on composting; covers the art and science of making a compost pile, how to improve its nutrient status, and how to use compost most effectively in the home garden.)

Shoji, K. 1977. Drip irrigation. *Scientific American* 237 (5):62–68. (This article discusses the merits of the drip irrigation process [trickle irrigation] in agriculture, the basic equipment needed for such systems, and the effects on soil structure and moisture and on plant growth and crop yield.)

Sprague, H. B., 1964. *Hunger signs in crops: A symposium.* New York: David McKay Co. (Covers the major signs of nutrient deficiency in plants; color plates illustrating nutrient deficiencies in different crops are superb.)

# Houseplants 5

Growing plants indoors is becoming a national pastime. There are many reasons for this renewed interest in plants. For example, plants add a touch of nature to our homes and brighten our indoor surroundings during the long winter months. Some large houseplants may actually become part of the interior design. Many people acquire plants as they would members of any group of such diverse and appealing objects, and certain plants may constitute status symbols as collector's items. Growing and caring for plants may also provide great horticultural therapy as these activities make one feel better and teach patience and responsibility. Anticipating the appearance of a spectacular flower gives one a sense of excitement. Some folks like the challenge of trying to get as many members of a certain group of plants as they can. Whatever the reason, the houseplant industry is booming like never before, and many new types of plants are being introduced to satisfy the demand.

Indoor gardening has many forms. Some people specialize in bonsai or dwarf plants. Others become addicts of African violet culture or orchid growing. Still others prefer foliage plants such as ferns, aroids, and begonias. Growing miniature flowering plants indoors under artificial lights has become a popular way of decorating a room. Growing plants in various kinds of hanging pots has become a widespread fad. To bring a little of the spring season indoors early, we can force bulbs such as hyacinths, narcissus, and crocus in the middle of winter. When spring finally comes, we often plant vegetable seeds indoors to get an early start on the outdoor growing season.

You can see that indoor gardening is one of the most fascinating aspects of practical botany.

Why take up indoor gardening? In the first place it can provide food; for example, herbs can be grown on a kitchen windowsill. In addition, plants "soften" harsh corners in kitchens, living rooms, and bedrooms by providing color and adding a sparkle of interest. Indoor gardening can be a hobby, as people collect groups of specialized plants. Perhaps best of all, psychologically, is the fact that *living* indoor plants are far better to have around than the plethora of plastic plants that many confront us in shopping centers, business places, doctors' offices, and service stations. Can one love a plastic philodendron? The answer for most people is a resounding NO!

## Getting acquainted with houseplants

### NATURE OF HOUSEPLANTS

The first thing to know about houseplants is that they are very different from the wildflowers, ferns, and garden plants that we grow outside in the temperate zones. Most of our houseplants come from tropical climates and are adapted to warmer conditions year around than native plants of the temperate regions which normally need to experience very cold temperatures in the winter while they are dormant. Tropical houseplants like the same year around temperature regime as humans, between about 18° and 24° C (65° to 75° F). Likewise, tropical plants are used to very little seasonal fluctuation in day length and light intensity. Many temperate plants, in contrast, may be triggered into periods of growth, flowering, or dormancy by changes in day length (see Chapter 3). Fortunately, tropical plants are not adversely affected by our seasonal extremes when grown under controlled conditions in the home or greenhouse.

Typical houseplant conditions may be defined as follows:

| | |
|---|---|
| Temperature | 21° C (70° F)-plus or minus 4° C (7° F) |
| Light intensity | Low to medium-75 to 500 foot-candles |
| Day length | About 8–12 hours light per day year around |
| Humidity | Medium to high-40 to 70 percent relative humidity |
| Soil | Evenly moist potting soil |
| Fertilizer | Regular addition of a balanced fertilizer (10-10-10) |
| Growth period | About even growth year around, slightly more in summer; flowering occurs anytime there is sufficient light |

Of course there are many tropical plants which depart from these typical conditions, and their special needs must be recognized. In the following discussion four such groups are described.

First are those plants which are indigenous to tropical latitudes but which grow at high elevations in the mountains and usually require cooler than normal conditions. It may be that a cooler nighttime temperature is necessary for sustained health, or a period of significantly cooler temperatures may be required to trigger the flowering period. Many orchids (especially *Cymbidium)* and peperomias fit this pattern and may not thrive under our extremely warm summer conditions. Such plants are naturally rare in cultivation and may be grown successfully only in certain parts of the country where their requirements can be easily met. Other foliage plants may adapt to these conditions and can be grown "cool," a situation which can save heating costs in the greenhouse.

A second group of tropical plants which departs from the norm is the desert **succulents**. These are plants which thrive in the hot, dry, semiarid environments of tropical regions in the Northern and Southern Hemispheres. Their requirements are characterized by higher temperatures during the growing season, less water, and higher light levels than are normally provided for tropical foliage plants. These strange plants are usually recognized by their succulent, or water-storing, stems and leaves. They must be watered adequately during the growing season, but not overwatered during the winter when they are dormant. Ironically, most succulents also benefit from significantly cooler temperatures during winter dormancy. Many cacti are known to bloom profusely in the spring after a dry, cool winter rest.

In the third group are many of the familiar smaller tropical jungle plants known as **epiphytes** (see Chapter 10). Epiphytes are plants which grow upon other plants, usually on the branches or trunks of trees in mats of accumulated organic matter. While their requirements for light, warmth, and humidity may be satisfied by typical houseplant conditions, they differ in one important way. As epiphytes, their roots are always found in moist, not wet, well-aerated loose organic matter. Overwatering and saturated soil can be the limiting factors preventing healthy growth for these plants growing in confining pots. Epiphytes can be a challenge to grow because they like higher than the minimum atmospheric humidity and fresh air movement at all times. Most orchids, gesneriads, ferns, bromeliads, and peperomias are in this important category.

Finally, some tropical plants such as thick-stemmed orchids, tuberous-rooted gesneriads and begonias, and especially bulbous plants need a dry, cool rest period after flowering. In this sense they are like succulents. This period of dormancy in nature is a response to a relatively brief dry period during their otherwise seasonless tropical year. In other words, these plants escape the unsuitable dry period by going dormant, like temperate plants escape the cold winter. Many such plants readily adapt to indoor cultivation when we recognize their need for dormancy and, ideally, allow it to coincide with our own winter season. Since our cool, short days of winter are generally unsuitable for optimal growth anyway, it is best if we force dormancy during this period on plants that require a rest. This is usually done by deliberately withholding water after flowering. Some plants cannot be induced to enter a rest period during our winter. They would normally be growing during their winter (our summer) in their native Southern Hemisphere or Mediterranean habitats, and they sometimes cannot be forced to change this regime. These plants must be treated carefully during their winter growing seasons because conditions may not be optimal and we may not be used to taking care of them properly during this time of year. It is as if these plants are responding to an internally controlled stimulus to grow during an absolute time of year without consideration for the relative quality of the growing conditions that normally trigger plants to go dormant. Examples of such "winter growers" are succulents from southwestern Africa like *Sedum multiceps, Conophytum, Fenestraria,* and *Bowiea volubilis.*

It is important to know something about the biology and life cycles of your favorite houseplants in order to treat them properly. While many plants in cultivation have adapted to various deviations from their natural environmental conditions (and this is indeed why they have become successful houseplants), others are more demanding and require careful treatment. As indicated in our discussion of the four groups of tropical plants, orchids, succulents, gesneriads, ferns, and peperomias are the large groups of houseplants most often containing species which deviate from the normal conditions. It is these groups that are the most challenging to grow, and they are the groups with some of the most unusual and rewarding types.

Despite the many exceptions just mentioned, it is still no accident that the plants best suited to typical indoor conditions have turned out to be tropical jungle plants, adapted to low light, moist soil, warm temperatures, and little fluctuation in conditions throughout the year. As you begin to recognize the degrees of success you have with different types of plants in your home, you will begin to see how matching their cultural requirements is important for sustained plant health over the long run.

## TYPES OF HOUSEPLANTS

One of the major ways to classify houseplants is to divide them into those which are primarily foliage plants and those which are flowering plants. This classification is certainly artificial, botanically speaking, because while we grow many nonflowering plants such as ferns (which reproduce solely by spores and vegetative means) and conifers (whose reproductive structure is a cone), the majority of our houseplants are true flowering plants (Angiosperms) which ultimately will produce flowers for reproductive purposes. In practice, however, **foliage houseplants** are recognized as those which are grown because they have particularly interesting and showy *leaves,* and their flowers, when produced, are either unattractive or inconspicuous (or both). **Flowering houseplants** are those which have large and attractive flowers, but whose leaves are not necessarily appealing. Also, in general, foliage plants are better suited for low-light situations; and thus, most of our large decorative interior plants are foliage plants.

Some reasons why plants which are capable of flowering usually do not include: (1) not enough light; (2) not enough humidity; (3) not enough phosphorus fertilizer; (4) lack of some particular environmental factor needed by a particular plant, such as a cold period; and (5) not mature enough to bloom, such as the larger species of *Ficus, Dracaena, Schefflera,* and palms. These latter plants usually do not reach flowering size in the confines of a home or average size greenhouse. They do make nice foliage houseplants as "youngsters."

The flowering houseplants may be conveniently classified according to their growth cycles (Table 5–1). The first group consists of **annuals** (or plants treated as annuals) which complete their vegetative growth, flowering, and seed production within one year, or one growing season. These would include marigolds, petunias, pocket-book plant, cineraria, and some impatiens. Then, there are **perennials** which are plants that live and bloom year after year. Among these perennial flowering plants we have several categories: (1) those which bloom continuously given adequate light and feeding, such as the African violet; (2) those which bloom seasonally and are kept year around as foliage plants, such as Christmas cactus; and (3) those which bloom seasonally, but which we usually throw away after blooming (onto the compost pile!) because they are too much trouble to rebloom in the home, such as the poinsettia and forced bulbs.

## CHOOSING A HEALTHY HOUSEPLANT

How do you know which is the best plant to select from among the many plants on display in a shop? The first thing to look for is a plant of the proper color, density, shape, and desired appearance. Select one that is well branched, if that is the growth habit, and is not "leggy" with spindly stems. Check to see if it has a good root system, but is not pot-bound, by tugging gently at the plant to see if it resists. Often, large woody-stemmed plants will be stuck as cuttings into pots of soil and sold before they have even rooted. Look for signs of new growth on your plant. If it is a flowering plant, look for young flower buds. Check to see if there will be buds coming into bloom over a long period of time, not just for the immediate future. For some types of pot plants, such as chrysanthemums, get them just as the flower buds are opening so they may be enjoyed for weeks.

If you are purchasing a foliage plant, look to see that the new leaves are approximately the same size, texture, and proportion as the older leaves. This means that the plant has been grown under the same optimal conditions throughout its development. Do not buy large indoor plants that have just recently been trucked to your region from some distant growing point. Plants experience shock when they change climates too quickly. Foliage plants grown under high light and humidity in Florida and brought north to be sold in a few days at bargain prices will usually show their lack of gradual adjustment by dropping leaves when put in a low-light situation in the home. Try to obtain a plant that has been acclimated for several weeks under conditions similar to those in which it will be positioned at home. Buying from a reputable dealer, one who has been around for a while, has built up a clientele of satisfied customers, and can answer your plant questions, is worth the little extra cost of a plant.

*TABLE 5–1*
*Types of flowering houseplants*

**I. Plants treated as Annuals**

*Begonia semperflorens*
Cineraria
German violet *(Exacum)*
Impatiens
Marigold
Petunias
Pocket-book plant *(Calceolaria)*
Primrose
Schizanthus

**II. Perennials**

A. Continuous bloom (under proper conditions)

African violet
Some begonias
Crown of thorns
Geraniums
*Hibiscus*
Some impatiens
*Spathiphyllum*

B. Seasonal bloom; keep as foliage plant

Amaryllis *(Hippeastrum)*
Some bromeliads
Cacti
Christmas cactus
Orchids
*Streptocarpus* (Some gesneriads go dormant after flowering)
Succulents
Thanksgiving cactus
Kaffir lily *(Clivia)*

C. Seasonal bloom; throw away (difficult to rebloom)

Azalea (may plant outside in the South)
Bulbs (except amaryllis)
Cyclamen
Easter lily (may plant outside in South)
*Fuchsia*
Gloxinia *(Sinningia)*
Hydrangea (may plant outside in the South)
Kalanchoe
Poinsettia
Zebra plant *(Aphelandra)*

Be sure to look for healthy foliage and disease- and pest-free specimens. Most pest infections enter the home from newly acquired plants. You might consider a quarantine period for all new plants, away from your other prized specimens, until you are sure there will be no outbreak of insects or fungus. Table 5–2 indicates some common houseplant problems to help you evaluate plants both before acquiring them and after they have lived in your home. Finally, beware of plants marked down drastically for quick sale; something is probably wrong with them!

*TABLE 5–2*
*Common houseplant problems*

| *Symptoms* | *Probable Causes* |
|---|---|
| Tips or margins of leaves turn brown | 1. Too much fertilizer<br>2. Soil has been allowed to dry out<br>3. Cold injury<br>4. Wind burn<br>5. Low humidity |
| Wilting, yellowing of leaves, and soft growth | 1. Too much heat<br>2. Damage to root system by disease<br>3. Too much or too little moisture |
| Small leaves and long internodes | 1. Too high a temperature<br>2. Insufficient light |
| Weak growth, light green yellow leaves | 1. Lack of fertilizer (especially nitrogen)<br>2. Damage to root system<br>3. Too high a light intensity |
| Yellowing and dropping of leaves from base up | 1. Overwatering<br>2. Poor drainage<br>3. Poor soil aeration<br>4. Gas fumes or air pollution<br>5. Not enough light |
| Smaller leaves than normal | 1. Lack of fertilizer<br>2. Soil mix too heavy<br>3. Not enough moisture |
| Large brown blotches on leaves | 1. Sunburn<br>2. Severe cold injury (stems turn mushy if frozen)<br>3. Fungal infection |
| Failure to bloom, plant otherwise healthy | 1. Not mature enough<br>2. Not enough light; photoperiod improper<br>3. Not enough phosphorus fertilizer<br>4. Not enough humidity—buds dry up or drop<br>5. Improper temperature—may cause bud drop or prevent their formation<br>6. Not a flowering plant (fern or conifer) |

## Some important groups of houseplants and their culture

The best way to get to know houseplants and their general culture is to study them by major group: ferns, aroids, arrowroots and spiderworts, cacti and succulents, bromeliads, orchids, gesneriads, and large foliage plants. Special plants for special places may also be considered, such as plants for low-light situations or plants for terraria. In the following pages we will discuss these major groups of plants, briefly describing some of the important cultivated types commonly used as houseplants. This listing, and the general cultural requirements given, is by no means complete. It is important that you consult specialized reference books on any of the groups of plants in which you become interested. (Some of the best ones are listed at the end of the chapter.) Talking to experienced growers and joining local chapters of the different plant societies will enable you to learn more and to solve problems, and will also be an enjoyable experience in social interaction. (See Appendix II) We have found plant lovers to be interesting people, always willing to share their knowledge (and plants!) with others.

### FERNS

Ferns are perennial plants and are widely distributed in temperate and tropical regions. The temperate species make excellent outdoor specimens for the moist woodland garden, while the tropical types are familiar as houseplants in situations where shade and adequate humidity (40 to 60 percent) are available. Ferns are best recognized as having delicately divided leaves, though there is a great deal of diversity among them (Figure 5–1). Ferns produce leaves *(fronds)* from a stem *(rhizome)* which may take several distinct growth forms: (1) a creeping underground rhizome producing many individual leaves, forming a mass, such as *Adiantum;* (2) strong clump-forming types like most pot ferns such as *Asplenium, Nephrolepis,* and *Pteris;* (3) types with distinctive hairy or scaly, creeping "rabbit-foot" rhizomes such as *Davallia* and *Polypodium;* and (4) the interesting epiphytic (tree-dwelling)

*FIGURE 5–1*
Representative genera of cultivated ferns. Characteristic patterns of sporangia (spore cases) are arranged into clusters called *sori* (singular, sorus). Some enlargements of sori are shown. a = sorus, b = indusium, a covering over the cluster of sporangia (sorus), c = leaflet margin, d = rhizome. Examples shown are *Nephrolepis* (Boston fern)—kidney-shaped sorus and indusium on the underside of a frond (leaf). *Asplenium* (bird's-nest fern)—linear arrangement of sporangia (sori) along the main veins. *Adiantum* (maidenhair fern)—tip of leaflet margin folded under, covering sporangia. *Pteris* (brake or table fern)—side of leaflet margin folded under, covering sporangia. *Platycerium* (staghorn fern)—masses of sporangia on underside of leaf tips. Note two types of leaves. *Cyrtomium* (holly fern)—round sorus with indusium. *Davallia* (rabbit's-foot fern)—pocketlike sorus with indusium. Note slender scaly rhizome. *Polypodium* (deer's-foot fern)—round sorus without indusium. Note large scaly rhizome.

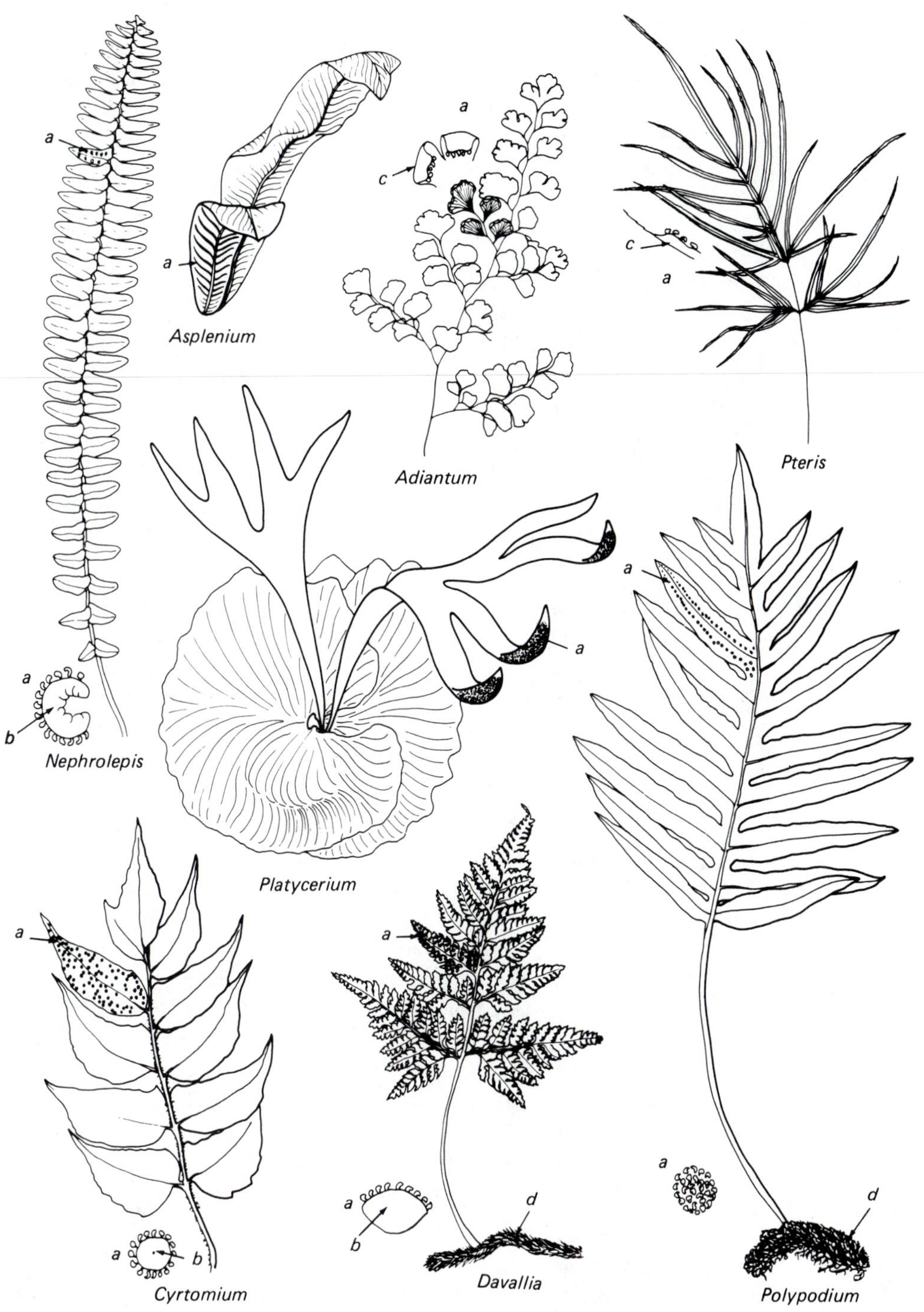
a
a
Asplenium
a
c
c
a
Adiantum
Pteris
a
a
b
Nephrolepis
a
Platycerium
a
a
a
a
b
d
a
b
Cyrtomium
Davallia
d
Polypodium

species, which usually produces two-form (dimorphic) types of leaves, such as the staghorn ferns, *Platycerium.* Epiphytes are plants which grow upon other plants in nature. In cultivation epiphytic plants are grown on clay pots, cork bark, and various other porous surfaces. Different ferns are identified by observing a combination of growth form, leaf shape, and position and arrangement of the spore-case clusters *(sori)* on the underside of the leaves. The common "fruit dots" on the underside of mature fern leaves are the sori.

## *General cultural requirements*

1. *Humidity* Most ferns like high humidity (at least 50 percent), especially the ones with thin, delicate leaves. Some, such as Boston fern, can take drier air and less water. On warm, sunny days it is useful to mist-spray fern foliage with water.
2. *Water* Provide evenly moist soil. Standing water or soggy soil is definitely to be avoided.
3. *Temperature* Night temperatures should range between 13° and 18° C (55° and 65° F) with a rise of 5° to 8° C (10° to 15° F) by day.
4. *Soil* Preferably fibrous soil with ample drainage. Some epiphytic types may grow on tree-fern root or osmunda fiber.
5. *Fertilizer* Bone meal in the potting soil is useful. Commercial fertilizers should be used at one-fourth strength on ferns, as they are very sensitive to overfertilization. Organic fertilizers such as fish emulsion (5–1–1) are ideal. Ferns do need regular fertilizing to keep them lush; and high nitrogen fertilizers help keep them dark green.
6. *Light* The amount of light depends on the type of fern. Winter sun is not too bright for most ferns, but it is best to avoid direct exposure, as in south windows. In summer, ferns must be kept in the shade.
7. *Propagation* Division of clumps from the crown, rhizome cuttings (scaly, creeping types), plantlets from the runners of Boston ferns, and plantlets from the fronds of some species are used. Ferns may also be propagated sexually from spores. Place a fertile leaf (one with mature sori) in a dry envelope for a few days to catch the brown dustlike spores as they are released. Sprinkle the spores evenly on the surface of sterile potting soil (African violet soil works well, or any peaty soil) in a plastic container with a lid.

   The classic way to germinate spores is by sowing them on the surface of a small inverted clay pot kept moist by placing on a saucer of water and covering with a glass jar. In this humid environment the spores will germinate in a few weeks to form small green heart-shaped plants (prothalli) on the surface of the medium. Spray (or mist) with water occasionally. In a few months, new fern leaves will appear from the prothalli and grow into larger plants after being transplanted to fern soil. Wait until baby ferns are large enough to have roots before transplanting. They must be kept very humid (near 90 percent) at all times.

*Cultivated genera of ferns useful as houseplants*

| | |
|---|---|
| 1. *Adiantum*<br>Maidenhair fern | A delicate-leaved fern requiring high humidity and shade. |
| 2. *Asplenium*<br>Bird's-nest fern | An unusual fern with broad, entire-margined leaves that form an open clump. Requires high humidity. |
| 3. *Cyrtomium falcatum*<br>Japanese holly fern | A tough, coarse fern which can tolerate drier air and less water. |
| 4. *Davallia*<br>Rabbit's-foot fern | A creeping-rhizome fern for hanging baskets and pots. Leaves are delicate. Good house fern if misted occasionally. |
| 5. *Nephrolepis*<br>Boston fern | Tough, old-fashioned fern with many cultivar types. Tolerant of most home conditions. |
| 6. *Platycerium*<br>Staghorn fern | Requires humid air and a thorough soaking at least once a week. May be grown as an epiphyte or in a pot. |
| 7. *Polypodium*<br>Golden-polypody fern<br>Deer's-foot fern | A coarse, very large fern with a large hairy rhizome. |
| 8. *Pteris*<br>Table fern<br>Brake fern | A medium-textured fern with many cultivated forms. Many have green and silver variegated leaves. Does well under most conditions. |

## AROIDS

Largely tropical herbs, the aroids are in the Arum family (Araceae) and have fleshy or semiwoody stems (Figure 5–2). Some aroids are tree-climbing vines. Aroids have varied leaf forms. Flower clusters *(inflorescences)* are composed of a dense spike, called a *spadix*, with a leaflike bract, called a *spathe*. The spathe is sometimes brightly colored and showy. Some are very easy to grow as houseplants because they can tolerate a wide range of growing conditions, especially low light.

*General cultural requirements*

1. *Humidity* Most prefer at least 30 to 40 percent humidity, but some will survive at lower values.
2. *Water* Most like evenly moist soil.
3. *Temperature* Average house temperature, but do not allow to fall below 13° C (55° F). Most like 15° to 21° C (60° to 70° F).
4. *Soil* Normal potting soil except as noted.
5. *Fertilizer* Fertilize flowering plants every two to four weeks when plants are actively growing; fertilize foliage plants every three to four months.
6. *Light* Only one genus, *Zantedeschia* (calla lily), likes direct sunlight; most other genera do best in bright indirect light, but several can tolerate lower light intensities.
7. *Propagation* Division, stem segment cuttings, or air layering for the larger species; stem-tip cuttings for the smaller species.

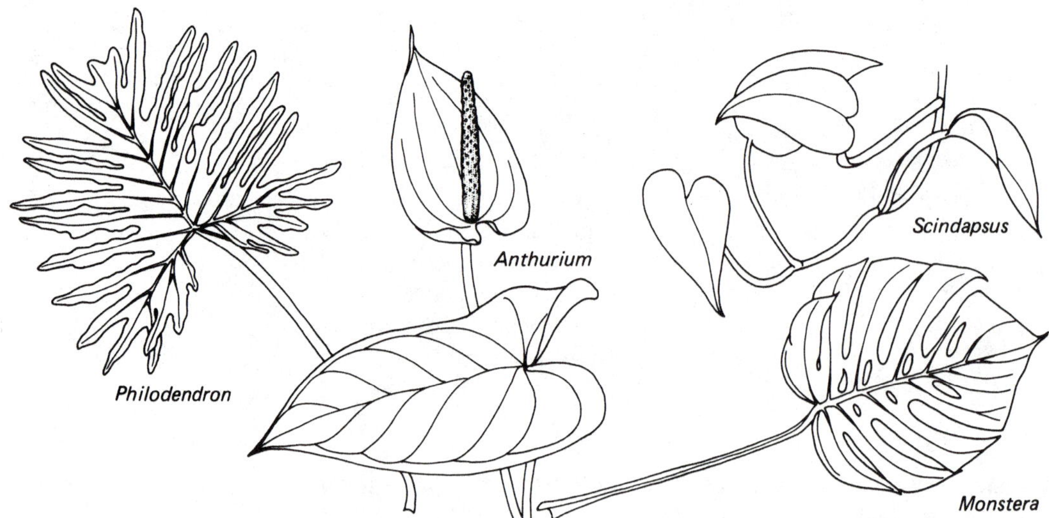


*FIGURE 5–2*
Some important cultivated aroids: *Philodendron* (philodendron); *Anthurium* (flamingo flower); *Scindapsus* (devil's ivy, pothos); *Monstera* (split-leaf philodendron).

Cultivated genera of the Arum family

| | |
|---|---|
| 1. *Aglaonema* Chinese evergreen | Very tough plant; best in shade with barely moist soil; can tolerate low humidity and very low light. |
| 2. *Anthurium* Flamingo flower | Showy red spathe with spadix. Requires high humidity (over 50 percent) and a loose soil (e.g., fir bark and peat moss); grows best in bright indirect sunlight and an evenly moist soil. |
| 3. *Caladium* | Colorful foliage. Prefers a rich soil, fairly bright light, and good humidity; likes evenly moist soil and requires a four- to five-month dormant or rest period after foliage dies back. Often used outdoors as a summer bulb for colorful foliage in shady places. |
| 4. *Colocasia* | Requires high humidity and a thoroughly moist soil. |
| 5. *Dieffenbachia* Dumb cane | Likes to have soil on the dry side and to have bright light (but can tolerate low light). |
| 6. *Monstera* Split-leaf philodendron | Grows best in bright, indirect sunlight and barely moist soil. |
| 7. *Philodendron* | Will survive in shade and low humidity; for cutleaf-climbing types provide a support for roots to cling to. |
| 8. *Spathiphyllum* Peace lily | Do not put in direct sunlight; likes thoroughly moist soil and warm temperatures (18° C, 64° F). May flower continuously. |
| 9. *Scindapsus* Pothos or Devil's Ivy | Will survive in low light but may lose colorful markings. |
| 10. *Syngonium* | Will survive in low light. |
| 11. *Xanthosoma* Tannia | Needs fresh air; likes high humidity and evenly moist soil. A large-leaved plant similar to *Caladium*. |
| 12. *Zantedeschia* Calla lily | Grow in sun and keep very moist and well fertilized during active growth period. |

## ARROWROOTS AND SPIDERWORTS

Plants in the arrowroot family (Marantaceae) and spiderwort family (Commelinaceae) are tropical herbaceous monocots. Many members of these two families are easily grown as houseplants (Figure 5–3). Although the flowers of these plants are not conspicuous, the foliage is attractively patterned and colored. The flowers on some are interesting.

### *General cultural requirements*

1. *Humidity* Most can survive in average house humidity, but prefer at least 40 percent relative humidity. Low humidity will cause browning of leaf edges.
2. *Water* Most prefer to be kept evenly moist; those that like to be kept on the dry side are noted in the list that follows. Do not overwater.
3. *Temperature* Average house temperature is sufficient, but not above 24° C (75° F) in winter.
4. *Soil* Normal potting soil for spiderworts. Arrowroots prefer a richer soil, so add 1 part leaf mold.
5. *Fertilizer* Fertilize every four to six weeks when plants are actively growing.
6. *Light* Most spiderworts will tolerate a wide range of light conditions but prefer bright indirect light. Arrowroots prefer semishade.
7. *Propagation* Arrowroots: division at the beginning of the growing season. Spiderworts: stem-tip cuttings or layering for trailing types; division for *Rhoeo* and clump-forming *Tradescantia*.

*FIGURE 5–3*
Popular representatives of spiderworts (Commelinaceae) and arrowroots (Marantaceae). *Zebrina pendula* (wandering Jew) is a beautiful creeping member of the spiderwort family with leaves that are silver and green above and purple on the underside. *Maranta leuconeura* (prayer plant) has two-tone green leaves with red veins above and purple on the underside.

**Arrowroots** (Marantaceae): tropical American herbs, many having tuberous roots; leaves large, with patterned feather design and sheathing bases; flowers surrounded by bracts. These are good plants for low-light situations (east and north windows).

| | |
|---|---|
| a. *Calathea* | Needs high humidity (50 percent or more); good for large terraria. |
| b. *Maranta*<br>Prayer plant<br>Rabbit-track plant | Leaves fold upward at night. Tolerates low light. |

**Spiderworts** (Commelinaceae): semisucculent herbs and creepers; leaves alternate, parallel veined, and with sheathing bases. Trailing types make nice, full hanging baskets if kept pinched back.

| | |
|---|---|
| a. *Cyanotis*<br>Teddy-bear plant<br>Pussy-ear plant | Leaves covered with dense hair. |
| b. *Gibasis multiflora*<br>Tahitian bridal-veil | Small leaves and tiny white flowers; makes an excellent rounded basket plant. |
| c. *Rhoeo discolor*<br>Moses-in-the-cradle<br>Moses-in-the-boat | Upright clump-forming herb with leaves dark green above, purple below. Flowers appear from between large bracts. |
| d. *Setcreasea purpurea*<br>Purple-heart | Leaves purple. Keep on dry side. |
| e. *Tradescantia blossfeldiana*<br>*Tradescantia fluminensis*<br>Wandering Jew | Green- and white-striped creepers. |
| f. *Tradescantia navicularis*<br>Succulent wandering Jew | A succulent-leaved creeper; smaller than other wandering Jew types. Has showy purple flowers. |
| g. *Tradescantia palludosa*<br>Spiderwort | Upright clump-former with narrow leaves and large blue flowers. |
| h. *Zebrina pendula*<br>Wandering Jew | Green- and purple-striped leaves; excellent hanging basket plants. Bright light intensifies colors. |

## SUCCULENTS, INCLUDING CACTI

Succulents are plants which have developed specialized water-storage tissues in their leaves, stems, or roots. This type of growth habit is found primarily in desert-inhabiting species but is also prevalent in species that grow in dry rocky mountain habitats. There are more than forty plant families that contain at least some succulent species, the total of which numbers in the thousands. The cactus family (Cactaceae) and the stonecrop family (Crassulaceae) are made up entirely of succulent species. The carpetweed family (Aizoaceae) and the spurge family (Euphorbiaceae) contain many important succulent

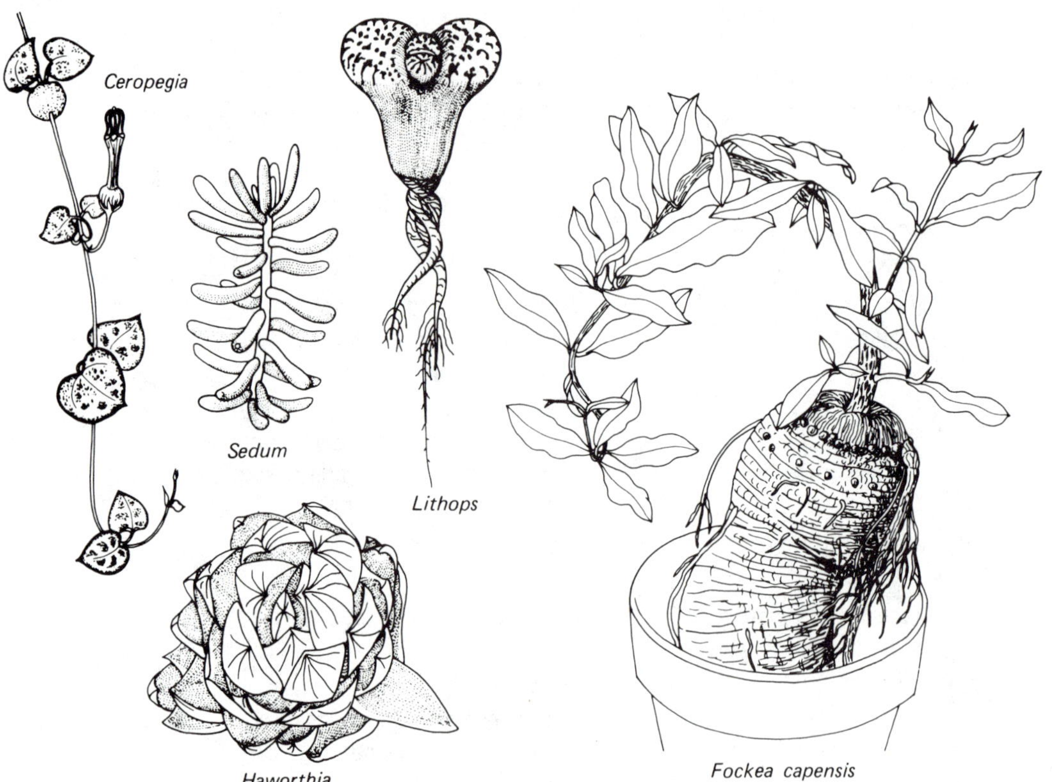


*FIGURE 5–4*

Representative leaf and caudiciform succulents. Milkweed family (Asclepiadaceae): *Ceropegia woodii* (string of hearts, rosary vine); *Fockea capensis* (fockea)—this is a caudiciform succulent with an enlarged, water-storing, root called a *caudex*. Stonecrop family (Crassulaceae): *Sedum pachyphyllum* (stonecrop or jellybean sedum). Lily family (Liliaceae): *Haworthia cymbiformis*. Carpetweed family (Aizoaceae): *Lithops* sp. (stone-plant, living stone, and stoneface).

species as well as species of more typical herbaceous growth habits. In general, succulents are relatively easy to care for and will survive if grown under less than ideal conditions.



Succulents may be categorized as either **leaf succulents** or **stem succulents** (Figures 5–4 and 5–5). Leaf succulents store water in modified leaves and range from the fat-leaved, jellybean sedums and crassulas and the rosette-forming haworthias and aloes to the highly unusual and very juicy mimicry succulents from South Africa such as the stone-plants *(Lithops* and *Conophytum).*

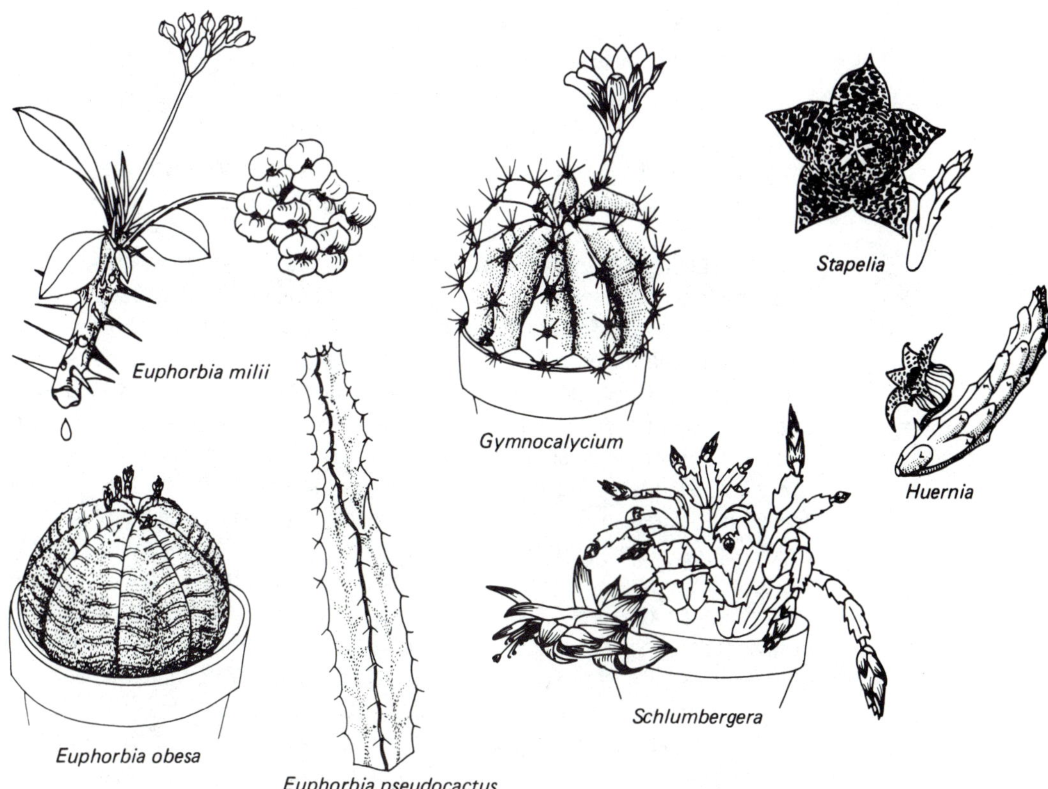


*FIGURE 5–5*
Representative stem succulents. Spurge family (Euphorbiaceae): *Euphorbia milii* (crown of thorns); *Euphorbia obesa* (basketball euphorbia); and *Euphorbia pseudocactus* (candelabra euphorbia). Cactus family (Cactaceae): *Gymnocalycium* sp. (chin cactus) and *Schlumbergera truncata* (Thanksgiving cactus). Milkweed family (Asclepiadaceae): *Stapelia variegata* (starfish flower) and *Huernia* sp. (a stapeliad).

Stem succulents are usually leafless and have enlarged stems which may bear spines. The most famous examples are the true cacti and the cactus-like euphorbias. Cacti are found natively only in the New World deserts of North, Central, and South America. Succulent euphorbias come mainly from southern Africa. Both groups may have short, rounded, globular plant forms which may produce many offsets (Figure 5–6); or they may be tall, columnar, branched treelike forms (Figure 5–5). True cacti may be distinguished from succulent euphorbias by the following characteristics:

| *Cacti* | *Euphorbias* |
|---|---|
| Aereoles present (cushionlike areas at nodes that bear bristles, spines, and flowers). | Aereoles absent. |
| Flowers large and showy, with many petals and stamens. | Flowers small, inconspicuous; petals absent; stamens few. |
| Fruits fleshy with many seeds. | Fruits dry, ovoid, with three seeds. |
| Sap watery (a few *Mammillarias* have milky juice). | Sap milky and abundant; may be poisonous. |

Succulents in general provide a diverse array of forms that make them a fascinating and popular group. They are interesting and attractive even when not in bloom. The stapeliads (or starfish flowers) along with *Ceropegia* and *Hoya* in the milkweed family (Asclepiadaceae) from southern Africa produce relatively showy flowers (Figure 5–7). Succulent asters like *Kleinia,* succulent geraniums *(Pelargonium),* and succulent coleus *(Plectranthus)* are among the more unusual types with well-known non-succulent relatives.

A most interesting type of succulent is the **caudiciform**. These plants are often bizarre in that they have a conspicuously enlarged stem or root which stores water; and delicate vines, stems, or leaves are produced from the top of the bulb-like swollen base. The base forms the caudex, and hence the name "caudiciform." The caudex may often be grotesquely twisted or disproportionately enlarged. Caudiciforms are generally slow growing and must be carefully watered to prevent rotting, especially in winter when they go into deep dormancy. (See *Fockea capensis* in Figure 5–4.)

One group of succulents that requires different care is the epiphytic cacti, *Epiphyllums.* They are often called orchid-cacti because of the unusually large and attractive flowers. They grow in the trees in warm, humid tropical jungles and hence are not desert plants. They require more moisture, higher humidity, filtered light, acidic soil and more fertilizer than their desert-dwelling succulent relatives.

Contrary to popular opinion, desert succulents thrive on good soil and frequent feeding during their growing period (which is usually March through October). Always use a low-nitrogen fertilizer for the slower growing succulents. The soil must be well drained, however, and allowed to dry

*FIGURE 5–6*
Hedgehog or sea urchin cactus (*Echinopsis* sp.) with flowers which are night blooming. (A) Habit view of plant with seven tubular flowers in bloom, all at the same time. (B) Close-up of two flowers, showing numerous stamens. *(Photographed by Virginia Ryan)*

*FIGURE 5–7*
*Hoya bella* (dwarf wax plant), a hanging or trailing species. ($\times 1\frac{1}{2}$)

out thoroughly between waterings; but the succulents may be watered as often as they dry out. Smaller cacti and seedlings should not be allowed to dry out as long as the larger ones. Watering once a week would not be unusual for a succulent houseplant if it is getting warm temperatures in a sunny window. Water the plants well when you do water. Avoid overwatering the highly juicy South African mimicry succulents (like *Lithops*) as they will absorb so much water that the fat leaves may swell and literally burst from the internal pressure.

Most succulents go dormant in the winter (a few go dormant in summer) and should be kept fairly dry and cool (12° to 15° C or 55° to 60° F) from October through March, watering just enough to keep the leaves and stems from shriveling. Give them plenty of air movement in the summer. They will even benefit from a good overhead spraying with water to periodically remove dust. If properly cared for, giving regular feeding and paying attention to growth cycles and water needs, succulents can remain healthy for years, blooming annually. Some growers have succulent plants which are older than they are!

### *General cultural requirements*

1. *Humidity* Most survive well in low humidity (as low as 5 percent).
2. *Water* Water well when you water; allow plants to dry out thoroughly between waterings; water less in winter or during dormant periods. Watch for new growth in the spring to begin watering again.
3. *Temperatures* Average house temperature is sufficient. Cacti may take temperatures as low as 10° C (50° F) in winter; while southern African succulents should be kept at 15° C (60° F) in winter.
4. *Soil* Rich but sandy soil is ideal; good drainage is important.
5. *Fertilizer* Fertilize every month when plants are actively growing; not at all when plants are dormant. Use a high phosphorus fertilizer (10-30-10) as too much nitrogen stimulates lush growth that is susceptible to rot. Apply at the rate of 2 tablespoons per gallon.
6. *Light* At least six hours of direct sun per day is best for most succulents; they can be scorched if placed in direct, hot sun (or if placed outside in direct sun after being indoors all winter), but they do need bright light.
7. *Propagation* Stem-tip cuttings or seeds for many; some can be propagated by leaf cuttings, and some can be divided. Succulents are very easy to propagate vegetatively. If you make stem or leaf cuttings, let the cut end dry off before sticking into dry (or just slightly damp) sandy soil. Do not water until new roots have formed; the plants will have enough water to carry them through the rooting period if kept shady and in a relatively humid atmosphere.

| | |
|---|---|
| 1. Agave family (Agavaceae) *Agave* (Century plant) | Rosette-forming, very thick-leaved succulents. |
| *Yucca* | Thick-leaved plants, may be tall. |
| 2. Cactus family (Cactaceae) | |
| a. Desert cacti: *Mammillaria, Opuntia, Echinocereus, Echinocactus* (golden-barrel), *Gymnocalycium, Echinopsis,* and many others | Follow general cultural requirements. |
| b. Epiphytic cacti: *Epiphyllum, Rhipsalis, Rhipsalidopsis, Schlumbergera* (Thanksgiving and Christmas cacti) | Need more shade, more moisture, and cooler temperatures than desert types. |
| 3. Carpetweed family (Aizoaceae) | |
| The South African mimicry succulents or "Mesembs": | |
| *Lithops* (stone-plants)<br>*Faucaria* (tiger's-jaws) | Give plenty of light; do not overwater. Most go dry and dormant in winter. |
| *Conophytum*<br>*Fenestraria* (baby-toes) | *Conophytum* and *Fenestraria* are "winter growers." Do *not* overwater. |
| 4. Lily family (Liliaceae) | |
| *Aloe* (*Aloe vera* is the famous "burn plant".) (*Aloe variegata* is the tiger Aloe) | Rosette-forming leaf succulents. Some have showy flowers. |
| *Bowiea volubilis* (succulent onion) | Winter grower; withold water in summer. |
| *Gasteria*<br>*Haworthia* | Smaller, rosette-forming plants with attractive markings on leaves. |
| 5. Milkweed family (Asclepiadaceae) | |
| *Stapelia, Huernia, Caralluma* (stapeliads) | Stapeliads are from southern Africa. Need well-drained soil and bright light. |

(Continued)

| | |
|---|---|
| *Hoya carnosa* (wax plant) | Epiphyte; needs bright indirect light. |
| *Ceropegia woodii* (string-of-hearts) | Bright light and frequent watering will allow this interesting delicate vine to grow into a massive plant. |
| 6. Spurges or Euphorbias (Euphorbiaceae) | Follow general cultural requirements. |
| *Euphorbia milii* (crown-of-thorns)<br>*E. tirucalli* (pencil-tree)<br>*E. obesa* (basketball-euphorbia)<br>*E. pseudocactus* (candelabra euphorbia) | Leafy types require more water than highly succulent-stemmed ones. |
| 7. Stonecrop family (Crassulaceae)<br>*Crassula, Sedum, Kalanchoe, Echeveria* | These are leaf succulents, often rosette forming. Some have showy flowers. Good houseplants. |
| 8. Composite family (Asteraceae or Compositae)<br>*Kleinia*<br>*Senecio* (some species) | Follow general cultural requirements. |

## BROMELIADS

Most members of the bromeliad family (Bromeliaceae) are tropical epiphytes which are grown for their colorful leaves, bracts, flowers, fruits, or unusual silvery gray leaves. Some are known as **air plants** because they can be grown without soil or substrate; others are referred to as **tank plants** because they hold water in the "cup" or "tank" formed by the cluster of tightly fitting leaves. There are three basic types of bromeliads: (1) terrestrial species, such as pineapple *(Ananas)*, which grow rooted in the soil; (2) tank epiphytes, illustrated by *Aechmea, Billbergia, Neoregelia,* and *Vriesia,* which hold water in a stiff rosette of overlapping leaves; and (3) air plants, or atmospheric epiphytes, such as the large genus *Tillandsia,* which have stiff narrow leaves covered with hairlike silvery scales and which do not hold water (Figure 5–8). *Tillandsia usneoides,* or Spanish moss, has no roots and just hangs from the trees in the southeastern United States. It grows well in a shady greenhouse if it is thoroughly watered every morning and allowed to dry out overnight. Most bromeliads have scales on their leaves which absorb nutrients and allow gas exchange, and so it is an important cultural requirement that the leaves dry out at night so the plants can "breathe."

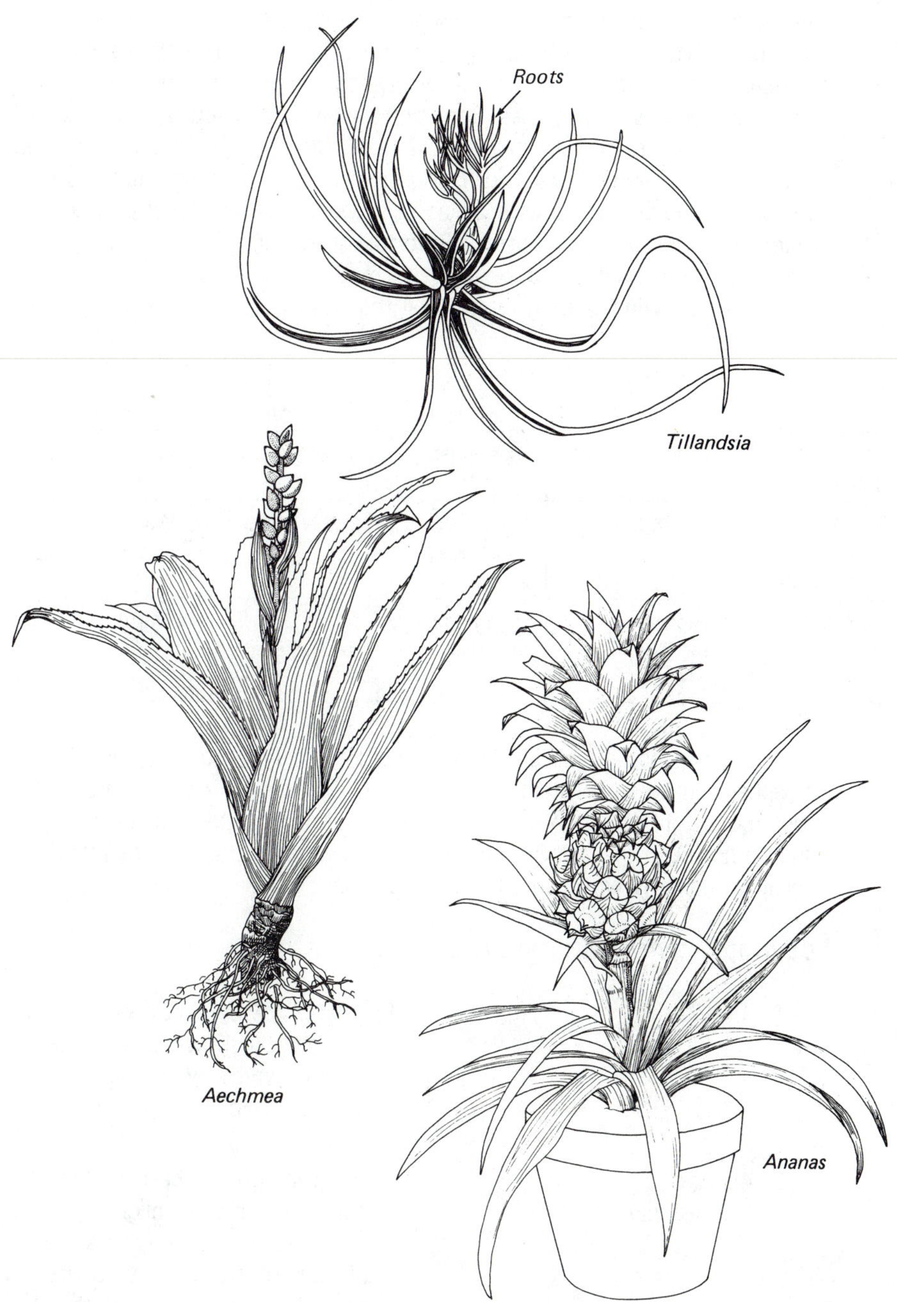


*FIGURE 5–8*
Three types of bromeliads. Terrestrial: *Ananas comosus* (pineapple). Tank-type epiphyte: *Aechmea* sp., fruiting stalk has colorful bracts. Atmospheric epiphyte: *Tillandsia schiedeana* (air plant).

The basic growth form of bromeliads is the rosette, or cluster, of leaves. The rosette forms on a creeping, horizontal rhizome (stem) which may be short, resulting in clump formers; or elongated, with the rosettes forming away from the pot. In any event, each rosette, or tank, blooms only once and then eventually dies. Sometimes many old rosettes stay together and new ones are formed around the sides. Old rosettes may be removed after blooming as new ones form. Rosettes formed on long rhizomes may be cut off and potted-up as soon as new roots develop near the base (for example, on *Aechmea* × 'Foster's Favorite'). Bromeliads may also be grown from seeds, but they take a long time to mature.

The following chart gives cultural requirements for the three types of bromeliads:

| | *Soil* | *Light* | *Moisture* | *Fertilizer* | *Temperature* |
|---|---|---|---|---|---|
| Type 1<br>Terrestrial | Loose, well drained | High | Just moist | Apply to soil | Warm; cool in winter |
| Type 2<br>Tank type | No soil; bark slab, fir bark, or osmunda | Bright Shade | Keep roots just moist; keep tank full | Apply to tank | Warm |
| Type 3<br>Air plant | No soil; bark slab or osmunda | High to medium | High humidity; mist daily | Apply as mist to leaves | Warm |

Bromeliads, like all epiphytes, benefit from high humidity and an atmosphere that has constant air movement. Do not grow bromeliads under continuous light (their "breathing pores," or stomata, open only in the dark) and do not keep their leaves wet at night when gas exchange is taking place. Fertilize regularly with dilute balanced fertilizer.

### *Cultivated genera of bromeliads*

1. *Aechmea* — Epiphytes with flowers surrounded by showy, spiny, pink bracts which will last for months. Foliage is often banded.

2. *Ananas*
   Pineapple — Terrestrial; grows best in high humidity and sunlight. To propagate, slice leafy rosette off top of fruit, allow to dry and form callus tissue; then plant in normal potting soil.

3. *Billbergia*
   Queen's tears — Very durable epiphytes, having leaves often variegated and

| | |
|---|---|
| | usually forming tubular or urn-shaped rosettes. Inflorescence drooping with brightly colored bracts. |
| 4. *Cryptanthus* Earthstars | Dwarf, low-growing terrestrials. Leaves are often banded or striped with silver, bronze, and/or red. High humidity preferable. |
| 5. *Neoregelia* | Mostly epiphytes with spiny-edged leaves. Very easy to grow. Flowers form down inside the tank. |
| 6. *Tillandsia* Air plants | Small epiphytes which need bright light, high humidity, and good ventilation. More difficult to grow. |

## ORCHIDS

The orchid family (Orchidaceae) is the largest of all flowering plant families, containing about 750 genera and over 25,000 species, not to mention the thousands of artificial hybrids. In nature, many orchids are tropical epiphytes (plants that grow on limbs and trunks of trees). There are numerous orchids, however, which are terrestrial, and some of these do grow in temperate areas (such as our lady's slipper). Many tropical genera can be grown successfully as houseplants; the temperate ones cannot.

Orchids owe much of their appeal to their unusual and attractive flowers which may last for several weeks. While the interesting and complex flowers of orchids may look very different from each other, they are all basically alike in structure, having three sepals and three petals with one of the three petals modified into a lip (Figures 5–9 and 5–10). In addition, the fusion of the stamens with the style and stigma to form the column is a diagnostic characteristic of all orchid flowers.

Orchids have two basic growth forms which relate to techniques of vegetative propagation. The most commonly encountered form is called **sympodial** (literally "feet together") where a single stem (often swollen into a pseudobulb) produces only one flower cluster; then a new shoot (stem) must grow from the base in order for another bloom to form. These new shoots may be produced on short or long rhizomes so that the plant appears clustered or sprawling, respectively. Vegetative propagation is accomplished by dividing the rhizome so that three to four pseudobulbs remain together as a unit. New roots and shoots will form from "eyes" at the base of these old pseudobulbs. Common sympodial orchids include *Cattleya, Cymbidium, Den-*

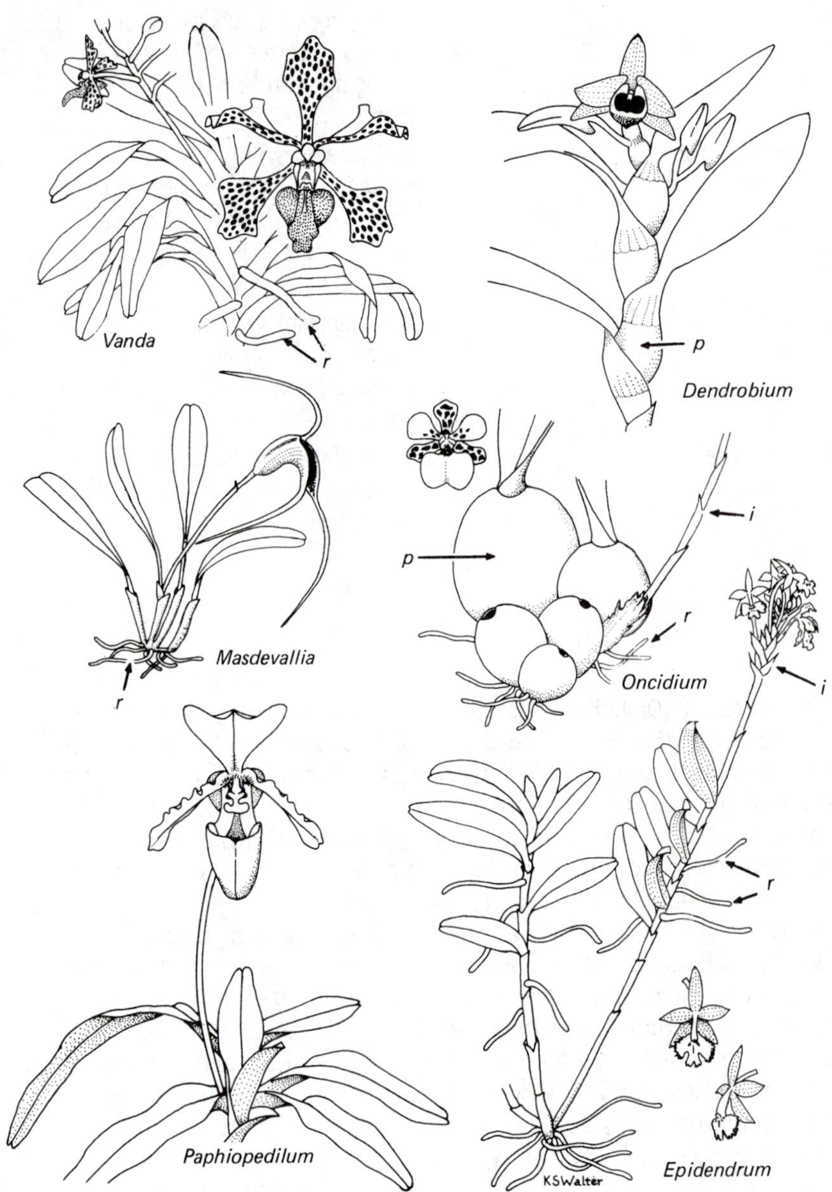


*FIGURE 5–9*
Representative genera of Orchidaceae (i = inflorescence; r = roots; p = pseudobulb). *Vanda tricolor* (habit × ⅛; flower × ⅜), monopodial. *Dendrobium nobile* (× ⅜), sympodial, with pseudobulbous stems. *Masdevallia calura* (× ⅜), sympodial, no pseudobulbs. *Oncidium ampliatum* (× ⅜), one large leaf per pseudobulb; multibranched inflorescence. *Paphiopedilum spicerianum* (× ½), sympodial, no pseudobulbs; single flower per bud; lower lip is "pouch"; slipper orchid. *Epidendrum radicans (ibaguense)* (habit × 3/16; flower × ⅜), elongated stems without pseudobulbs; flower clusters at tips. *(Drawings by Kerry S. Walter)*

*drobium, Encyclia, Epidendrum, Masdevallia, Oncidium,* and *Paphiopedilum* (Figure 5–11).

The other growth form is called **monopodial** (literally "one foot") and has a single main stem that grows taller throughout the life of the plant. The leaves come off the sides of the stem like steps of a ladder, roots are produced on the lower parts of the stem, and flowers are formed from the upper portions (Figure 5–9, *Vanda*). Vegetative multiplication must be from offsets or plantlets as you cannot cut the main stem into pieces and root them.[1] The monopodial habit is less common than the sympodial form, but there are common examples such as *Phalaenopsis* and *Vanda.*

Sexual reproduction in most orchid species and hybrids can be accomplished by germinating the minute seeds which may take many months to mature in the seed pods. Since the seeds are so small and contain virtually no stored food, they are best grown on special nutrient media under sterile conditions (see Chapter 2). As with other flowering plants, orchid hybrids can be produced only by growing the seeds resulting from a successful cross between two distinct parents. More than any other group, orchids are notorious for being able to hybridize freely (even between different genera), and many spectacular man-made hybrids have been produced. Since 1856, there have been over 57,000 horticulturally produced orchid hybrids registered with the Royal Horticultural Society (London).

More recently, tissue-culturing (or mericloning) techniques have been perfected to enable growers to mass-produce many orchids asexually (see Chapter 6). As many as four million specimens have been produced per year from a single tissue explant (original tissue from the parent plant). Practically, however, commercial tissue culture labs seldom produce more than 1,000 to 2,000 plantlets from an explant because of the chances of deleterious mutations occurring during the process. The benefit of such mass production is to make available to the average home orchid grower highly desirable specimens of outstanding quality at lower prices.

The two most adaptable houseplant orchids are *Phalaenopsis* (moth orchids) and *Paphiopedilum* (slipper orchids) (Figures 5–10 and 5–11). They like low to medium light (never full sun), warm temperatures (15° to 27° C or 60° to 80° F), evenly moist growing medium, and regular feeding. Never put an orchid or any other houseplant or greenhouse plant out in full summer sun when it has been growing indoors protected by a glass or plastic covering. The strong ultraviolet rays will burn the leaves unless the plant has been gradually acclimatized to the sun, just like humans who need gradual exposure to avoid sunburn. Some orchids, however, need full summer sun to do well; and plants like *Vanda* and *Cymbidium* may be gradually moved out in early spring to remain in full sunlight until just before frost in the fall.

---

[1]Many orchid genera regularly produce vegetative plantlets on their older flowering stems or pseudobulbs. These plantlets are referred to as *kikis* (pronounced "key-keys") by orchid hobbyists and are ready sources of new plants for your collection.

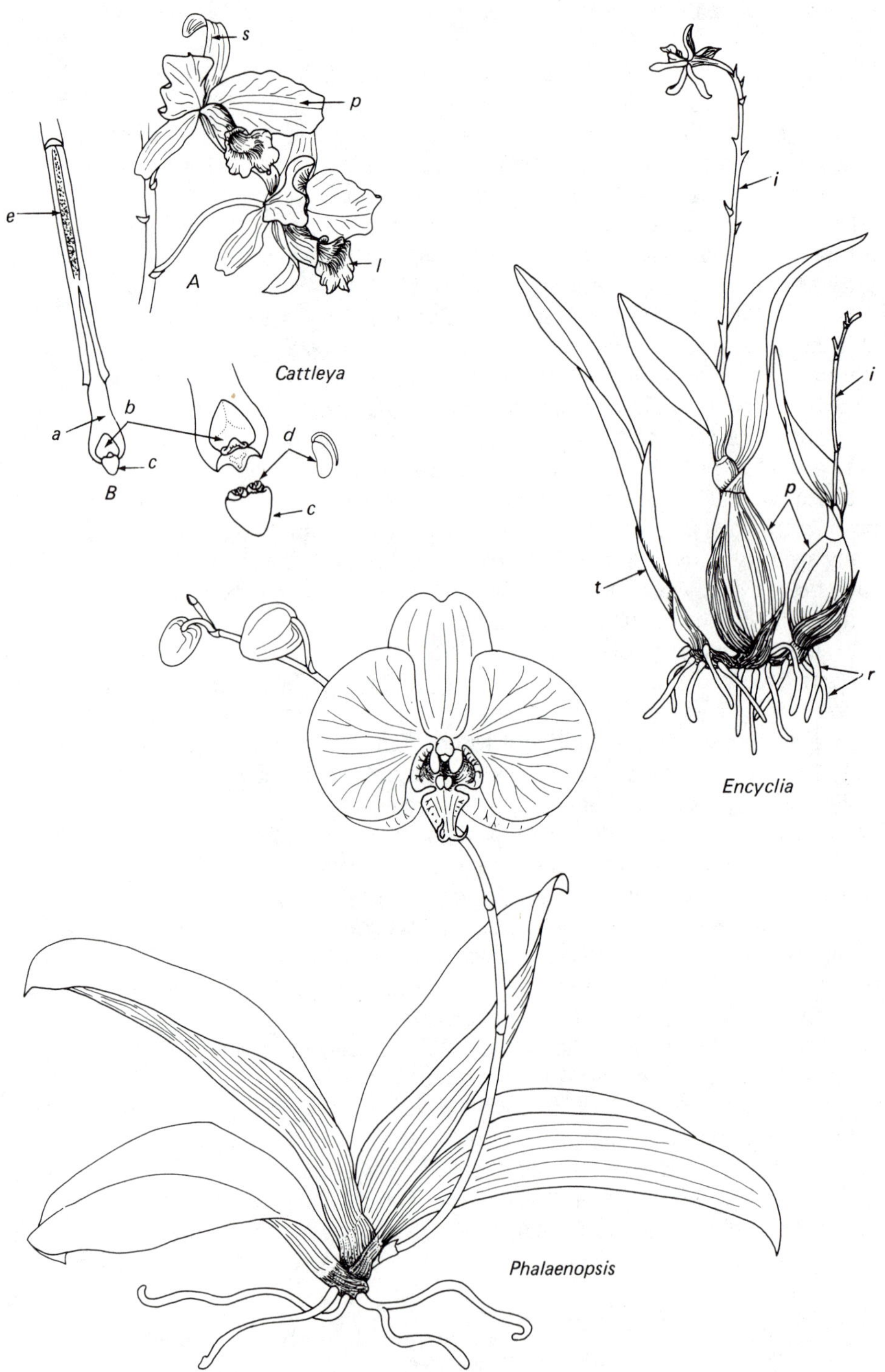
s
p
e
l
A
Cattleya
a
b
c
B
d
c
i
i
p
t
r
Encyclia
Phalaenopsis

*FIGURE 5–10*
Representative genera of Orchidaceae. *Cattleya:* (A) flower showing sepals (s), petals (p), and lip (l); (B) reproductive structures showing column (a) which includes the stigmatic surface (b), the anther cap (c) which contains four pollinia (pollen sacs) (d). The ovary (e) enlarges after fertilization and contains the numerous small seeds. Plant habit with conspicuous pseudobulb similar to those of *Encyclia* shown below. *Encyclia:* epiphyte showing pseudobulbs (p) (enlarged stems), epiphytic roots (r) and old flowering stalks (i). Note the new shoot (t) forming from the short rhizome. *Phalaenopsis* (moth orchid): popular for its very long-lasting flowers, ability of the flower stalk to branch and produce more buds, and adaptability to indoor conditions.

*FIGURE 5–11*
*Paphiopedilum* (slipper orchid) in flower.

## *General cultural requirements*

1. *Humidity* Requirement varies with genus, but in general, 50 percent humidity is sufficient. Good ventilation is also very important, but do not place plants in hot or cold drafts.
2. *Water* Epiphytes: allow growing medium to become almost dry between waterings. Terrestrials: keep moist but not wet.
3. *Temperature* Depends on genus. A 5° to 8° C (10° to 15° F) drop from day to night is very desirable.
4. *Soil* Epiphytes: fir bark or osmunda fiber (should be very porous and allow air circulation). Terrestrials: finely chopped fir bark, soft brown osmunda, or any mixture which drains well but retains moisture.
5. *Fertilizer* Depends on genus.
6. *Light* Depends on genus, but in general, as much light as possible, without injuring the plant. Too much light will cause yellowing; too little will cause foliage to turn dark green and plant will not bloom.
7. *Propagation* By division or by seeds started in sterile culture; mericloning from small portions of tissue near the growing tip, also requiring sterile culture techniques. Both practices are difficult to perform under ordinary home growing conditions. Propagation from kikis (pronounced "key-keys") is a readily adaptable procedure however.

## *Selected orchid genera*

1. *Cattleya* — A pseudobulb-producing epiphyte. A common corsage flower and quite tough. Needs bright light but not full summer sun. Temperature: days 18° to 24° C (65° to 75° F); Nights 13° to 16° C (55° to 60° F). Feed a dilute solution of 18-18-18 every two to four weeks. Not generally recommended as a houseplant unless good light can be provided.
2. *Cymbidium* — Terrestrial with very long-lasting flowers, long grasslike leaves, and characteristic large egg-shaped pseudobulbs. These plants come in standard and miniature sizes, the smaller ones making the best houseplants. They like very bright light (grow outdoors in sum-

| | |
|---|---|
| | mer) and an evenly moist growing medium. Fertilize regularly as these plants are heavy feeders. Need 5° C (40° F) temperatures for three to four weeks in the fall to set flower buds, which bloom in the winter indoors. |
| 3. *Dendrobium* Especially, *D. nobile* | Epiphyte with elongated pseudobulbs; many need bright light and cool temperatures to bloom well. Temperatures to 13° C (55° F) or lower are tolerated, and a cool, dry rest is usually required after blooming. |
| 4. *Epidendrum* Especially, *E. radicans* | Epiphytes with very long flowering stems and clusters of small flowers. Blooms regularly and over a long period, then produces plantlets (kikis) on the flowering stem. Tolerates a wide range of light and temperature. |
| 5. *Masdevallia* | A miniature orchid, excellent under artificial lights. Requires average home conditions and medium light; likes higher humidity. |
| 6. *Oncidium* | A medium to large orchid with distinctive pseudobulbs and multibranched inflorescences. Requires good light. |
| 7. *Paphiopedilum* Slipper orchid | Excellent houseplant; average conditions; feed regularly as you would an African violet. Mist leaves if air is too dry. Blooms once a year but makes an attractive foliage plant, especially the mottle-leaved types. Temperature: 21° to 29° C (70° to 85° F) during the day; 13° to 18° C (55° to 65° F) during the night. |

8. *Phalaenopsis*
Moth orchid

Monopodial epiphytes with long arching sprays of flowers; may stay in bloom for months. Do not remove old flower stalks as lateral buds often develop which can continue blooming. Grows best under average conditions. Temperature: 18° to 29° C (65° to 85° F) during the day; 15° to 21° C (60° to 70° F) at night. A nighttime temperature drop to 13° C (55° F) for several weeks in the fall may help to produce flower buds. Keep the very loose growing medium evenly moist.

9. *Vanda*

A monopodial orchid that may be medium to very large. Long roots may get tangled in other pots. Needs high light to bloom well. Keep warm (15° to 18° C or 60° to 65° F at night). Smaller types are best suited to indoor growing; keep outside in summer.

## GESNERIADS

Gesneriads (Gesneriaceae) are primarily tropical plants having handsome, usually velvety leaves with tubular, or bell-shaped flowers (Figure 5–12). The flowers are usually large and quite showy. In most cases a diagnostic characteristic of gesneriads is that the anthers (not the filaments) are fused to each other.

Gesneriads generally make good houseplants as they prefer the warm conditions of the home year around. Because many of them are small plants, they make excellent specimens for indoor light gardening. While the African violet is one of the best old-fashioned gesneriads, many new varieties are constantly being produced; and many easy-to-grow genera of Gesneriaceae are finding their way into the trade.

Gesneriads are divided into groups based on the nature of their underground organs: fibrous rooted, rhizomatous, and tuberous. Gesneriads are a marvelous group to collect because of the diversity of flower color, plant size (from miniature-sized terrarium plants to large and robust specimens), growth form (pot plants and hanging basket types), and propagation techniques. They are fairly adaptable to home conditions.

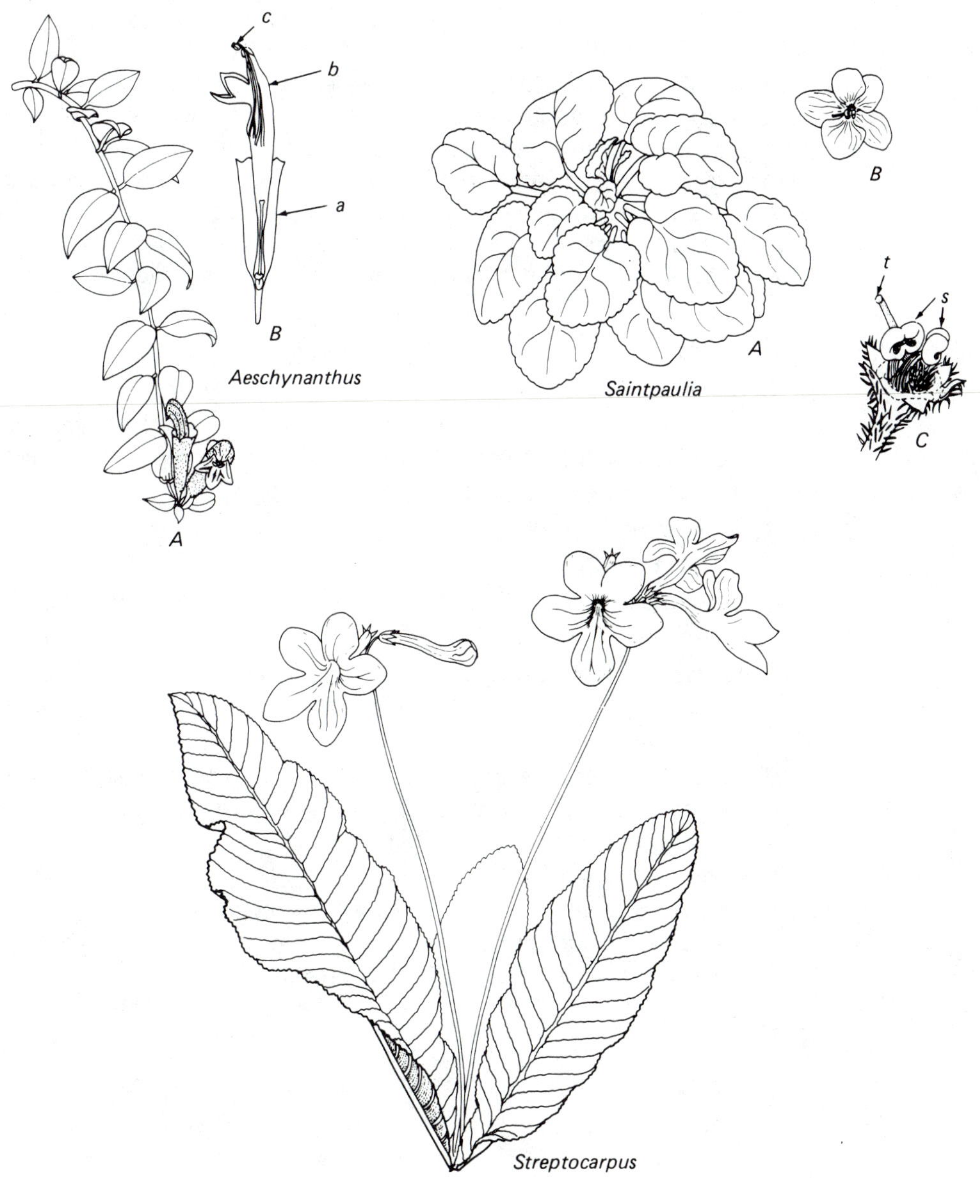


*FIGURE 5–12*
Representative genera of Gesneriaceae. *Aeschynanthus* (lipstick plant): (A) a hanging basket plant with long tubular red flowers; (B) cutaway view of flower (a = sepal tube, b = petal tube, c = stamens protruding from "mouth" of flower). *Saintpaulia* (African violet): (A) rosette-forming pot plant with flat flowers; (B) flower; (C) flower with petals removed (stamens = s and stigma = t). *Streptocarpus* (cape primrose): a large-leaved pot plant with colorful tubular flowers.

One special point to remember is that many of the rhizomatous and tuberous types need a rest period following the blooming period, when water and fertilizer are withheld. The thickened underground parts may then be unpotted and stored in vermiculite or peat moss to await signs of new sprouts before repotting into fresh medium and returning to normal watering routines. Gesneriads are very sensitive to overwatering and drying and should be kept just evenly moist at all times; it is better to keep them on the dry side than too wet. Avoid sudden changes in temperature or humidity, as the leaves are easily damaged by cold water or cold air in winter.

### *General cultural requirements*

1. *Humidity* Thirty to 50 percent.
2. *Water* with room temperature water from top or bottom; keep soil slightly moist. Cold water or sunshine on water droplets may spot foliage.
3. *Temperature* Daytime 21° to 27° C (70° to 80° F) with 5° C (10° F) drop at night.
4. *Growing medium* Basic potting soil mixture or special commercial African violet mix.
5. *Fertilizer* Use weak solution every two to four weeks when plants are actively growing. Use a high phosphorus fertilizer for best flowering.
6. *Light* Most need plenty of bright light. Direct summer sunlight is not recommended, but diffuse light through a south window is ideal. Many thrive under 14 to 16 hours of artificial light per day.
7. *Propagation* Stem cuttings, divisions, rhizomes, or tubers depending on type of plant. Many genera may be propagated by leaf cuttings. Many may also be started from seeds. A most satisfactory group to work with because of the interesting propagation techniques.

### *Gesneriad genera*

1. **Fibrous rooted:** no thickened underground storage organs.

| | |
|---|---|
| a. *Aeschynanthus*<br>Lipstick plant | Make showy hanging baskets. May be pinched or pruned to desired shape. |
| b. *Chirita* | Prefer moderate temperature, light shade, and high humidity. |
| c. *Columnea*<br>Goldfish plant | Mostly pendent or trailing. Prefer to be slightly pot-bound. Like sunniest window (except in summer) and high humidity for bud formation. |

| | |
|---|---|
| *d. Episcia* | Many episcias have leaves with contrasting veins and ever-blooming flowers. Make attractive hanging baskets. Propagate from runners. |
| *e. Hypocyrta* | Everblooming under good conditions. Make nice hanging basket plants. |
| *f. Saintpaulia*<br>African violet | Eleven described species, countless cultivars; not related to garden violet. Must have good light to flower. Prefer fresh air but not drafts. |
| *g. Streptocarpus*<br>Cape primrose | Tolerate lower temperatures, lower humidity, and stronger light than African violets. Some need a two- to three-month dormant period after flowering. |

2. **Rhizomatous:** scaly rhizomes are pinecone-shaped, food-storing structures formed from underground stems. Single scales may be planted like seeds.

| | |
|---|---|
| *a. Achimines* | Require medium light and evenly moist soil. Need a four-month dormant period after flowering. |
| *b. Kohleria* | Attractive foliage. Need staking for upright growth. Cut back after flowering to encourage new growth. |
| *c. Smithiantha*<br>Temple bells | Need a three-month rest period after blooming. |

3. **Tuberous:** tubers are enlarged underground stems.

| | |
|---|---|
| *a. Rechsteineria*<br>Cardinal flower | Need a three- to four-month dormancy period after blooming. |
| *b. Sinningia*<br>Includes florist's gloxinia | After blooming, decrease water and store in a dark place until new growth appears (check frequently). A very large and diverse group, ranging from miniatures to standard sizes. |

## LARGE HOUSEPLANTS

The following are miscellaneous larger plants that are often grown in the house. They are grown in normal potting soil (2 parts loam, 1 part sand, and

1 part leaf mold or peat moss). Other cultural requirements are listed as follows:

1. *Araucaria*
   - *A. excelsa:* Norfolk Island pine
   - *A. bidwillii:* monkey puzzle tree

   Bright indirect light. Keep humus-rich soil evenly moist. Propagate by seed.

2. *Brassaia actinophylla*
   Umbrella tree, Schefflera

   Direct sunlight or bright indirect light. Allow soil to become dry between waterings. Propagate from seeds or layering.

3. *Citrus*
   - Lemon, lime, orange, grapefruit
   - *C. mitis:* calamondin orange

   Bright light necessary. Keep soil evenly moist (especially when fruits are forming). Likes cool temperatures, particularly during winter rest period. Propagate by seeds or cuttings of young growth.

4. *Coffea arabica*
   Arabian coffee tree

   Bright indirect light. Humus-rich soil should be kept barely moist. Propagate by seeds or cuttings of young growth.

5. *Dracaena*
   - *Dracaena fragrans* 'Massangeana': corn plant
   - *D. godseffiana:* gold-dust Dracaena
   - *D. sanderiana:* ribbon plant

   Bright indirect light. Keep soil moist. Propagate by air layering. Some can tolerate low light.

6. *Fatsia*
   Japanese aralia

   Bright light, and cool temperatures. Keep humus-rich soil evenly moist. Propagate by rooting young shoots.

7. *Ficus*
   - *F. benjamina:* weeping fig
   - *F. carica:* edible fig
   - *F. elastica:* rubber plant
   - *F. lyrata:* fiddle-leaf fig

   Bright indirect light. Keep barely moist. Avoid drafts which may cause leaf drop. Propagate by air layering.

8. *Grevillea*
   Silk oak

   Direct sunlight and cool temperatures. Allow to dry between waterings.

9. *Persea americana* Avocado — Bright indirect light; keep evenly moist. Naturally grows as a young tree with widely spaced branches. Prune large specimens as necessary to fit room space. Requires two different specimens for cross-pollination to produce fruit.

10. *Phoenix* Date palm — Bright indirect light. Keep soil evenly moist. Propagate from seeds. Other palms have similar cultural requirements.

11. *Podocarpus* Podocarpus — Bright light and cool temperatures are best. Keep soil barely moist. Propagate from stem cuttings in fall.

12. *Punica granatum* Pomegranate — Bright light. Keep soil evenly moist. Propagate by cuttings or air layering.

## Guide to interior foliage plants

Interior foliage plants are those that have been grown and selected or specially acclimatized to live in the reduced conditions of lighting, humidity, and so forth found in the interiors of modern homes and offices. Climate control in modern buildings is usually designed to meet the comfort requirements of humans; likewise, offices are designed for the convenient location of desks, traffic flow, and specific business activities, not necessarily for the ease of growing living plants that need light, water, humidity, fresh air, and a minimum night temperature. More recently, efforts have been made to make architects and designers aware of the needs of interior plants so that buildings may be designed with these in mind.

Light is usually the most limiting environmental factor encountered when growing foliage plants indoors. Minimum light requirements must be met for a plant to subsist even if other conditions are optimum. As shown in Table 5–3 plants can be categorized as to their light requirements.

*TABLE 5–3*
*Lighting-level categories*

**Low** 75 to 150 foot-candles (fc) of light per 12- hour day[1] (minimum for subsistence 58 fc)
**Medium** 200 fc/12-hour day (min. 75 to 150 fc)
**High** 500 fc/12-hour day (min. 200 fc)
**Very high** 1,000 fc/12-hour day (min. 500 fc)

[1]Values per 12-hour day, 7 days a week.
*Source:* Adapted from Gaines 1977.

These values indicate the amount of light generally found to be necessary to maintain plants in a healthy state with minimal growth. The amount of light indicated is for a 12-hour day, 7 days a week. If only a shorter time for exposure is available, then the light intensity must be increased. For example, if a plant receives 200 fc of light for 12 hours per day, it will be receiving a total of:

$$200 \text{ fc} \times 12 \text{ hours} = 2400 \text{ fc hours in a 24-hour day}$$

If only 100 fc of light are provided, the light must be on for 24 hours to provide the same total amount:

*TABLE 5–4*
*Estimated amounts of light (in foot-candles) available under various situations in the home*

| *Light Source* | *Distance*[1] | *Foot-Candles (fc)*[2] |
|---|---|---|
| Daylight | 1 foot from north window | 220 to 500 fc |
| | 3 feet from north window | 100 to 180 fc |
| | 1 foot from south window | 500 to 900 fc (shade)<br>5600 fc (direct sun) |
| | 1 foot from east or west window | 250 to 400 fc |
| | 2 feet from east or west window | 150 to 250 fc |
| Incandescent bulb[3] | | |
| 75 watt | 1 foot away | 150 fc |
| 100 watt | 3 feet away | 40 fc |
| 150 watt flood | 3 feet away | 90 fc |
| 300 watt spot | 3 feet away | 180 fc |
| Fluorescent tube | | |
| 40 watt 1 tube | 1 foot away | 120 fc |
| | 2 feet away | 75 fc |
| 40 watt 2 tubes | 1 foot away | 240 fc |
| | 2 feet away | 120 fc |
| 40 watt 4 tubes | 1 foot away | 550 fc |
| | 2 feet away | 320 fc |

[1]A quick and fairly reliable technique for estimating foot-candles is to use the built-in light meter in a single-lens reflex camera according to the following formula:

$$\text{Foot-candles} = \frac{20\,(f)^2}{TS}$$

where f= the f. stop number, T= the shutter speed in seconds, and S= the film speed in ASA units. First, take a meter reading 12 inches from a large sheet of white paper (8½″ × 11″) under the lighting situation to be measured and adjust the camera until the meter needle indicates a proper exposure. Then plug the values obtained into the formula and calculate the approximate foot-candle number.

[2]A foot-candle is a measure of the amount of light falling on a one-foot square surface held one foot away from a standard tallow candle. This is an old designation, and more modern ways of expressing irradiance are used; but this can still be useful for indicating relative light intensities.

[3]Incandescent bulbs produce sufficient heat to burn plants placed closer than one foot.

*Source:* Adapted from Graf 1978.

$$100 \text{ fc} \times 24 \text{ hours} = 2400 \text{ fc hours in a 24-hour day}$$

If only 8 hours of light are provided, the intensity must be increased to 300 fc:

$$300 \text{ fc} \times 8 \text{ hours} = 2400 \text{ fc hours in a 24-hour day}$$

In Table 5–4, estimated average amounts of light are given for various situations in the home. In Tables 5–5 through 5–8, selected plants are listed in their light-requirement categories. In addition, desirable temperature ranges are given to help in plant selection.

## *TABLE 5–5*
## *Plants requiring low light*

*75 to 150 foot-candles/12-hour day (min. 50 fc for subsistence)*

| *Plant* | *Temperature Range*[1] |
|---|---|
| *Aglaonema commutatum*<br>Chinese evergreen | warm |
| *Aspidistra elatior*<br>Cast-iron plant | cool |
| *Chamaedorea elegans*<br>Parlor palm | warm |
| *Dracaena deremensis* cv. 'Janet Craig'<br>Janet Craig dracaena | warm |
| *Dracaena deremensis* cv. 'Warneckii'<br>Warneckii dracaena | warm |
| *Dracaena fragrans* cv. 'Massangeana'<br>Corn plant | warm |
| *Dracaena godseffiana*<br>Gold-dust dracaena | warm |
| *Epipremnum aureum (Scindapsus aureus)*<br>Devil's ivy; golden pothos | warm |
| *Maranta leuconeura* cultivars<br>Prayer plant | intermediate |
| *Philodendron scandens*<br>Common philodendron | warm |
| *Philodendron selloum*<br>Selloum philodendron | intermediate |
| *Pteris* cultivars<br>Table fern | intermediate |
| *Spathiphyllum* cv. 'Mauna Loa'<br>Peace lily | warm |
| *Zebrina pendula*<br>Wandering Jew | intermediate |

[1]Temperature designations: *warm*—21° to 23° C (70° to 75° F) day and 16.5° to 18.5° C (62° to 65° F) night; *intermediate*—15° to 21° C (60° to 70° F) day and 10° to 13° C (50° to 55° F) night; *cool*—10° to 15° C (50° to 60° F) day and 4.5° to 7.5° C (40° to 45° F) night.
*Source:* Data adapted from Gaines 1977 and Hawkey 1974.

## TABLE 5–6
## Plants requiring medium light

| *200+ foot-candles/12-hour day (min. 75 to 100 fc for subsistence)* Plant | Temperature Range[1] |
|---|---|
| *Aphelandra squarrosa* Zebra plant | warm; humid |
| *Aucuba japonica* Gold-dust tree | cool |
| *Araucaria heterophylla* Norfolk Island pine | intermediate |
| *Brassaia actiniophylla* Schefflera | warm |
| *Chamaedorea erumpens* Bamboo palm | warm |
| *Chlorophytum comosum* Spider plant | intermediate |
| *Cissus rhombifolia* Grape ivy | intermediate |
| *Coffea arabica* Common coffee | warm |
| *Cordyline terminalis* var. *minima* Baby-doll dracaena | warm |
| *Crassula argentea* Jade plant | intermediate |
| *Cycas revoluta* Sago palm | intermediate |
| *Cyrtomium falcatum* Japanese holly fern | intermediate |
| *Dieffenbachia amoena* Dumb cane | warm |
| *Dracaena marginata* cv. 'Tricolor' Striped dragon tree | warm |
| *Fatsia japonica* Japanese fatsia | cool |
| *Ficus benjamina* Weeping fig | warm |
| *Ficus lyrata* Fiddle-leaf fig | warm |
| *Fittonia argyroneura* Mosaic plant | warm |
| *Gardenia jasminoides* Common gardenia | intermediate |
| *Hedera helix* cultivars English ivy | intermediate |
| *Howea forsteriana* Kentia palm | intermediate |
| *Nephrolepis exaltata* cv. 'Bostoniensis' Boston fern | intermediate |
| *Nephrolepis exaltata* cv. 'Fluffy-Ruffles' Fluffy-ruffles fern | intermediate |

## TABLE 5–6 (Continued)

| | |
|---|---|
| *Peperomia* species<br>Peperomia | intermediate |
| *Plectranthus australis*<br>Swedish ivy | intermediate |
| *Polyscias balfouriana*<br>Balfour aralia | warm |
| *Polyscias fruticosa*<br>Ming aralia | warm |
| *Tolmiea menziesii*<br>Piggyback plant | intermediate |

[1]Temperature designations: *warm*—21° to 23° C (70° to 75° F) day and 16.5° to 18.5° C (62° to 65° F) night; *intermediate*—15° to 21° C (60° to 70° F) day and 10° to 13° C (50° to 55° F) night; *cool*—10° to 15° C (50° to 60° F) day and 4.5° to 7.5° C (40° to 45° F) night.

*Source:* Data adapted from Gaines 1977 and Hawkey 1974.

## TABLE 5–7
## Plants requiring high light

*500+ foot-candles/12-hour day (min. 200 fc for subsistence)*

| *Plant* | *Temperature Range*[1] |
|---|---|
| *Aloe vera*<br>Burn plant, med-plant | warm |
| *Asparagus densiflorus* cv. 'Sprengeri'<br>Sprengeri asparagus fern | intermediate |
| *Beaucarnea recurvata*<br>Pony-tail palm | warm |
| *Begonia* species and cultivars<br>Begonia<br>Foliage types can take less light<br>Flowering types may need more light | intermediate |
| *Bryophyllum (Kalanchoe) daigremontiana*<br>*Bryophyllum tubiflora* (produces plantlets on leaves)<br>Mother-of-thousands | intermediate |
| *Caryota mitis*<br>Fish-tail palm | warm |
| *Chamaerops humilis*<br>European fan palm | intermediate |
| *Cibotium chamissoi*<br>Tree fern | warm; humid |
| *Coleus blumei*<br>Coleus | intermediate |
| *Cordyline terminalis* cv. 'Bicolor'<br>Hawaiian ti plant | warm |
| *Dizygotheca elegantissima*<br>False aralia | warm |
| *Epiphyllum* species<br>Orchid cacti (jungle plants, epiphytic) | intermediate |

*(Continued)*

*TABLE 5–7 (Continued)*

| Plant | Temperature Range |
|---|---|
| *Euphorbia millii*<br>Crown of thorns | intermediate |
| *Gynura aurantiaca*<br>Velvet plant, purple-passion plant | warm |
| *Hoya carnosa*<br>Wax plant | intermediate |
| *Kalanchoe blossfieldiana*<br>Kalanchoe | intermediate |
| *Peperomia* species<br>Peperomia | intermediate |
| *Podocarpus macrophyllus*<br>Podocarpus | intermediate |
| *Sansevieria trifasciata*<br>Snake plant | intermediate |
| *Sansevieria trifasciata* cv. 'Golden Hahnii'<br>Golden hahnii sansevieria | intermediate |
| *Yucca elephantipes*<br>Spineless yucca | intermediate |

[1]Temperature designations: *warm*—21° to 23° C (70° to 75° F) day and 16.5° to 18.5° C (62° to 65° F) night; *intermediate*—15° to 21° C (60° to 70° F) day and 10° to 13° C (50° to 55° F) night; *cool*—10° to 15° C (50° to 60° F) day and 4.5° to 7.5° C (40° to 45° F) night.
*Source:* Data adapted from Gaines 1977 and Hawkey 1974.

*TABLE 5–8*
*Plants requiring very high light*

| *1000+ foot-candles/12-hour day (min. 500 fc for subsistence)* | |
|---|---|
| *Plant* | *Temperature Range*[1] |
| Cacti<br>Desert cacti—tolerate dry air, dry soil | warm summer;<br>intermediate winter |
| *Citrus mitis*<br>Calamondin orange | intermediate |
| *Codiaeum variegatum* (many cultivars)<br>Croton | warm |
| *Euphorbia tirucalli*<br>Pencil tree | intermediate |
| *Euphorbia* species<br>Succulent desert euphorbias | warm, summer and winter |
| *Hibiscus rosa-sinensis*<br>Japanese hibiscus | warm |
| *Jacobinia carnea*<br>Jacobinia | warm |

Most flowering houseplants need very high light and intermediate temperatures.

[1]Temperature designations: *warm*—21° to 23° C (70° to 75° F) day and 16.5° to 18.5° C (62° to 65° F) night; *intermediate*—15° to 21° C (60° to 70° F) day and 10° to 13° C (50° to 55° F) night; *cool*—10° to 15° C (50° to 60° F) day and 4.5° to 7.5° C (40° to 45° F) night.
*Source:* Data adapted from Gaines 1977 and Hawkey 1974.

## Growing plants in hanging containers

One of the nicest ways to utilize the vertical space available in a greenhouse, in an office, or at home is to grow plants in hanging pots or baskets. It helps to soften harsh walls and ceilings, provide focal points in a room, and better utilize the space available in a greenhouse above the benches or floor beds. To help you get started in the fine art of growing plants in hanging pots or baskets, we shall explore first the kinds of containers that may be used.

### CONTAINERS FOR SUSPENDED PLANTS

Several types of hanging containers are useful. The most common is the wire basket. Such containers must be lined with suitable material to retain the potting soil mixture in the basket. The advantage of a wire basket is its light weight compared to a wood basket. Another type of hanging basket is made of plastic. These tend to be a little less expensive than wire baskets and have the advantage of not rusting; however, wire baskets may be repainted periodically to avoid rust. Plastic baskets, like wire baskets, must be lined with a material such as plastic film that will retain the soil mix in the basket. A third type of hanging basket is made of wood. Shapes may vary, but they are made of slats of wood to reduce weight. To prevent rotting of the basket itself, the wood must be treated with a fungicide-type wood preservative such as Cuprinol which contains copper. These baskets are very attractive but have a much shorter "life expectancy" than wire or plastic baskets.

Hanging pots may be made of glass, plastic, or ceramic materials. Coupled with macramé hanging devices, such containers can be unusually artistic. Saucers are necessary for pots with holes in the bottom.

### PREPARING A HANGING CONTAINER

The basic directions we present here apply to both hanging pots and to hanging baskets. However, with hanging baskets, we must add one additional step, that of lining the basket.

For hanging baskets, we can use presoaked sphagnum moss, sheet moss, partly decomposed leaves, or plastic film for lining. If leaves or sphagnum moss are used, the layer should be about one inch thick to assure that the soil mixture is retained in the basket, especially after watering.

The next step, that of preparing the soil mixture, applies to both hanging baskets and pots. One of the best formulas is as follows: 1 part loam: 1 part leaf mold: 1 part perlite: 1 part peat moss. This potting mix is then added to your container until it is half-filled. Now the container is ready for plants.

Decide whether you want a pendent species, an upright one, or a combination of both. In combination, upright species are usually planted in the middle of the container, and pendent species are planted around the periphery of of the container. A number of plants can be alternatively planted along the sides of a hanging basket.

The final step is to suspend your pot or basket on an appropriate hook or support, making sure that the environmental conditions are favorable for the plants you have selected.

## SOME PLANTS USEFUL FOR HANGING CONTAINERS

There are literally hundreds of species of plants that can be grown in hanging baskets and pots. The following lists include some of those species that in our experience have proved to be particularly successful in these containers.

### *Plants to grow upright in the center of a basket or pot*

1. *Fuchsia*
2. *Begonia* (upright cultivars)
3. *Browallia*
4. *Coleus blumei*
5. *Epiphyllum* (orchid cactus)
6. *Pelargonium* (upright cultivars) (geraniums)
7. *Asparagus plumosus* (asparagus fern)
8. *Nephrolepis exaltata* cv. *'Bostoniensis'* (Boston fern)
9. *Schlumbergera truncata* (Thanksgiving cactus)

### *Plants to grow pendent at the periphery of a basket or pot*

1. *Gibasis* (Tahitian bridal-veil)
2. *Pelargonium* (pendent cultivars)
3. *Chlorophytum* (spider plant)
4. *Senecio mikanioides* (German ivy)
5. *Cissus rhombifolia* (grape ivy)
6. *Zebrina pendula* (wandering Jew)
7. *Tradescantia* (spiderwort)
8. *Commelina*
9. *Fittonia*
10. *Petunia hybrida*
11. *Tropaeolum majus* (nasturtium)
12. *Columnea*
13. *Episcia*
14. *Sedum morganianum* (donkey's tail)
15. *Setcreasia*
16. *Saxifraga sarmentosa* (strawberry begonia)

Hanging pots are relatively easy to suspend. Generally, they are hung from screw eyes or hooks screwed into wood. Ceilings pose another problem. Fortunately, special hooks have been devised to take care of this problem. These hooks (toggle bolts) have special wings on them that slip up through the hole you make in the ceiling, then spread out on the upper side to suspend the hook.

Hanging baskets are also suspended from hooks or screw eyes. Many baskets already have a hook made from the three wires that suspend the basket. This hook can be slipped over a pipe, branch, a hook, or a screw eye.

## *Maintaining plants in hanging containers*

1. Water the plants well whenever they need it, usually at least once a week. Cacti may be watered less frequently, while types like petunias, wandering Jew, and nasturtium may have to be watered more often. If your houseplants should dry out thoroughly, it may be difficult to rewet the soil unless you soak it, pot and all, in a tub of water for 30 minutes or more. The use of a wetting agent in the original soil mix may help in rewetting peat-based soil mixes.
2. Periodically remove dead leaves because they harbor disease and are unsightly. Pinch off old flowers after bloom to encourage new buds or vegetative growth to form.
3. Repot your plants periodically. Some slow-growing plants may last for several years in a hanging pot, while fast growers like Swedish ivy and wandering Jew may need to be restarted by rooting cuttings every year to keep them looking full and healthy. Repotting allows you to divide an overgrown plant or to restart new cuttings in fresh soil.
4. Fertilize your plants as often as necessary to keep them looking healthy. Some may need fertilizing several times a month, especially flowering plants. Use a balanced (10-10-10) fertilizer for foliage plants, and a high phosphorus mix (10-30-10) for flowering plants. Use the fertilizer at the recommended strength, or even less, so as not to burn the roots. It is better to fertilize *weakly,* and do it *weekly*! Unless plants are grown under optimal conditions it is better not to force growth by constantly fertilizing. Fish emulsion (5-1-1) and other organic fertilizers are good for foliage plants. They will not burn the roots or cause salt buildup. Flush hanging baskets with fresh water occasionally and let the water drip through copiously to help remove excess fertilizer salt buildup.
5. Watch for pests like mealybug, red spider mite, whitefly, and aphids. (See Chapter 8 for methods of pest control.)
6. As with any houseplants, never place them unprotected in full summer sun after they have been growing indoors, in the shade, or in a green-

house. Sunburn may result if the plants have not had a chance to protect themselves from the burning ultraviolet rays. Plants grown in full sun generally develop a reddish tinge to the leaves; this is a natural sun screen pigment and is analagous to tanning in humans.

## Reblooming seasonal houseplants

You will often receive gift houseplants such as azaleas, poinsettias, cyclamens, lilies, hydrangeas, gloxinias, and Christmas cactus. Remember that all of these plants should be kept cool (18° C or 65° F or less) and out of full, direct, hot sunlight that might burn the leaves. (For example, they should not be set against a hot, sunny window.) The plants should be kept evenly moist, but not wet; and they would probably benefit from moderate humidity (30 to 50 percent). Since these plants will have been produced for special occasions by forcing them into bloom under optimal controlled conditions in a greenhouse, it will be difficult to rebloom them satisfactorily in the home, where the precise temperature, light, humidity, and fertilization schedule cannot be maintained. The best thing to do is to throw them away after the period of enjoyment and not worry about them. If you have a greenhouse, or would just like to try your luck at reblooming them (realizing that most of today's gift plants are selected for their performance under ideal conditions and careful attention to a regimented forcing schedule), then consider the following guidelines.

All of these plants need a period of "drying off" and resting sometime after they bloom. Keep them cool and give them just enough water to keep them from wilting. The bulbous plants (lilies, cyclamen, and gloxinia) need full sun and regular feeding while they "ripen" their leaves and store food in their bulbs, corms, or tubers, respectively. Then, keep them cool and dry until time to start the new growth cycle, usually anytime during the year after a two- to four-month rest period. You may even take the "bulbs" out of the pot and store them separately. The other plants can be kept under average houseplant conditions while they rest, watering only enough to keep them from severely wilting or shriveling. The azalea and hydrangea will bloom in the spring after several weeks of cool (10° to 15° C or 50° to 60° F) temperatures and regular weak feeding. Azaleas should be pruned back to induce new growth immediately after blooming. Hydrangeas should be pruned back to lower buds on old stems just before new growth is desired. Christmas cactus should not be pruned unless you want to make cuttings. Poinsettias should be kept cool and dry after blooming. Then, in mid-August take six-inch cuttings, root them with bottom heat, and start the forcing process by keeping them warm (18° to 21° C or 65° to 70° F) at night, in full sun, and apply a balanced fertilizer regularly. Operators of commercial greenhouses fertilize their poinsettias two or three times a week with the equivalent of 30 milliliters (2 tablespoons) per 3.75 liters (one gallon) of 10-20-10 fertilizer. Old-fashioned poinsettias needed short days (or long nights, greater than 12½ hours) to induce flower buds; and this had to be started in September by covering the plant benches with dark cloth to artificially provide long nights. The cuttings took 14 to 16 weeks of growth to bloom and have colorful bracts (leaves) by Christmas. Modern poinsettia cultivars (which are protected from unauthorized propagation by plant patents), however, are selected to bloom on schedule under normally existing lighting regimes and optimal greenhouse conditions. The science of forcing seasonal plants is highly exacting, and environmental conditions and watering schedules must be monitored closely all along the way. A few degrees difference in temperature can radically change (especially delay) a forcing schedule, and a greenhouse full of poinsettias is useless the day after Christmas! Christmas and Thanksgiving cacti are the easiest seasonal plants to induce to rebloom. Low temperatures and long nights in the fall, along with a dry period in September and October, are enough to induce healthy plants to bloom on schedule. Many modern hybrids have been produced which may bloom at any season under average home conditions, however.

## REFERENCES

### *General*

Ball, Inc. 1972. *The Ball red book.* Chicago: George Ball. (A commercial grower's manual containing much useful information.)

Baumgardt, John P. 1972. *Hanging baskets for home, terrace, and garden.* New York: Simon & Schuster. (An interesting and useful book.)

Crockett, James U. 1971. *Flowering house plants.* New York: Time-Life Encyclopedia of Gardening. (Good for beginners.)

———. 1967. *Foliage plants for indoor gardening.* New York: Doubleday and Co. (Good for beginners.)

Cruso, Thalassa. 1969. *Making things grow: a practical guide for the indoor gardener.* New York: Alfred A. Knopf.

Elbert, George and Virginie. 1974. *Plants that really bloom indoors.* New York: Simon & Schuster.

Fitch, Charles Marden. 1972. *The complete book of house plants.* New York: Quadrangle Books.

Graf, Alfred B. 1976. *Exotic house plants.* 10th ed. East Rutherford, N.J.: Roehrs Co. (Compact identification manual.)

———. Graf, Alfred B. 1978. *Exotic plant manual: exotic plants to live with.* 5th ed. East Rutherford, N.J.: Roehrs Co. (Large size identification manual.)

———. Graf, Alfred B. 1978. *Exotica III: pictorial cyclopedia of exotic plants.* 9th ed. East Rutherford, N.J.: Roehrs Co. (Standard tome for houseplant identification—you should learn to use it, even if you can carry it.)

Hay, Roy, and Synge, Patrick M. 1969. *The color dictionary of flowers and plants for home and garden.* New York: Crown Publishing Co. (All color identification reference.)

Herwig, Rob, and Schubert, M. 1979. *A treasury of houseplants.* New York: Macmillan. (Color pictures for identification.)

Kaufman, P. B., and LaCroix, D. 1979. *Plants, people, and environment.* New York: Macmillan. (A diversity of topics about plants, their uses, and their importance to people.)

Larson, Roy. 1981. *Introduction to floriculture.* New York: Academic Press. (Growing plants on a commercial scale; good information for the home gardener, too.)

Menage, R. H. 1974. *Growing exotic plants indoors.* Chicago: H. Regnery Co. (Many useful tips.)

Northen, Henry T. and Rebecca T. 1973. *The complete book of greenhouse gardening.* New York: John Wiley and Sons.

Readers Digest. 1979. *Success with house plants.* New York: W. W. Norton Co. (An excellent general book.)

Wright, Michael, ed. 1979. *The complete indoor gardener.* 2d ed. New York: Random House. (A general houseplant reference.)

***Ferns***

Foster, F. Gordon. 1971. *Ferns to know and grow.* New York: Hawthorn Books. (Mostly hardy outdoor ferns.)

Hoshizaki, Barbara J. 1975. *Fern growers manual.* New York: Alfred A. Knopf. (Best book for indoor ferns.)

***Succulents***

Chidamian, Claude. 1958. *The book of cacti and succulents.* Garden City, N.J.: Doubleday and Co. (A good book for beginners.)

Jacobsen, Hermann. 1974. *The lexicon of succulent plants.* England: Blandford Press. (Important reference.)

Rowley, Gordon D. 1978. *Illustrated encyclopedia of succulents, including cacti.* New York: Crown Publishing Co. (A must for succulent enthusiasts on the biology of succulents.)

***Bromeliads***

Padilla, Victoria. 1973. *The bromeliads.* New York: Crown Publishing Co. (The best general guide.)

Rauh, Werner. 1978. *Bromeliads: for home, garden, and greenhouse.* England: Blandford Press. (An advanced reference.)

***Orchids***

American Orchid Society (AOS). 1976. *Handbook on orchid culture.* (Available from AOS, Inc., 84 Sherman St., Cambridge, Mass. 02140.) (For beginners.)

Northen, Rebecca T. 1970. *Home orchid growing.* 3d ed. New York: Van Nostrand Reinhold Co. (The most complete reference on home orchid growing.)

______. 1976. *Orchids as house plants.* New York: Dover. (A smaller version of the above.)

Oregon Orchid Society. 1974. *Your first orchids and how to grow them.* Portland: Oregon Orchid Society. (A handy booklet for beginners; available from AOS.)

Williams, B. et al. 1980. *Orchids for everyone.* New York: Crown Publishing Co. (An excellent book on orchid culture with many color pictures and "how to" items.)

### Gesneriads

Elbert, Virginie F. and George A. 1976. *The miracle houseplants.* New York: Crown Publishing Co. (An excellent book covering all aspects of gesneriads.)

### Interior Plants

Gaines, Richard. 1977. *Interior plantscaping.* New York: Architectural Record Books. (An important book on interior uses of plants.)

Hawkey, William S. 1974. *Living with plants.* New York: W. Morrow and Co. (A textbook on interior uses of plants.)

### Indoor Light Gardening

Elbert, George A. 1975. *Indoor light gardening book.* New York: Crown Publishing Co.

McDonald, Elvin. 1974. *The complete book of gardening under lights.* New York: Popular Library.

### Encyclopedias and Series

Brooklyn Botanic Gardens Handbooks:

- #40 *House plants*
- #42 *Greenhouse handbook for the amateur*
- #43 *Succulents*
- #53 *African violets and their relatives*
- #54 *Orchids*
- #59 *Ferns*
- #62 *Gardening under artificial lights*
- #70 *House plant primer*
- #75 *Breeding plants for home and garden*
- #81 *Bonsai for indoors*

Time-Life Encyclopedia of Gardening:

- *Flowering house plants*
- *Foliage house plants*
- *Cacti and succulents*
- *Ferns*
- *Orchids*
- *Miniature gardens and terraria*
- *Gardening under artificial lights*

# Cloning plants 6

Cloning plants is a way to increase the number of plants without using seeds. It is done by vegetative propagation, also called asexual reproduction. Whereas sexual reproduction results in seeds, vegetative propagation results in genetically identical individuals (clones). This is because several genetic mechanisms for recombining traits have been avoided. A *clone* is thus a group of genetically identical individuals produced by vegetative propagation.

Several reasons for propagating plants vegetatively include: (1) it is usually *quicker*—larger plants can be obtained earlier than by starting from seeds; (2) the exact same kind of plant can be produced in *large quantity;* (3) it provides a way of obtaining *disease-free plants;* (4) it is a way to *maintain unique characteristics* of a cultivar whose seeds will not result in an alike plant; and (5) it provides *plants from seedless specimens,* such as sterile hybrids or plants whose flowers are self-incompatible and thus produce no seeds.

It is important to know the major methods for propagating plants vegetatively. Vegetative reproduction may involve specialized organs or it may involve cuttings of typical roots, stems, and leaves which have the ability to regenerate a complete plant. Our discussion will focus on (1) bulbs, corms, and other food storage structures; (2) runners, offsets, and other stem structures; (3) cuttings; (4) layering; (5) grafting; and (6) mericloning.

## Bulbs, corms, tubers and tuberous roots, rhizomes, and bulblets

You can learn to multiply and force your own bulbs. For example, you can obtain new bulbs from lily scales or mature hyacinth bulbs, start tubers of Jerusalem artichoke as well as the ordinary Irish potato, and culture bulbs of amaryllis and tubers of begonia. All of these specialized reproductive structures are food-storage organs.

### BULBS

A bulb is a stem axis (the basal plate) which has shortened internodes and bears fleshy leaf bases (bulb scales) at the nodes (Figure 6–1). Nodes represent regions of a shoot where leaves are attached, and internodes are stem axis regions between the nodes. Bulbs represent one means of vegetative reproduction and overwintering. During the growing season lateral buds in the axils of the bulb scales develop into new bulblets. These eventually separate from the parent bulb, and after several seasons of storing food manufactured by the leaves, they become mature enough to flower. Thus, removal of the green leaves after flowering prevent development of the next season's bulbs from the parent bulb and subsequent food storage. Bulbs are all characterized by fleshy leaf bases which permit the storage of a considerable quantity of carbohydrates and nutrients, thus allowing an early start on the next season's growth. In fact, when mature, all the bulbs contain complete

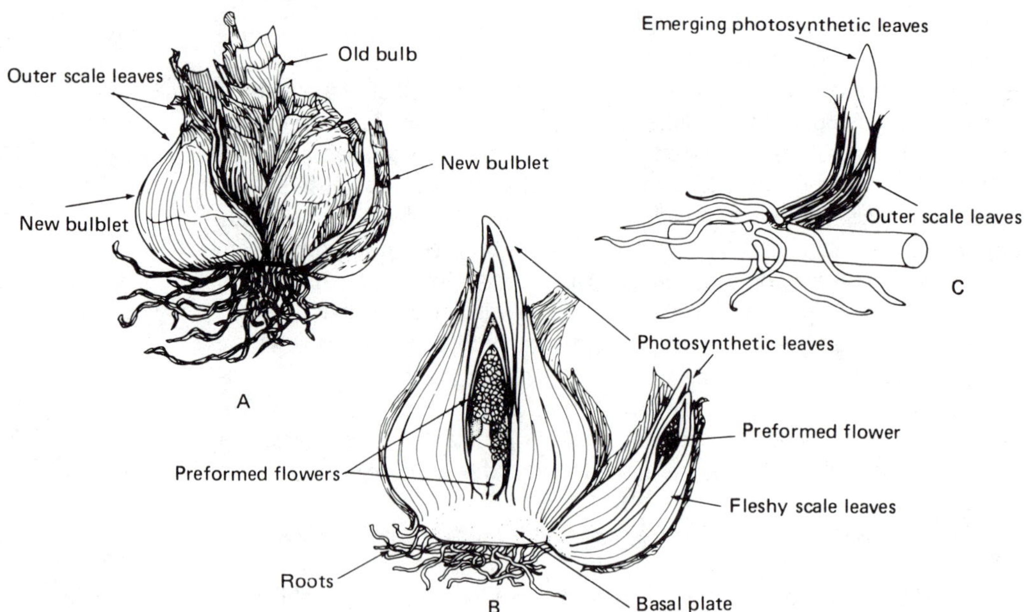


*FIGURE 6–1*
Bulbs and bulblike structures: (A) external view of a tunicate bulb (narcissus); (B) Longitudinal section through a tunicate bulb (hyacinth); (C) bulblike lateral bud ("pip" on lily of the valley rhizome).

plants with enough carbohydrate and nutrients to carry the plant through its blooming period. For a subsequent season, however, the new green leaves must again store new supplies of carbohydrates and re-form storage leaves in the new bulb. This is why it is important to allow the leaves to remain after flowering, until they turn yellow or brown and begin to wither. This is called ripening.

*TYPES OF BULBS.* The two basic types of bulbs are *tunicate* and *scaly.* Tunicate bulbs, such as hyacinth, tulip, daffodil, and amaryllis, have a compact structure of tightly fitting, overlapping, fleshy, nonphotosynthetic leaf bases and a looser, dry outer papery covering that helps resist drying and tends to retard decay and mechanical injury. Scaly bulbs, such as those of lily, lack the outer protective dry papery covering and compact structure of tunicate bulbs. Consequently, these bulbs are easily injured by drying and mechanical abrasion. In storage they need to be kept slightly damp. They fare best when kept in soil, vermiculite, or other loose material such as wood shavings, which serves to prevent drying.

*PROPAGATION OF TUNICATE BULBS.* Tunicate bulbs are progagated vegetatively by separating the bulblets that develop by the end of the growing season from the parent bulb. Hyacinth bulbs offer a special type of vegetative propagation. Invert the bulb and cut out the basal plate with a knife. Then make V-shaped cuts around the periphery of the bulb where the basal plate was removed. Leave the bulb exposed, and in a few days callus will form. *Callus* is undifferentiated parenchymal tissue, the basic living tissue of a plant. Then, half-bury the bulb upside down in moist sand. New bulblets will regenerate from the wounds made in the base of the bulb.

Tunicate bulbs such as tulip and daffodil are easily divided at the end of the growing season, after they have had time to store food and develop young bulblets. The entire bulb cluster is dug up, the parent bulb scales removed, and young bulblets separated from the old withered parent bulb and flowering stalk. The small bulblets are planted in the autumn. During their first year they produce only leaves. After two or three years, they will initiate flower shoots (inflorescences).

*PROPAGATION OF SCALY BULBS.* Scaly bulbs are propagated by removing the fully turgid, healthy scales one at a time, dusting the bases of each with a rooting hormone powder, such as Rootone, to help induce formation of new bulblets from the bases of the scales. Then, the scales are half-buried in moist sand. They will produce adventitious roots and buds from the scales. These "budded" scales can be planted directly in the garden for the hardy types of lilies. Nonhardy ones are planted indoors.

*FORCING BULBS TO FLOWER IN WINTER.* Certain cultivars of almost all spring bulbs (tulip, hyacinth, bulbous iris, Roman and grape hyacinths, and narcissus) can be induced or forced to bloom indoors in the winter (Figure 6–2). It is important to select bulb varieties which are recommended for

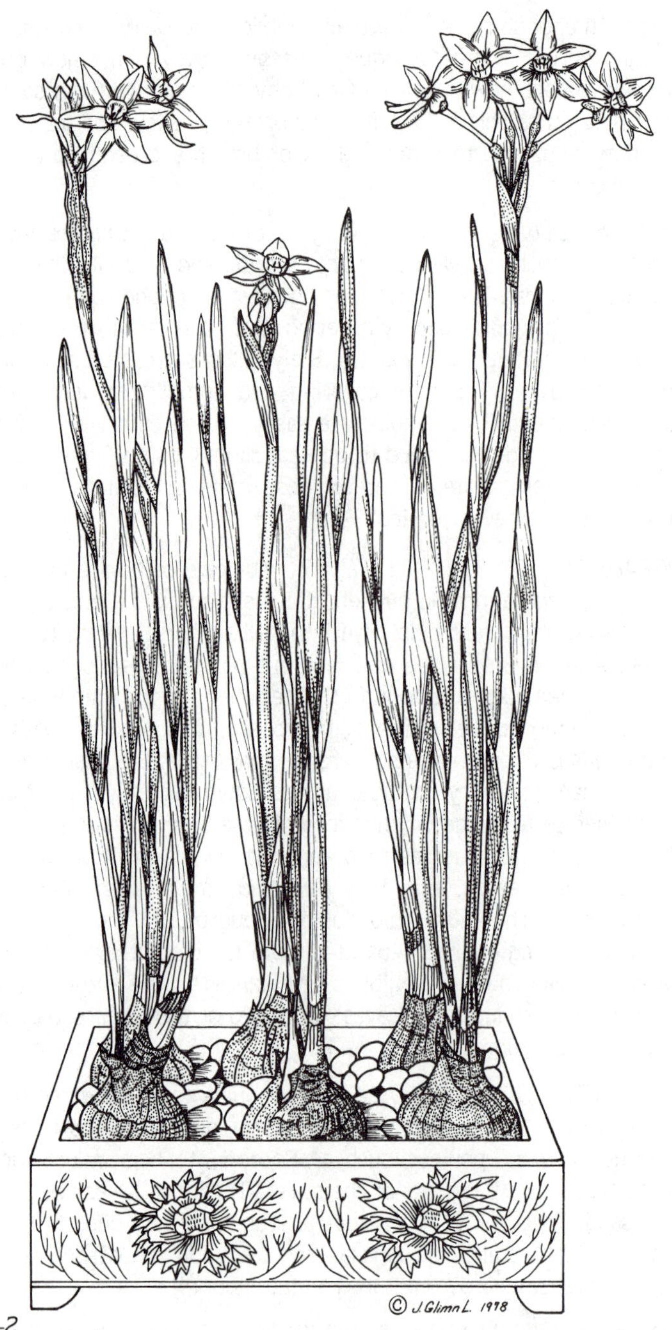


*FIGURE 6–2*
Forced paper white narcissus bulbs. To force, the bulbs are held in place by pebbles in a container and kept in a refrigerator for six weeks, moved to a cool, dark closet for one week, and then into a warm, lighted room where flower shoots emerge.

forcing. Forcing bulbs requires special media, planting, and care. For media, a mix providing good drainage and organic matter to hold moisture is desirable. A standard potting mix with a soil base of a peatlike mix may be used. (See Chapter 4, Table 4–4.) Because bulbs contain sufficient nutrients for forcing, fertilizers are not necessary.

Hyacinths and narcissus can be grown in water and pebbles or pearl chips. A special hourglass-type container is available for forcing hyacinth bulbs; it supports the bulb while allowing roots to grow in water (Figure 6–3). Select a container which will hold water and is deep enough to allow

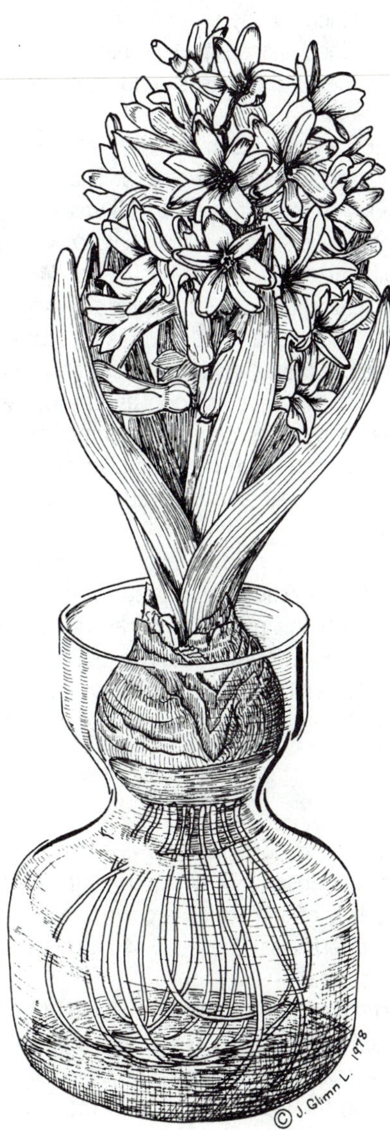


*FIGURE 6–3*
Forced Roman hyacinth bulb. To force, scrape the bulb base clean with a knife or spoon and place in an hourglass jar, water-fill to base of bulb. Refrigerate at 4° C (about 40° F) for six weeks, maintaining the water level. Then remove and maintain at room temperature, out of direct sunlight, until the flower stalk emerges.

adequate room for root growth; otherwise, the bulb is likely to "heave"; that is, to be pushed up and topple over as its roots develop. In a shallow container pebbles should be at least one-half inch below the pot rim and yet cover the shoulder of the bulb. If several larger bulbs are planted in the same container, they should be at least 1.25 centimeters (one inch) apart, the smaller ones one-half inch apart. Keep the water level at the base of the bulb.

If a soil mix is used, place a piece of pot shard over the pot's drainage hole and add a base layer of mix. Plant the bulb tops level with the pot rim and keep the soil level one-half to one inch below the pot rim. The number of bulbs per pot will be determined by the size of the bulbs and the size of the pot. Space large bulbs one inch apart, smaller ones one-half inch apart. When planting tulips the flat side of the bulb should face inward. Firm the soil in the pot.

After watering, the bulbs are placed in the dark at about 10° C (50° F) for 6 to 8 weeks until the roots are well developed. Then the temperature should be lowered to just above freezing for 13 to 15 weeks. A refrigerator or a cold, but not freezing, garage is suitable. After one to two inches of leaf growth are showing, the bulbs are moved to a cool room (10° C or 50° F) with indirect light until the leaf shoots are four to six inches tall. When brought out for floral display, a cool, sunny situation is recommended. Paper white narcissus do not require prolonged cold treatment and can be brought out of the dark, cool room into the light as soon as one or two inches of leaf growth appear. If facilities with some degree of temperature control are not available, the bulbs can be placed in an outdoor pit or cold frame in the garden. The entire pot is placed in peat and covered with a layer of hay or straw for insulation.

Summer bulbs, which are tender, can be grown for indoor as well as outdoor display. These bulbs do not require cold pretreatment so are started in the light at normal greenhouse or room temperature. Some commonly planted ones include amaryllis and crinum. Amaryllis bulbs are potted in a container just an inch or two larger than the bulb. Fill in the potting soil up to midlevel of the bulb. Plant crinum so that the tip of the bulb is barely covered.

## CORMS

A corm is a type of enlarged underground stem. It is essentially the swollen base of a solid stem axis that is enclosed by dry papery leaf bases. In contrast to a bulb, which consists primarily of leaf scales, a corm is a solid stem structure with distinct nodes and internodes (Figure 6–4). Crocus and gladiolus are examples of plants with corms.

Corms are treated much the same as bulbs. Development of new corms takes place during the summer months after flowering, so it is essential not to remove the leaves after flowering for the best possible corm development. This is the time when food from the leaves is stored in the newly developing

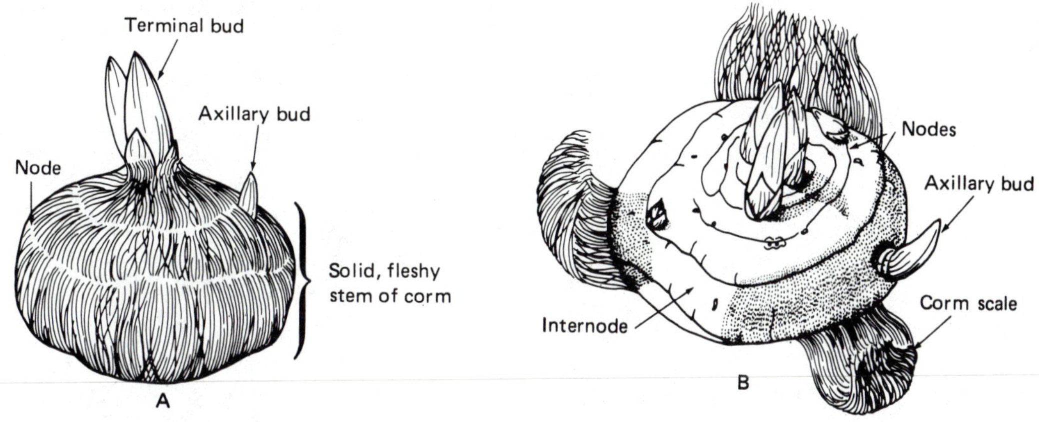


*FIGURE 6–4*
Crocus corms. A surface view shows scales (A) intact and (B) pulled back to expose stem structure of the corm.

corms. In the fall, the corms can be dug, divided, and the new cormels (small corms) of tender species such as gladiolus stored over winter to be planted in the spring. The cormels of hardy species such as crocus may be discarded after dividing as it will take a year or so before these corms are large enough to produce flowers. Large, well-developed hardy corms are planted in the fall. Root growth occurs during late fall and winter, followed by flowering in the spring. Tender corms such as gladiolus should be planted in late spring for summer flowering.

Corms, like bulbs, can also be forced. Following their planting in gravel, stones, or soil, they may be placed in the refrigerator or under straw in a protected place such as a cold frame, so that they can be removed or dug up in midwinter to be forced. Non-hardy summer flowering corms such as gladiolus are not usually forced, but hardy crocus corms are frequently forced.

## TUBERS AND TUBEROUS ROOTS

It is important to distinguish between tubers and tuberous roots because they are very different structurally and can be treated differently in propagation. Since tubers are enlarged stems with nodes and buds, they can be cut into pieces and planted directly. Tuberous roots, on the other hand, must be cut so that a piece of stem is attached or allowed to sprout new stems from one end without cutting it up.

Tubers are the swollen ends of underground stems or rhizomes; in contrast, tuberous roots are swollen roots. The most familiar example of a tuber is that of the white or Irish potato *(Solanum tuberosum)* (Figure 6–5). Less familiar examples are the edible tubers of Jerusalem artichoke *(Helianthus tuberosus)*, arrowhead *(Sagittaria)*, and nutgrass *(Cyperus esculentus)*. Tuberous roots are classically seen in dahlia, in tuberous-rooted be-

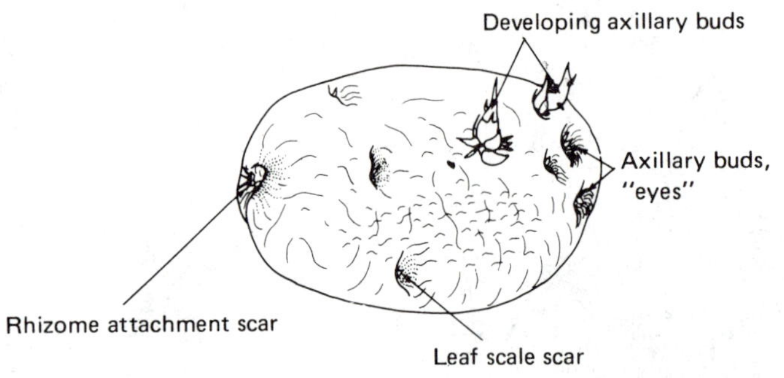


*FIGURE 6–5*
Mature Irish potato tuber.

gonias and sweet potato (Figure 6–6). We also encounter tuberous roots in indoor gardening with many members of the gesneriad family (Gesneriaceae) such as *Sinningia* and *Gesneria.* One of the largest tuberous roots known is that of manroot *(Echinocyctis fabacea),* a common weed in California; roots of this plant can grow to almost a meter in diameter!

In temperate regions, tubers must be stored in a cool, dry place during the winter. Tubers of white potato can be placed in a box, basket, or large can, since they are used throughout the winter as food. Dahlia and begonia tuberous roots can be packed in dry peat moss to prevent desiccation over the long winter months. In the spring, the tuberous roots are placed in a flat of moist peat moss for several weeks to induce bud initiation and development. Then, begonias can be planted in baskets, pots, tubs, or garden beds and left outdoors after danger of frost has passed. Dahlia tuberous roots with young shoots on them are planted outdoors, again after the last spring frost. In tropical climates, and in the deep South, dahlias can be left outdoors year-round, the tuberous roots being allowed to "rest" in the soil during the cool or dry season of winter.

Most tuberous plants can be propagated by division. The familiar tuber of Irish potato is cut into several pieces, each piece with two or three "eyes" (axillary buds) present. Each piece is planted with the eye up, 10 to 15 centimeters deep. They develop roots and shoots in about 10 to 15 days. With dahlias and day lilies, large clumps of tuberous roots can be divided with a knife. Each piece should be divided so that there will be sufficient stem tissue at the upper end for adventitious bud production. The tuberous root alone, without a stem piece, usually will not develop a new plant. Dahlia tuberous roots are best separated in the spring. Day lilies are frost-hardy and can be divided anytime except during flowering. The tuberous roots of sweet potato can be placed in containers with water to half-cover them. Thus, adventitious root and sprout formation are induced. The sprouts with roots can then be separated and planted in the garden after danger of frost is over.

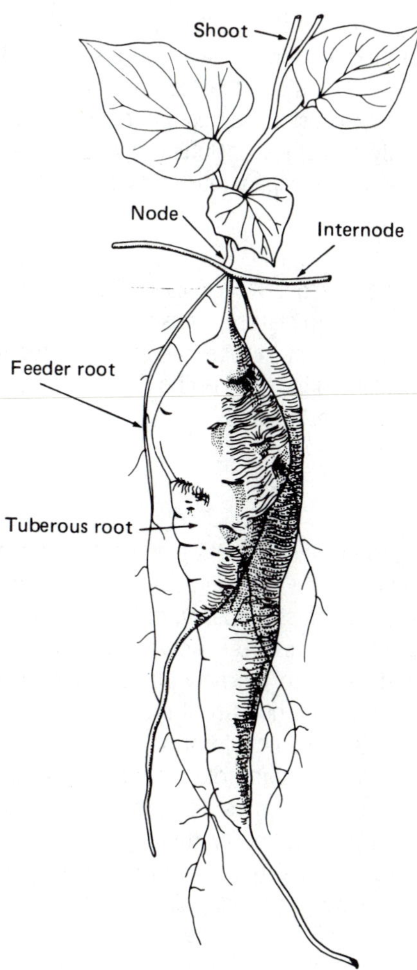


*FIGURE 6–6*
Sweet potato plant with two tuberous roots.

The gloriosa lily (*Gloriosa* spp) is a tuberous-rooted climbing vine that has spectacular red and yellow flowers. It is suitable for forcing in the greenhouse or growing outdoors in the summer.

## RHIZOMES

Rhizomes are enlarged underground stems. Some examples of plants with rhizomes are iris (a cultivated garden perennial), bamboo (a landscaping screen plant with edible shoots), cattail (an edible wild plant), lily of the valley (a fragrant-flowered garden ground cover), and Solomon's seal (a common woodland wild flower). Even the tubers of potato and Jerusalem artichoke represent enlarged tips of rhizomes. Rhizomes are simply propagated by cutting into pieces, each piece having one, or preferably two, nodes. Each node has an axillary bud which has the potential for developing into an above-ground shoot. Iris rhizomes have such shortened internodes (with nodes, therefore, close together) that it is easier to propagate them by breaking the rhizome at constricted regions or where they branch. Several houseplants have scaly rhizomes, including the gesneriads, *Achimenes* and *Kohleria*. These, too, can be propagated by division.

### BULBLETS

Bulblets are borne above ground such as in lily leaf axils and in bunching onion inflorescences. These miniature bulbs normally drop to the ground and propagate themselves. But they can be harvested, stored in a cool, dry place for later planting, or planted in the greenhouse, cold frame, or garden in new sites away from the "mother" plants.

## Runners, offsets, rootstocks or crowns, and plantlets

### RUNNERS

Runners or stolons produce new plants along creeping stems or at the ends of long, arching stems and can provide us with new plants annually. Runners typically develop each year on strawberry, myrtle, ajuga, mint, and on a number of other perennial plants after flowering. Some vigorously growing strawberries can produce up to ten new plants per runner! After each plant on the runner has developed a good root system, it can be cut from the parent plant and transplanted to a new location. This is especially useful in starting new strawberry beds, renewing old strawberry beds, and in establishing new beds for ground covers such as dichondra, myrtle, pachysandra, sedum (some species), and ajuga. Strawberry runners typically develop between the rows of the parent plants and can be used to establish strawberry beds over three- to four-year rotating periods. To encourage runner development, plants must be well watered and mulched.

Mulching around the plants with straw or plastic sheeting helps to suppress competing weeds and to conserve moisture in the soil. It also prevents, in large measure, end-rot of the ripening fruit. Care should be taken that such aggressive perennials as ajuga and the various other mints (peppermint, spearmint, gill-over-the-ground, and dead-nettle) do not extend beyond their desired boundaries and crowd other, less "gregarious" plants out of the garden.

In the greenhouse, runners are produced by the familiar spider plant *(Chlorophytum)*, Strawberry begonia *(Saxifraga sarmentosa)* and the gesneriad episcia (*Episcia* spp). This characteristic makes these plants ideal for hanging baskets, as their runners hang over the edge and form cascades of shoots.

### OFFSETS

Offsets, usually from adventitious buds or axillary buds, are small plants produced from the stem-root junction or crown of the parent plant. When rooted, offsets are cut from the parent plant and started as new propagules in much the same way as runners. For many plants, offsets replace the parent plant after it flowers. Examples are *Streptocarpus* (a gesneriad); *Vriesia, Tillandsia,* and other bromeliads; Spanish bayonet *(Yucca)* and century plant *(Agave);* and many kinds of sympodial orchids *(Cymbidium, Oncidium, Epidendrum*). Offsets of hen and chickens *(Sempervivum tectorum)* also continue to enlarge the colony and to provide new plants after the parent plants have flowered. Many globular cacti such as *Mammilaria, Echinopsis,* and *Rebutia* form clustered colonies made up of the original plant and its many offsets which remain in the pot and enlarge annually.

Orchid pseudobulbs are not really bulbs but offsets. They represent enlarged stems with the leaf blades above and roots emerging below. These pseudobulbs are best separated when a pot becomes crowded or the plants become pot-bound. In dividing them, it is usually best to keep three to four pseudobulbs in a cluster to assure that the new plant gets off to a good start. The pieces should be planted in the same type of mix as the original plant.

## ROOTSTOCKS OR CROWNS

One of the simplest and oldest means of propagating plants vegetatively is by division, or cutting apart. This can be done with plants which produce rootstock or crowns (as in clump-forming ferns and herbaceous perennials such as phlox, daisies, and asters), in many kinds of shrubs, and in mosses (used for terraria).

Practically all perennials can be propagated by division of their rootstocks or crowns. It is best done by lifting the parent plant from the soil with a garden fork or shovel. The clump can then be separated into smaller units with a sharp-edged shovel, a hatchet, or sometimes by hand. Each separated piece should have one to several leaf buds present in the crown (above the roots) so as to assure growth of the propagule. The pieces can be planted immediately or stored in a cool, moist place until planting. Place the pieces at the same depth as the original parent plant. Mulch with pine needles, compost or leaf mold for protection from summer drying and winter injury in cold climates. The best time to divide perennials is after they bloom, either in the fall when crown buds are well developed or in the spring before active shoot growth begins.

## PLANTLETS

A number of plants naturally produce plantlets along the margins of their leaves by epiphyllous (epi = upon, phyll = leaf) budding such as *Kalanchoë* and *Asplenium bulbiferum* (mother fern). Some plants produce plantlets along the central midrib of the leaf blade, as in the piggy-back plant *(Tolmiea).* These plantlets can be detached and planted in appropriate soil medium in the same way you might transplant young plantlets of annuals or perennials. Succulents such as *Kalanchoë* prefer a sandy soil, and *Tolmiea* and the ferns need a rich potting soil with compost or leaf mold. Some of the *kalanchoes* produce plantlets that drop to the soil from the mother plant. (See leaf cutting, Figure 6–14.) They can be removed and planted. *Kalanchoe* plants have been shown to secrete a chemical substance which may stunt the growth of other plants (termed *allelopathy*), so do not let them fall where they will become a nuisance.

## Cuttings

Cuttings represent one of the most widely used methods for propagating plants vegetatively. The reasons are that (1) cutting material is almost always available; (2) the method is simple; and (3) the chances of rooting cuttings are usually good. But it takes proper treatment and the knowledge of easily rooted species to be consistently successful with this technique. Some plant cuttings difficult to root include pines, spruces, firs, and many deciduous trees such as maple, cherry, hickory, and walnut. These plants are best propagated by seeds or grafting.

### ROOT CUTTINGS

Plants whose roots tend to produce adventitious buds can be propagated readily by root segment cuttings. This is true for perennial phlox, chrysanthemum, fork-leaved sundew *(Drosera binata)*, anemone, lilac, deciduous azalea, and roots of many woody plants that form colonies in nature.

The cut root pieces can be placed either horizontally in the rooting medium or vertically. If vertical, make sure the pieces are placed in the direction of original growth, as the shoot will form from the end nearest the mother plant and roots from the end furthest away.

### STEM CUTTINGS

The most common type of cutting is a stem cutting from herbaceous stems or "softwood." It has two or more nodes and a variable number of leaves (Figure 6–7). Common examples of plants from which stem cuttings are made include coleus, begonia, impatiens, geranium, and peperomia.

A variation on this theme is the stem midpiece cutting, used in the vegetative propagation of aroids such as *Philodendron, Monstera,* and *Dieffenbachia* (Figure 6–8) and the rubber tree (*Ficus* spp). The midpieces, 20 to 25 centimeters long with several nodes included, are cut from large stems. Both ends of each piece are dipped into hot, soft paraffin, a coating which prevents water loss and rotting caused by fungi. The cuttings are then placed horizontally and half-submerged into the rooting medium–a moist sand:peat moss, 1:1 mixture. In a few weeks, one to several shoots will develop from dormant axillary buds at the nodes, and roots develop below. Once this happens, the cutting can be transplanted to a large pot, keeping the original stem piece horizontal, buried halfway in soil. An alternative method of propagating aroids is air layering.

From woody plants, we can make either softwood or hardwood stem cuttings. New green shoots of woody plants are softwood cuttings before they become highly lignified. The best time to obtain softwood cuttings is during active growth in the spring or early summer for most temperate zone

*FIGURE 6–7*
Preparing a stem cutting of coleus *(Coleus blumei):* (A) remove lower leaves and half of large leaf blades; (B) root in pot of vermiculite as shown here, or a glass of water; (C) rooted cutting.

plants. In the tropics, the same period coincides with wet seasons. Examples of softwood cuttings from tropical plants include *Hibiscus, Allamanda, Bougainvillea, Passiflora,* and *Fuchsia* (Figure 6–9).

A hardwood cutting of a deciduous or broadleaf evergreen tree or shrub is made from a one-year-old shoot, taken during the dormant season when tissues are woody (fully lignified and mature). Some plants which can be propagated by this method are rose *(Rosa),* grape *(Vitis),* privet *(Ligustrum),* forsythia *(Forsythia),* willow *(Salix),* aucuba *(Aucuba),* spirea *(Spiraea),* snowberry *(Symphoricarpos),* holly *(Ilex),* and conifers such as *Juniperus* and *Chamaecyparis.* The best time to obtain deciduous hardwood cuttings is after one or two frosts, but before the tissues freeze, during late fall, or early winter. The cuttings are placed in rooting media and stratified over winter to satisfy dormancy requirements. By spring, callus tissue will have formed at the base of the cutting. Once callus forms, roots are initiated from the callus tissue, usually when the buds break dormancy. Well-rooted cuttings are potted, planted in a nursery, or placed in the garden (protected from herbivores). The best time to make coniferous hardwood cuttings is winter.

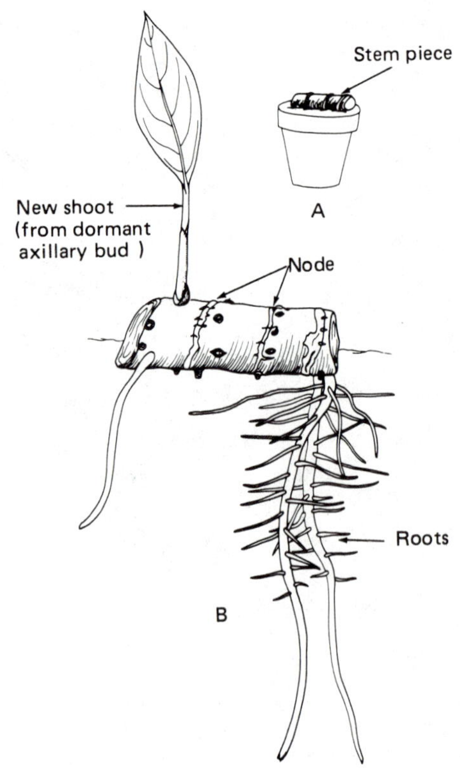


*FIGURE 6–8*
Stem midpiece cutting of *Dieffenbachia:* (A) in pot and (B) after shoot and root development.

## *Factors in successful rooting of stem cuttings*

1. Use a clean, new razor blade or knife to make a sharp, oblique cut at the base of each cutting to increase the root-forming surface area.
2. Treat the base of the cutting with a fungicide to curtail rooting.
3. At the same time, treat the base of the cutting with a synthetic auxin hormone preparation to stimulate callus development and rooting. Auxin treatment hastens rate of root initiation and increases the number of roots initiated. (See Chapter 3, Figure 3–3.) There is no need to use hormone preparations on easy-to-root plants such as coleus, wandering Jew, and willow. Hardwood cuttings may require a stronger hormone treatment than softwood cuttings to stimulate rooting.
4. For softwood cuttings from woody or herbaceous plants, retain approximately half of the number of leaves. Remove the basal leaves and cut off the tips of upper leaves to prevent excessive water loss by transpiration.

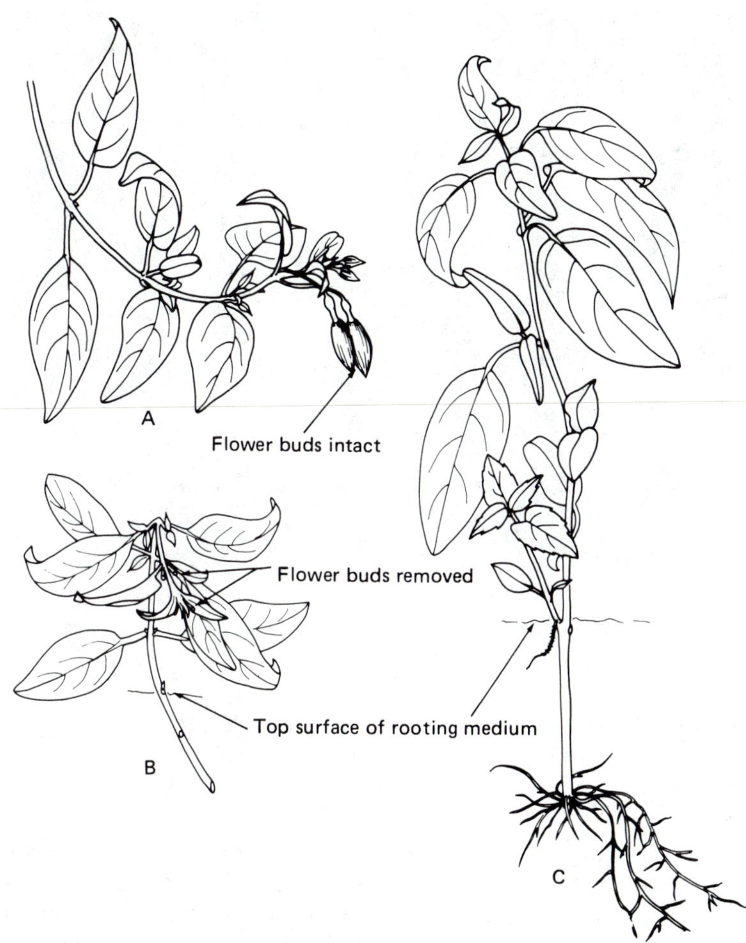


*FIGURE 6–9*
Softwood cutting of woody plant *(Fuchsia):* (A) shoot during flowering period; (B) shoot with flower buds and lower leaves removed, stem end treated with rooting hormone, and potted; (C) rooted cutting.

*CARE OF CUTTINGS.* Fleshy cuttings, including all cacti and succulents, need to develop scar tissue over cut areas to prevent rotting by fungi. Exposure to the air until scar tissue forms may take several days, perhaps up to a week. Then place the basal end of the cutting in sand or sand/fine gravel as a rooting medium, which also helps to prevent rotting (Figure 6–10). Do not water succulent cuttings until roots have formed. Place in shade, fully exposed to the air.

For nonsucculent-type cuttings, place in a rooting medium such as 1 part vermiculite to 1 part perlite, which allows good aeration plus high water-holding capacity. The container of cuttings can be watered via a small

*FIGURE 6–10*
Succulent dish garden. Cacti and succulents being grown in sand and fine gravel. These plants were rooted as stem cuttings.

clay pot embedded in the medium in the center of the larger "propagation pot" (Figure 6–11). The hole in the bottom of the small pot is plugged and filled with water to slowly seep into the propagation pot. A misting device over the cuttings, spraying 30 seconds every 30 to 60 minutes, helps to keep cuttings damp and curtails loss of leaves. In the absence of such a luxury, place plastic bags over the cuttings with provision for air entry (Figure 6–12). Glass "chimneys" are another substitute for misting devices. Avoid placing plastic bag-covered plants in bright, direct sunlight since the buildup of high temperatures in the bag may kill the cuttings.

For fastest rooting, both succulent and nonsucculent cuttings should be kept warm. Use bottom heat from a heating cable in the propagation bench at 21° C (70° F). Additional light to lengthen days in winter may help some cuttings to root faster. The faster a cutting roots, the less likely that it will be attacked and killed by disease-cause fungi.

## LEAF CUTTINGS

There are several types of leaf cuttings, varying from pieces of leaves to leaf bud and leaf blade types. Let us consider each type. A relatively simple cutting is prepared from pieces of a leaf blade, such as that of mother-in-law's tongue *(Sansevieria)* (Figure 6–13). The leaf blade is severed at the base, then cut into four- to five-centimeter pieces and inserted into moist sand with the *top end up*. In other words, maintain the direction of growth of the

*FIGURE 6–11*
Watering cuttings. A small clay pot set in the center of the container of cuttings facilitates the addition of water to the rooting medium; add water as needed to the small pot. Shown here are both herbaceous stem cuttings and a leaf cutting.

*FIGURE 6–12*
Rooting cuttings. Cuttings planted in a pot covered with a plastic bag to maintain high humidity and avoid wilting. Since direct sun causes temperature buildup under plastic, punch several holes in bag and keep pot in shade.

original leaf blade, as roots will form only from the morphological bases of the cuttings. Roots appear in a few weeks followed by leaf buds. When the leaf buds emerge from the sand, the new plants can be transplanted to potting soil. A 2 to 1, sand to loam mixture is preferred.

A second type of leaf cutting is represented by cuttings prepared from entire leaf blades. Plants such as crassula, sedum, jade plant *(Crassula argentea)*, rex begonia, and *Kalanchoë* are commonly propagated vegetatively by means of this type of cutting. For *Kalanchoë* or rex begonia, remove a blade and place it so that the lower surface is in contact with moist sand. Anchor in place with copper wires or hairpins stuck through the leaf. With leaf cuttings of *Kalanchoë daigremontiana*, new plantlets will form along the leaf margins at the notches *(sinuses)* (Figure 6–14). This type of epiphyllous budding is promoted by long days. Rex begonia leaves can also be cut through

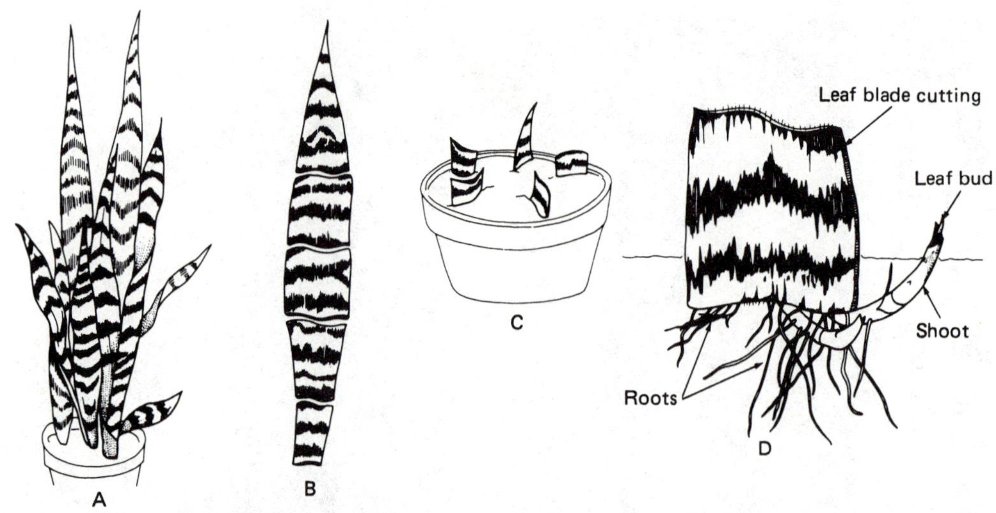


*FIGURE 6–13*
Leaf blade cutting of *Sansevieria zeylanica* (bowstring hemp or mother-in-law's tongue): (A) plant to be cloned; (B) leaf blade cut into sections; (C) leaf sections placed in sand—be sure to keep the leaf sections bottom end down in the rooting medium; (D) leaf section with roots and shoots.

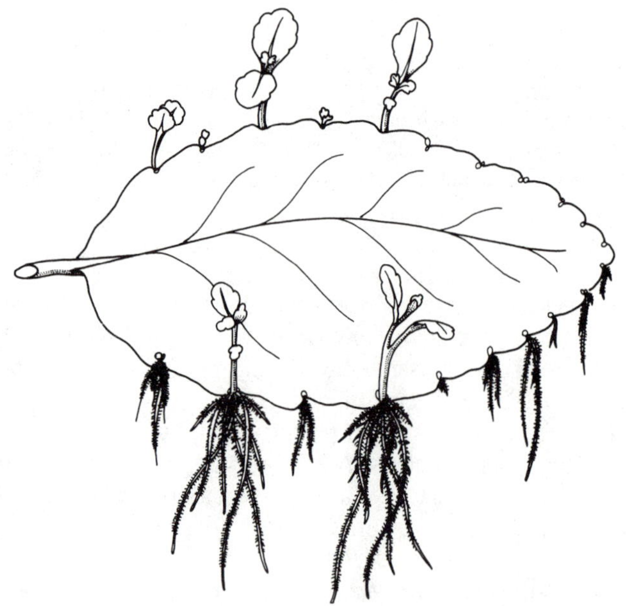

*FIGURE 6–14*
Leaf blade cutting of *Kalanchoë daigremontiana*. Plantlets develop from notches along the leaf margin. This is called *epiphyllous budding*.

the major veins to stimulate plantlet development from these wounded regions (Figure 6–15). For jade plants, other crassulas, and sedums, leaves are removed from the parent plant and inserted vertically, bottom third buried, into the rooting medium rather than horizontally.

A third type of leaf cutting is slightly more complex. It is called the *leaf blade-petiole cutting* and is best used for plants such as peperomia and African violet (Figure 6–16). The cuttings of blade with petiole are placed in a moist rooting medium such as 1 part vermiculite to 1 part perlite. Place the cut petiole into the medium so that the blade is above. Roots typically form first at the petiole base, followed by shoots after several weeks.

The most complex type of leaf cutting is exemplified by the leaf bud type. Such a cutting also includes a piece of stem where the leaf and axillary bud are inserted. This type of cutting is typically used with such plants as rubber plant *(Hevea brasiliensis)* and *Rhododendron.* In fact, leaf cuttings of *Rhododendron* must have the axillary buds present; otherwise, they will not root. Apparently, the axillary bud produces the auxin necessary to induce root development. The cuttings are inserted into a rooting medium, such as vermiculite or a vermiculite/perlite mixture, up to the leaf blade base.

Care of leaf cuttings is the same as care of nonsucculent stem cuttings.

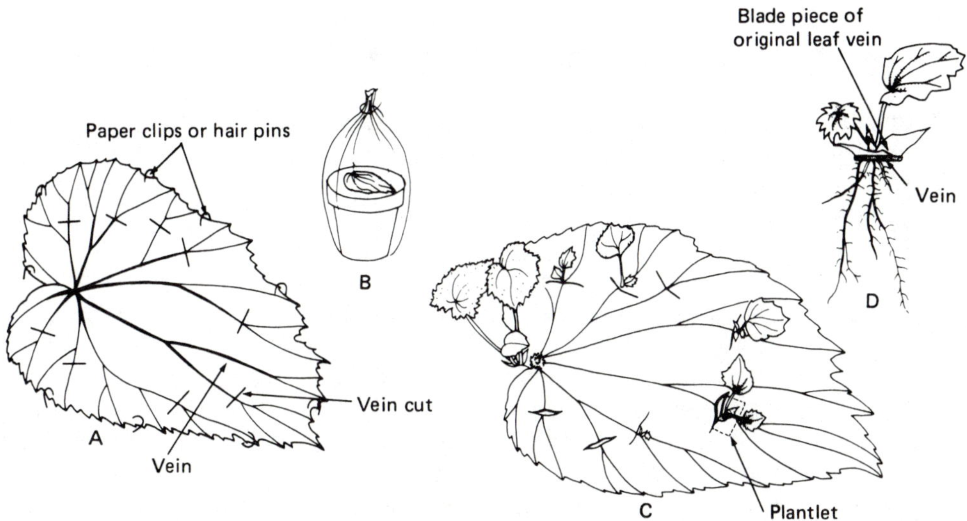


*FIGURE 6–15*
Leaf blade cutting of rex begonia: (A) detached leaf anchored on vermiculite, major veins cut; (B) propagation pot enclosed in plastic bag, placed out of direct sunlight; (C) plantlets developed on leaf at sites where veins were cut; (D) plantlet is ready for potting.

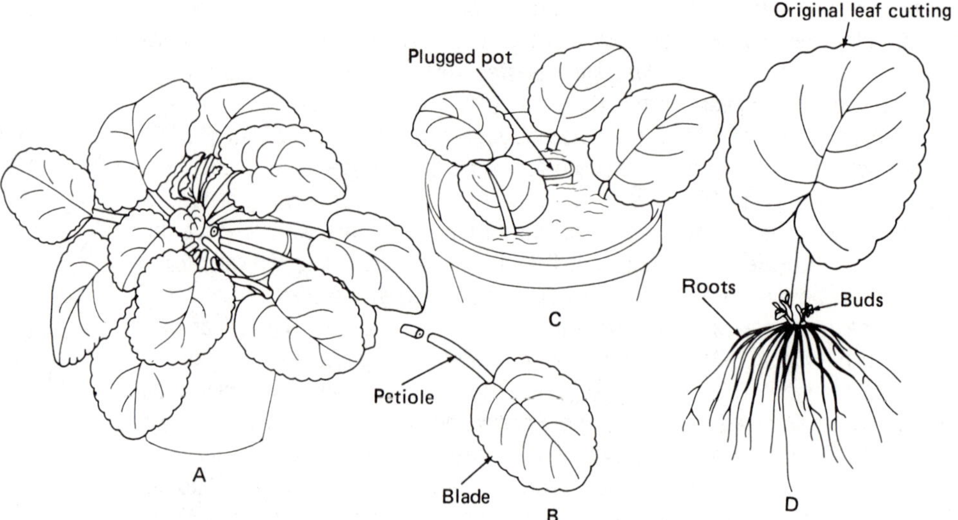


*FIGURE 6–16*
Leaf blade-petiole cutting of African violet *(Saintpaulia ionantha):* (A) original plant; (B) leaf blade-petiole cutting; (C) cuttings in rooting medium such as vermiculite with clay pot for watering in center; (D) rooted cutting.

## Layering

Layering plants, in contrast to preparation of cuttings, is often an easier and faster way of propagating small numbers of plants from a large mother plant. Layering takes several forms.

### TIP LAYERING

Tip layering involves rooting the tip of a shoot to produce a new plant, almost like a runner except that the shoot to be layered is the main stem of the plant, not a specialized structure. Several of the best examples include wild blackberries and black or purple raspberries (all species of *Rubus*). Because of their length and weight, the tips of the canes come in contact with the soil. The tips then take root followed by development of one or more new shoots; eventually, they will form a new crown of canes. This then can give rise to a proverbial bramble thicket. Tip-rooted canes may be severed from the parent plant and transplanted to a new planting site. Practically all commercial cultivars of blackberries, purple and black raspberries, and boysenberries are propagated by tip layering. Red raspberries, in contrast, do not tip layer and must be divided at the crown-root junction.

## SIDE LAYERING

Side layering occurs closer to the crown of the parent plant. In nature, we encounter lower branches, buried by fallen limbs or decomposed leaves, taking root in such plants as *Rhododendron* and winter-creeper *(Euonymus)*. This results in enlargement of the crown mass of the mother plant as the rooted branches grow up near the parent. We can do this artificially by bending a branch down, anchoring it in the soil, and then covering it with soil or compost. After several weeks rooting may have occurred.

The rooting process may be speeded up by wounding the buried portion of the branch on the lower side and treating with an auxin-type hormone powder such as Rootone. After rooting has occurred, sever the branch below the rooted portion and transplant it to a new location, making sure to take plenty of soil with the rooted branch. Side layering should be done with the current year's branches that are close to the soil. Many small fruits such as currants and gooseberries side layer naturally; but of course, you can promote this process to obtain new plants or to enlarge the crowns of the plants. Not all woody plants layer easily. If they do not, propagate by seeds, cuttings, or grafting.

## AIR LAYERING

Many houseplants such as the Chinese rubber tree *(Ficus elastica)*, monstera or Swiss cheese plant *(Monstera deliciosa)*, dumb cane *(Dieffenbachia)*, the fiddle-leaf fig *(Ficus lyrata)*, and other houseplants can be propagated by means of Chinese or air layering (Figure 6–17). This method of propagation is especially useful when a plant grows too tall for a room or when another plant from the parent plant is desired.

Air layering is simple. First, make three to four 3 millimeter (one-eighth-inch) deep and four 2.5 centimeter (one-inch) long longitudinal cuts parallel to the stem axis at the site where you want roots to develop. This is done to create wound tissue from which adventitious roots will regenerate. You may also cut and remove a notch of tissue from the stem; this prevents healing before roots form. After abrading or notching the stem, cover it with moist sphagnum moss; then, wrap it with plastic film, tying it at top and bottom. Check the moss every few days to help insure its moisture content. It may help to place some Rootone hormone powder on the abraded region before covering the stem with sphagnum moss and plastic film. This will help to stimulate root formation. After several weeks, when several roots are well developed, remove the plastic film, cut the stem below the rooted region, and pot the rooted stem piece. Now you have a plant whose genetic makeup is identical to that of the large parent plant left behind.

The remaining parent shoot may be cut in pieces for stem cuttings, layered again on a lower section, or left to develop new shoots from dormant buds. The lower buds will develop because of loss of the shoot tip. As

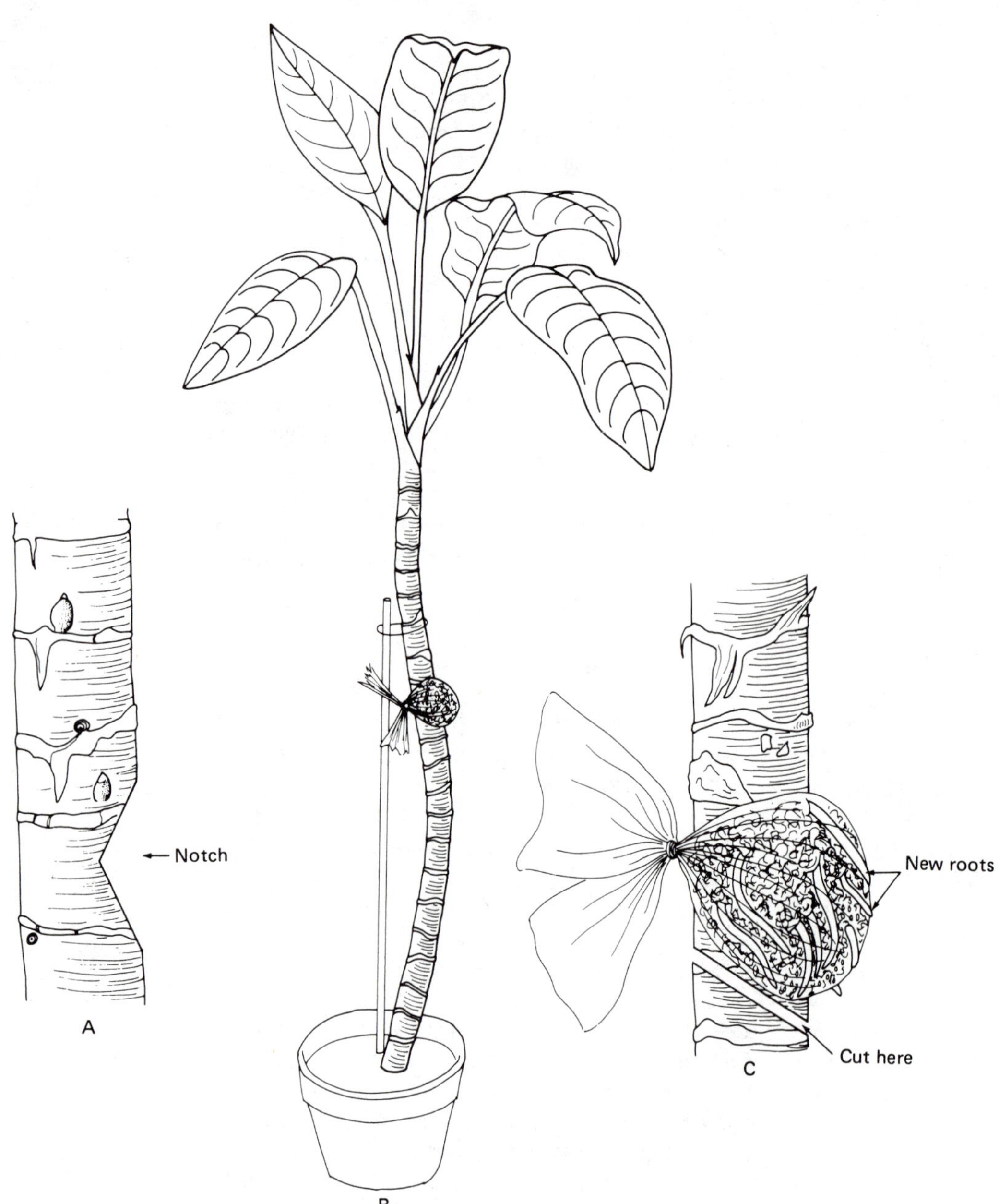


*FIGURE 6–17*

Chinese or air layering with a *Dieffenbachia* plant: (A) a cut is made in the side of the stem and rooting hormone applied; (B) the wounded area is covered with moist sphagnum moss, held in place with a piece of plastic, and tied with a "twist 'em"; (C) a closeup of rooting site with a clear plastic covering. Roots can be seen as they extend to the surface through the sphagnum moss. After extensive rooting, cut the stem below, remove the plastic, and pot in soil.

long as the shoot tip remains, it exerts an "apical dominance" by producing auxin hormone, inhibiting growth of lower buds. When the shoot tip is cut off, growth of lateral buds is stimulated by cytokinin hormones produced in the roots and moved upward in the xylem of the stem.

## Grafting

Almost all cultivars of walnuts, roses, apples, peaches, and grapes sold are shoot grafts on a rootstock of the same species. Grafting of plants is done for several reasons: (1) to obtain greater root or shoot winter hardiness; (2) to obtain disease- or insect pest-resistance; (3) to reduce plant size by dwarfing the shoot portion of a plant such as dwarf peach, apple, and pear trees; (4) to introduce new varieties of a desirable fruit on a rootstock of proven hardiness, such as five or more varieties grafted onto one pear or apple tree rootstock (this also eliminates the need of more than one tree for cross-pollination.); (5) to improve the branch patterns of older trees by restoring branches and increasing growth vigor; (6) to repair trunk injuries or wounds; and (7) to grow cuttings which are not easily rooted.

These advantages are derived from the fact that the lower part of the graft (called the rootstock) is often more cold-hardy, or insect- or disease-resistant, or able to confer dwarfness on the top portion (the scion). Likewise, the scion top portions can be more vigorous than or produce different varieties from the shoot branches of the mother tree onto which the scion branches are grafted.

In grafting, the two basic components, the *stock* (or root portion) and the *scion* (or shoot portion), are placed in contact with each other in various ways to promote their "knitting together." The knitting process takes place between the living vascular tissue of both portions so that new tissue forms a bond between the two. The vascular cambium tissue is located just under the bark. The success of any graft depends upon the lining up together of this living cambial tissue so that the cambium of the stock is in contact with the cambium of the scion. Usually the graft of scion onto stock is between two different cultivars of the same species, but sometimes from different, closely related species.

There are many different types of grafts. We shall examine only the basic types because many other grafts are simply variations of these.

### THE SPLICE GRAFT

The splice (or whip and tongue) graft is used to knit two pieces of similar diameter together, both usually no larger than a centimeter across. For the splice graft, make a long, oblique cut on the basal or root end of the piece to be grafted on (scion). Then, cut a matching angle slice on the rootstock stem (Figure 6–18). Next cut into the middle of each of the oblique areas, but not all the way through. The scion is then fitted onto the stock. Grafting wax is now applied to seal the region, or rubber grafting bands, tape, or strands of raffia can be wrapped around the graft site.

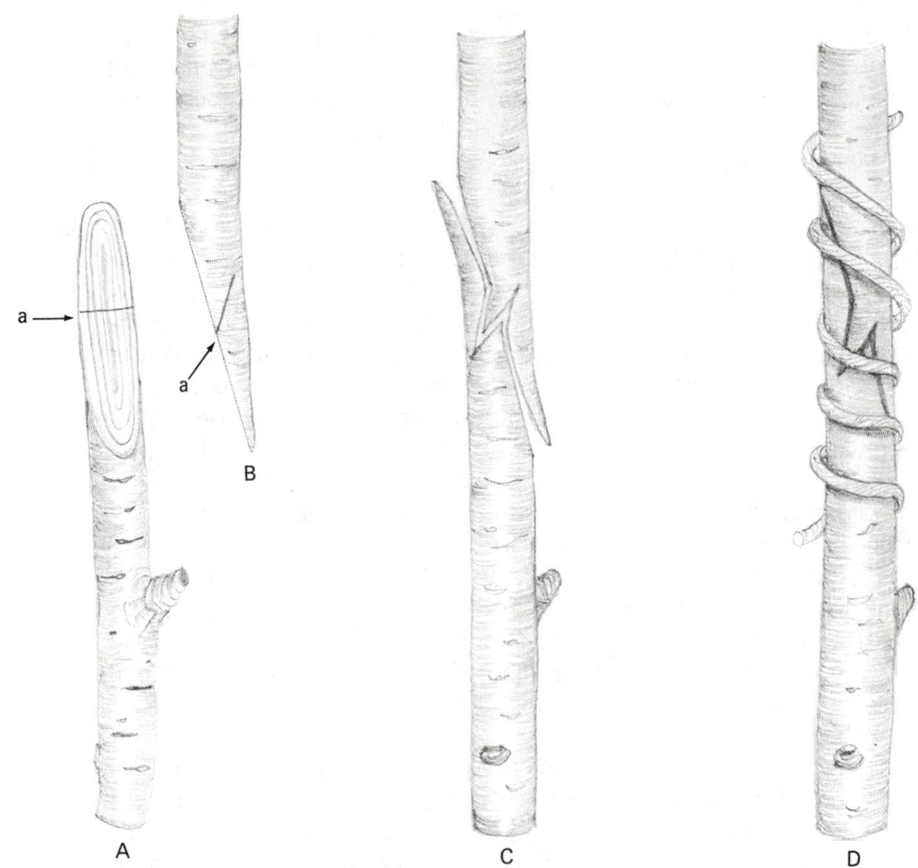


*FIGURE 6–18*
Whip and tongue or splice graft, using cherry *(Prunus):* (A) prepared stock with oblique cut and notch cut (a); (B) prepared scion with oblique cut and notch cut (a); (C) scion fitted on stock with oblique cuts face-to-face and notches interlocked; (D) graft site wrapped. (Root systems are not shown on stocks.)

## THE CLEFT GRAFT

The cleft graft is employed for "top-working" of fruit trees to improve the branching patterns or to graft different cultivars of the same species on a single tree. In this type of graft, the stock is the parent tree, and the scion is a much smaller branch piece that is grafted onto the parent tree. The stock is first cut across transversely, then split vertically in the middle for a distance of 8 to 12 centimeters. (This can be done with a grafting knife.) Two scion pieces are each cut obliquely at the base on opposite sides to form a long wedge-shaped base (Figure 6–19). Each scion piece is then inserted into

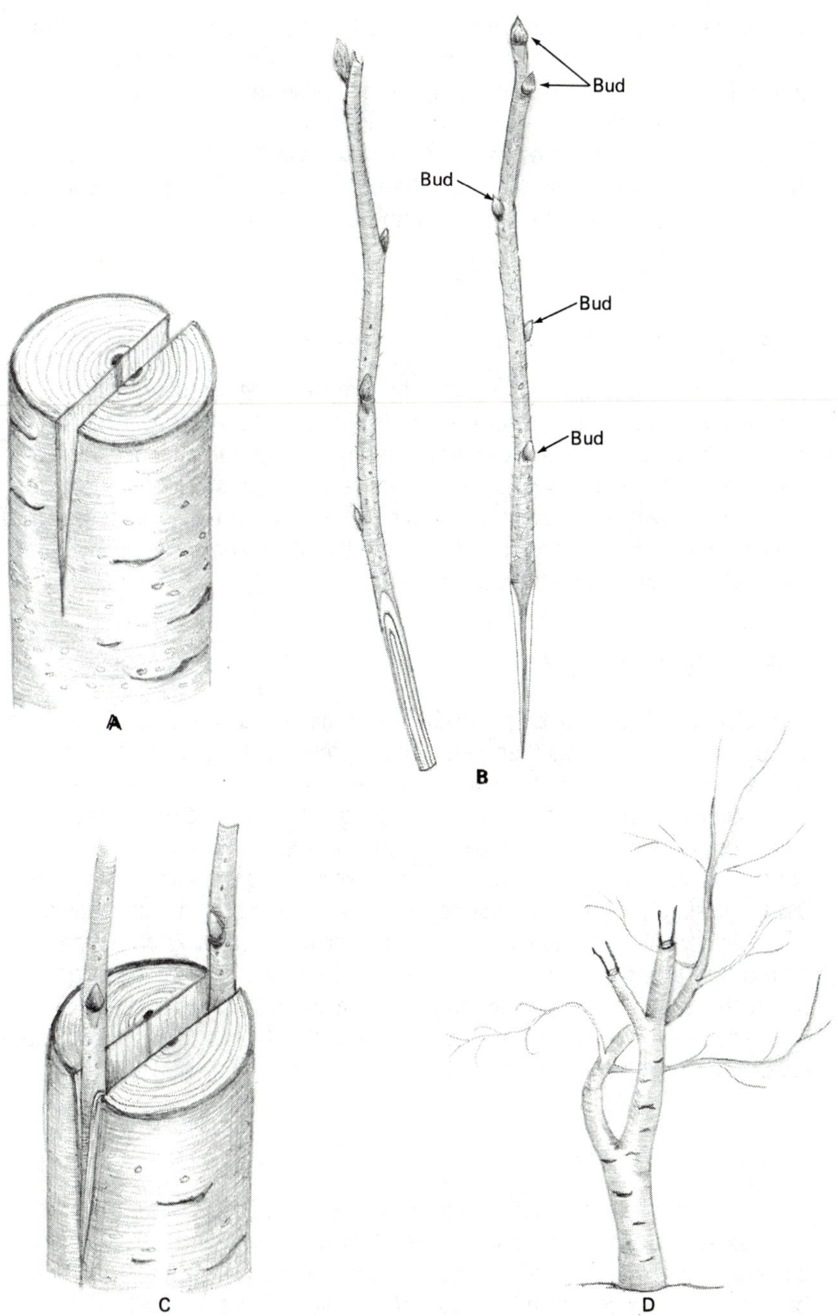


*FIGURE 6–19*
Cleft graft of apple *(Malus):* (A) stock prepared; (B) two scion pieces prepared; (C) scion pieces fit into stock cut; (D) two graft sites on stock. (Root system of apple tree is not shown.)

the vertical cut of the stock, making sure that the respective vascular cambium layers of both stock and scion are in contact with each other. If not, no graft will occur. After this operation, it is advisable to cover the entire transverse portion of the stock and basal portions of the scions with grafting wax. Make sure the scion pieces have several buds which can produce the new shoots. Finally, to prevent drying of the scions, cover with a plastic bag that is perforated with several pencil-sized holes to prevent the bag from becoming a "heat-trap."

## THE VENEER GRAFT

The side or veneer graft is by far the simplest one discussed so far. An oblique cut is made in the stock along its side. The base of the scion cut to match is fitted into the stock and then wrapped with raffia, rubber grafting bands, or tape (Figure 6–20). The respective cambia of stock and scion must be in contact with each other before taping the two portions together. After a successful graft has occurred, the top of the *stock* above the veneer graft is cut off and covered with grafting wax.

## THE APPROACH GRAFT

Two plants are set next to each other, in a manner that a shoot from each can be grafted easily together. One plant represents the stock to be retained for its root system, and the other plant contributes a scion for a new shoot system. The approach graft involves making a tangential cut on both stock and scion. A slightly more complex version involves making an oblique cut in both stock and scion and fitting the two pieces together. Whichever one is used, one simply fits the two cut regions of stock and scion together and ties them with raffia, grafting bands, or grafting tape. After grafting has occurred, the stock portion is cut off above the graft, and the scion portion below the graft. It is important to stress that both stock and scion portions must have their own separate root systems during the grafting process; otherwise, the scion will not make a successful graft.

## THE WEDGE GRAFT

Another type of graft is the wedge or saddle graft. It is similar to the cleft graft except that the scion is usually much larger than that of the cleft graft, and only one piece is inserted. If both pieces are of equal size, the scion must be cut as an inverted V and placed directly over the stock which is also cut in a V shape (Figure 6–21). Thus, the two pieces interlock with the respective cambia in contact with each other. Saddle grafts are particularly useful for grafting different species of cacti together.

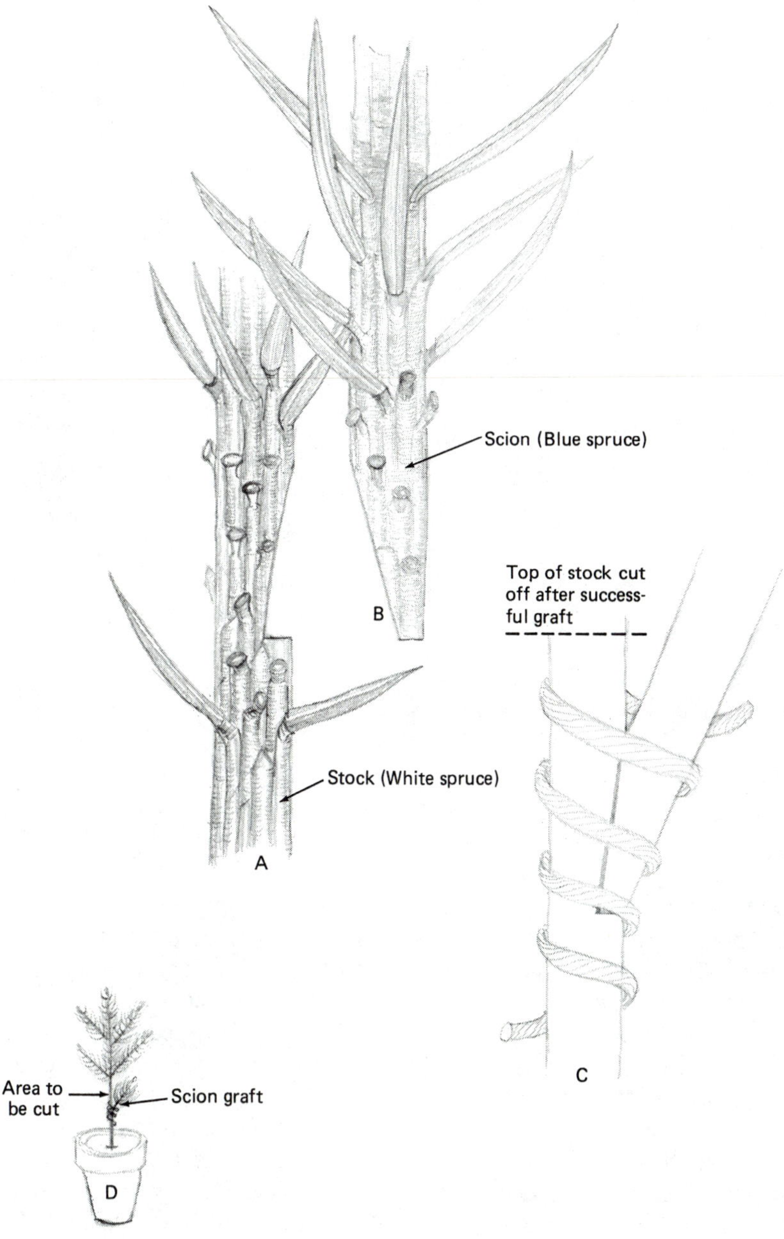


*FIGURE 6–20*
Side or veneer graft: (A) white spruce *(Picea glauca)* stock prepared; (B) blue spruce *(Picea pungens)* scion prepared; (C) scion fit on stock and wrapped; (D) shoot portion of stock to be cut off after scion graft is growing. [Root systems of stocks are not shown in (A), (B), and (C).]

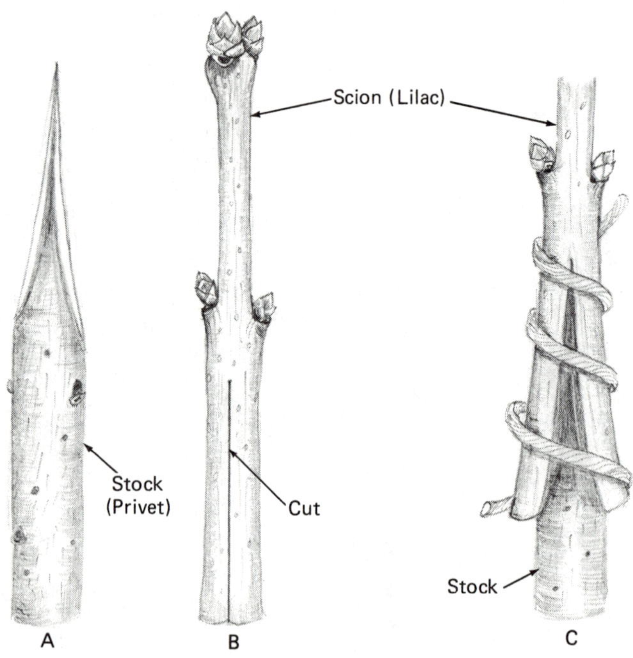


*FIGURE 6–21*
Wedge graft: (A) privet *(Ligustrum)* stock prepared; (B) lilac *(Syringa)* scion prepared; (C) scion fitted on stock and wrapped. (Root systems of stocks are not shown.)

## THE T-BUD GRAFT

Budding is a form of grafting, but instead of employing a branch piece for the scion, it involves the use of a bud with a lens-shaped piece of the stem attached to it (Figure 6–22). The bud is an embryonic shoot surrounded by protective bud scales. The stock, usually a young one-year-old plant called a *whip,* is cut through the bark to the cambium in a T-shaped configuration. The bud-heel scion piece is inserted into the T-shaped cut on the stock. The bud is left exposed. Grafting bands or raffia are wrapped around the T-shaped cut above and below the exposed bud. One point to remember with budding is that if you make the incision in the stock near the base of the seedling tree, after the bud starts to grow (a successful graft), you must then cut off the stock above the bud. Otherwise, the stock has a tendency to produce branches from its own buds.

Here are two final notes on grafting. First, if the graft does not take, the scion will die. This may be because the stock and scion are incompatible types or the two cambial regions fail to knit together properly and swell up, with the scion falling off. And second, if a branch grows out below the graft union from the stock portion, it must be removed soon or its vigor will out-

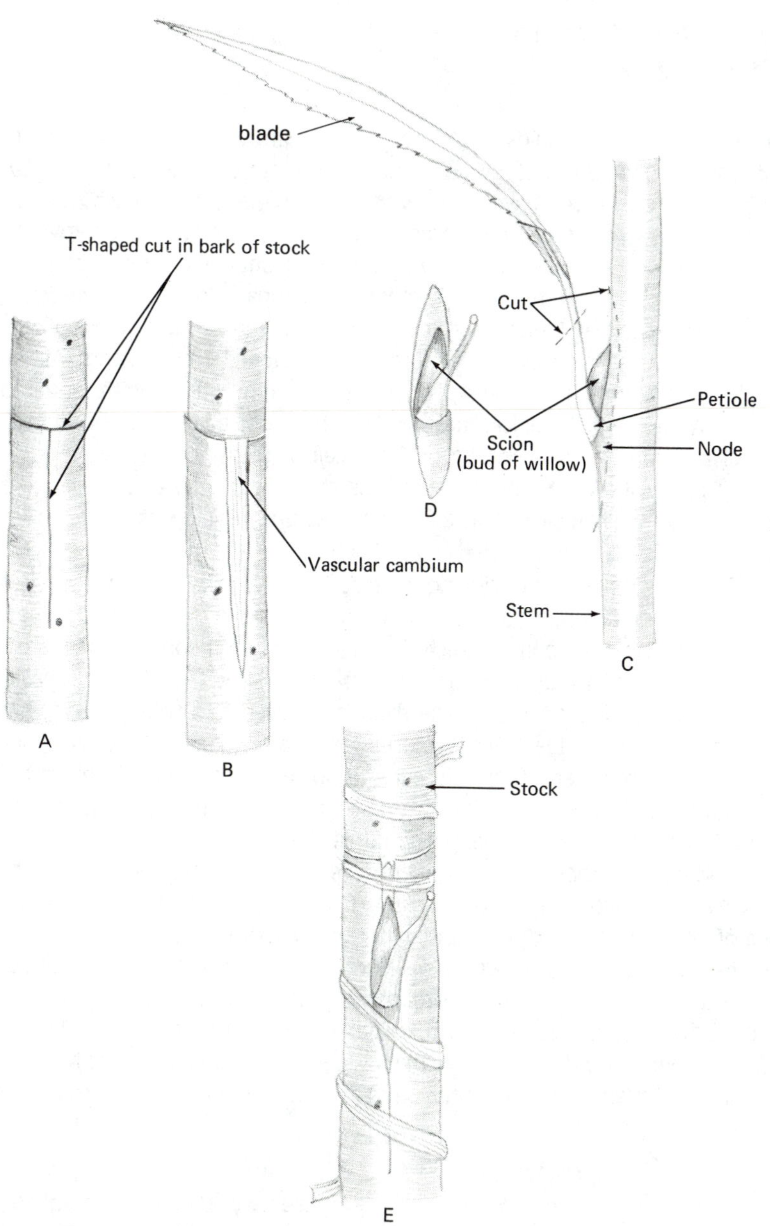


*FIGURE 6–22*
The T-bud graft using willow *(Salix):* (A) the bark of the stock is cut through in a T shape; (B) the cut bark of the stock is separated from the shoot to expose the living cambial tissue; (C) the T-bud, a portion of stem tissue with a leaf scar and axillary bud above it, is inserted into the T-cut of the stock; (D) after insertion, the T-bud scion is wrapped to close the T-cut on the stock. (Root system of stock is not shown.)

compete the weaker scion growth and your tree specimen will be that of the rootstock.

## Mericloning

Seed propagation of orchids can literally take years to produce a flowering plant, for example, up to seven years in *Cattleya*, but only two years in *Phalaenopsis*. Of course, you have to grow plants from seed if you wish to cross two plants for a hybrid having characteristics of both parents. Vegetative division of pseudobulbs yields only the number of plants as pseudobulbs available. However, another method of vegetative propagation, called *mericloning*, is now possible. Mericloning is the culture and multiplication of the vegetative shoot apical meristems from fully developed orchid plants. At the tips of orchid shoots and all vascular plants, are embryonic, undifferentiated cells which make up the apical meristem or the shoot apex. The bud in each leaf axil also has an apical meristem. Each cell of the apical meristem is capable of producing a new plant. It was discovered that the apical meristem cells of orchids, when grown in a culture medium, develop as though they were seed embryos.

The protocol for mericloning orchids is as follows. The leaves of a young orchid shoot are removed, often with the aid of a dissecting microscope in a sterile environment such as a laminar flow hood. The meristem (small mass of cells) thus exposed is cut off and placed in a flask of special nutrient solution (Figure 6–23). The flask is attached to a rotary shaker under constant light and ideal temperature. Within a few weeks protocorm-like tissue masses proliferate. (For reference to normal orchid seed development see Chapter 2, Figure 2–2.) These tissue masses are then cut into pieces, placed into separate flasks, given fresh nutrient solution, and returned to the shaker. Meristems that are not shaken develop only a seedlike germination pattern and produce only a single plant. Constant shaking prevents attainment of root-shoot polarity, inhibiting shoot formation. Therefore, the protocorm-like masses enlarge without differentiation into root-shoot plantlets. After successive cuttings of the protocorm masses and dividing into fresh nutrient flasks, they are placed (subcultured) on solid nutrient agar in bottles or flasks where they develop as normal root-shoot plantlets. Several hundred thousands of plants can be obtained from one original meristem in this manner!

*FIGURE 6–23*
Orchid propagation. (A) Asexual, by mericloning: (a) leaves removed, shoot apex exposed; (b) shoot apex, placed in nutrient agar; (c) callus formed; (d) clusters of cells; (e) protocormlike mass, to be cut into pieces; and (f) one of many new plants have developed with leaves, shoot, roots. (B) Sexual, by seeds: (a) green seed pod; (b) seeds removed ($\times 100$); (c) seedlings form green carpet on nutrient agar; and (d) eventually form new plants. (C) Asexual, by division: (a) crowded pot; and (b) rhizome cut into divisions, one pseudobulb per pot.

Asexual: by mericloning

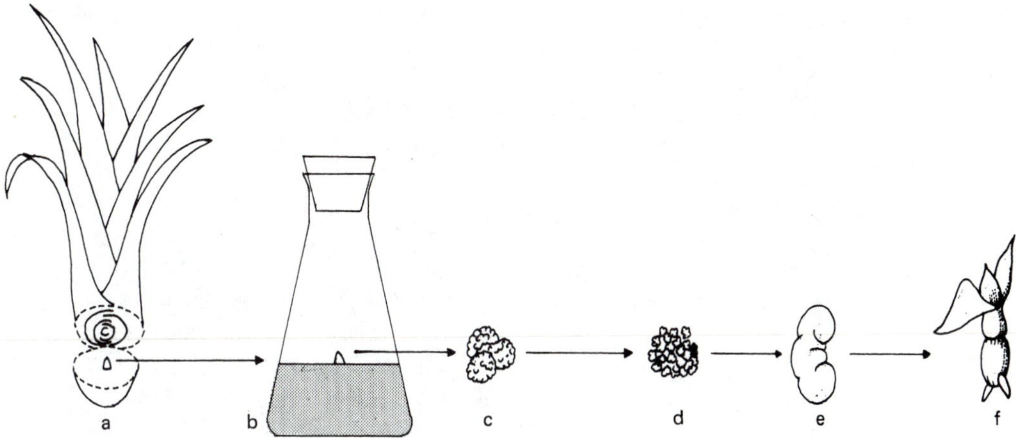


Sexual: by seeds

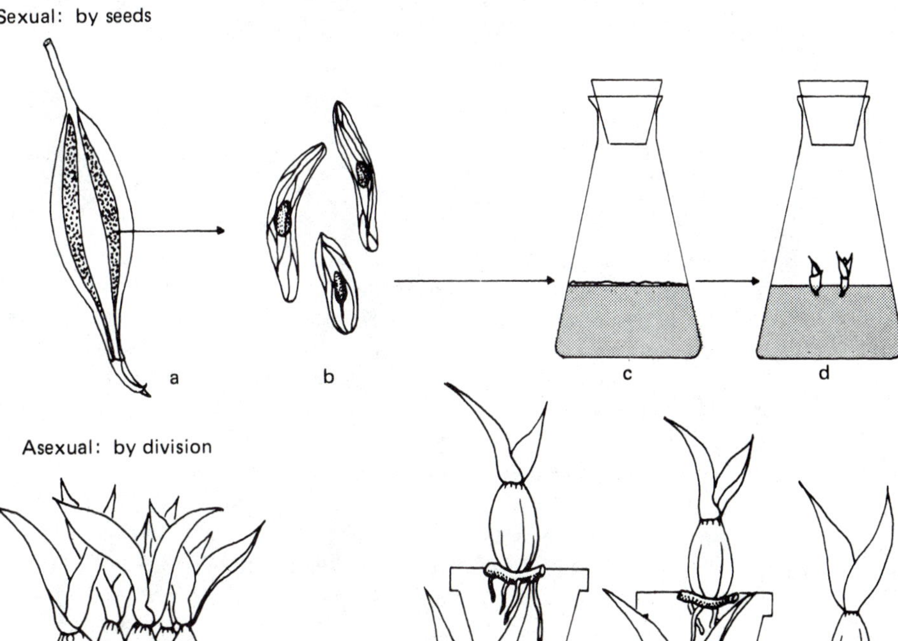


Asexual: by division

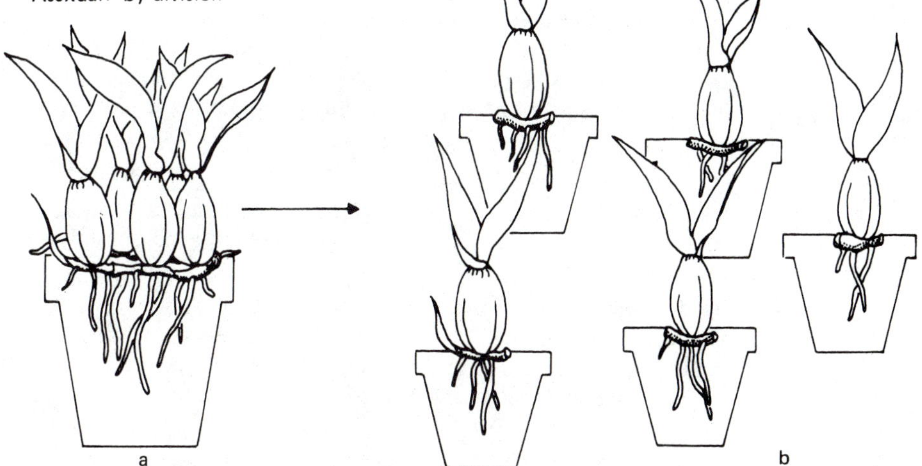

Care is taken throughout to prevent contamination of the original meristem by surface cleaning and sterilization of the parent plant and following sterile procedure in the subsequent nutrient solution changes and cutting procedures.

Cross-breeding can produce an unusual hybrid plant which can then be mericloned to result in thousands of genetically identical specimens. Mericloning provides faster orchid propagation, and there is no need to wait for a choice plant to develop mature seeds. Mericloning also makes possible the propagation of virus-free orchids even if the original plant is infected. This is because the apical meristem is usually virus-free. Now with cloning, rare and exotic orchids can be made available to the public at reasonable prices.

Mericloning is also being used to propagate African violets *(Saintpaulia ionantha)*, aroids such as anthuriums and philodendrons, geraniums, and many other houseplants as well as the important timber tree, Douglas fir *(Pseudotsuga menziesii)*. The main advantages of mericloning are speed, production of large numbers of plants, and ridding plants of disease-causing pathogens.

## REFERENCES

Crockett, J. 1970. *Bulbs. Time-Life encyclopedia of horticulture.* New York: Time-Life Books. (A treatise on both hardy and tropical bulbs: their propagation, forcing, times of flowering, methods of culture, and use in both indoor and outdoor landscaping.)

Free, M. 1957. *Plant propagation in pictures.* Garden City, N. Y.: American Garden and Doubleday and Co. (A very handy guide to the different methods of plant propagation by vegetative means for many different types of woody and herbaceous plants that are grown both indoors and outdoors—shown by means of photographs.)

Hartmann, H. T.; Flocker, W. H.; and Kofranek, A. M. 1981. *Plant science: growth, development, and utilization of cultivated plants.* Englewood Cliffs, N. J.: Prentice Hall. (A good basic text for gardeners and professional horticulturists on the principles of plant growth and development as applied to the growing of horticultural plants for home use, landscaping, and commercial production.)

Hartmann, H. T., and Kester, D. E. 1975. *Plant propagation: principles and practices.* 3d ed. Englewood Cliffs, N. J.: Prentice-Hall. (The standard reference)

Hutchinson, W. A. 1980. *Plant propagation and cultivation.* East Westport, Conn.: AVI Publishing Co. (A basic text on plant propagation methods, similar to Hartmann and Kester's book.)

Morel, G. M. 1964. *Tissue culture: a new means of clonal propagation of orchids.* American Orchid Society Bulletin 33, 473–78. (A now classic article from one of the pioneers in the field of plant tissue culture; this paper outlines the basic protocol for plant tissue culture as applied to orchids.)

Rockwell, F. F., and Grayson, E. C. 1977. *The complete book of bulbs.* Rev. ed. Philadelphia: J. B. Lippincott Co. (A classic; describes life cycles of bulbous plants)

Taylor, N. (ed.) 1948. *Encyclopedia of gardening.* Boston: American Garden Guild and Houghton Mifflin Co. (The "Bible" of gardeners for many years and still relevant to the basic principles of gardening; especially nice because it is in one volume. It is packed with a wealth of information on all major aspects of gardening.)

USDA. 1972. *Home propagation of ornamental trees and shrubs.* Home and Garden Bulletin No. 80. Washington, D.C.: U.S. Department of Agriculture. (A practical handbook, useful to the home landscaper and to those who grow nursery stock, covering the major methods of propagating ornamental trees and shrubs.)

Wimber, D. E. 1963. *Clonal multiplication of cymbidiums through tissue culture of the shoot meristem.* American Orchid Society Bulletin 32: 105–7. (A now classic paper, similar to the Morel one previously cited. Deals expressly with the mericloning methods as applied to *Cymbidium* orchids.)

# Home landscaping and maintenance 7

Most of us aspire to own our own houses and land. But whether we own or rent our homes, the time arrives when it becomes necessary for us to think about landscaping or keeping up the previously existing landscape. Most people take their yards for granted, never looking at them with a critical eye nor attempting to "decorate" them as they would the interior of their homes. Outdoor landscaping is really very similar to interior decorating: you take the individual pieces and place them aesthetically into a room of a certain size and use, to produce a functional and pleasing arrangement. Consider the yard as an extension of the indoor living area, with the same problems and potentials. You begin by becoming familiar with suitable plant material and its upkeep, just as you become familiar with wallpaper, wood furniture, and carpeting. As you learn more about plants and their uses, not only will you be able to enhance your surroundings, but you will come to admire the handiwork of others.

*Home landscaping* is the art and science of planning, designing, planting, and maintaining an attractive, functional, low maintenance environment. Sometimes this is referred to as landscape architecture, though such a term implies professional training. Professional landscape architects are more often employed to design urban and commercial landscape projects where larger scale and more expensive materials are involved. However, effective landscaping is possible for anyone willing to follow a few basic principles.

The *art* of landscaping involves the subjective creation of a plan, following simple rules of design, choosing from a vast array of living plant and structural material which combine forms, textures, colors, and shapes into a personally satisfying and functional design. The *science* of landscaping necessitates an awareness of the characteristics—potentials and limitations—of the plants and structural materials involved so that a successful long-term result is achieved. This means you need to know your plants, their correct names, their ecological requirements, and how to manipulate them to achieve the desired effect.

A *home landscape* (or a municipal landscape, for that matter) may be considered in contrast to the *natural landscape* as is seen when one views the terrain and native vegetation of the undeveloped countryside. Pristine natural landscapes are becoming rarer, but everyone should appreciate the fact that we are trying to recreate some semblance of the serene harmony in our home landscapes that is exhibited by the natural beauty found in the wild. In nature, ecological requirements are a major determining factor in deciding what grows where. We can learn much by observing the designs exhibited in nature as we seek to produce a more natural setting for our daily lives. This leads to an inner peace and satisfaction which helps us retain our sanity in a world of ever-increasing dehumanizing activities.

## Why landscape?

The two most obvious reasons for landscaping, as we have indicated, are (1) to complement your home to its best advantage in a unified setting and (2) to create a useful and pleasing environment around the home for the occupants and viewers. In order to accomplish this, you will have to put forth some effort to think about your own situation and to educate yourself by consulting various books (some of which are listed in the references at the end of the chapter) and asking questions of experienced people such as your knowledgeable neighbors or nursery personnel. Find someone who has a real love for plants and who has worked to produce a satisfactory landscape of his or her own, and you will have a ready and congenial source of information. We have never met a person who worked with plants who was not overly willing to share knowledge and experiences with others. You will find that your new found love and enthusiasm for landscape plants will eventually lead to rewarding personal relationships with your fellow humans. People who own a home with some land are fortunate to have the opportunity to experience the rewards of interacting with nature as they nurture their plants and watch them grow and change. Those who live in apartments must be satisfied with growing houseplants and dwarf container plants on the patio. They have to visit botanical gardens and arboreta to experience the joys of vicarious possession of fine specimens of trees and shrubs.

There are many other important reasons for landscaping. These include the following:

1. Expressing oneself artistically through the design and use of plant material in various settings (Figure 7–1).
2. Improving the mental well-being of people through the use of the color green in the landscape. It is a restful, peaceful color in contrast to the

*FIGURE 7–1*
A personalized landscape in a suburban development. Note the use of specimen plants (*left foreground:* weeping birch; *center background:* weeping spruce; *right foreground:* dwarf conifers) and the distinctive planting arrangement. While the forms and colors are rather diverse, the similar textures lend a degree of unity to the design. (Sterling Heights, Michigan)

concrete and stone of the asphalt jungle. Remember: "a plant can be a natural element in the middle of man-made arrangements."

3. Re-creating a natural landscape by the use of native plants in a residential setting. Not only is it low maintenance, but it is an effort to capture a part of our natural heritage.
4. Stopping erosion by planting ground covers on a steep hill or slope; or reducing mowing maintenance by planting ground covers in certain yard situations.
5. Reducing noise and visual pollution by creating a living barrier of tall and/or dense plantings between the home and the source of the pollution: a highway, factory, or rowdy neighbors! Plants can also be used to cut down glare from bright walls, bodies of water, and so forth.
6. Producing an environment that smells good—not just the seasonal scents of flowers in the spring, but that fresh smell of real plants. Certain plants, like hemlocks and pines, do provide distinct, natural odors.
7. Increasing humidity locally, even a little bit. This occurs because plants absorb water from the soil and evaporate it into the air. It can make a difference in how you feel and how the surrounding vegetation reacts if a large broadleaf tree is positioned in the landscape.
8. Attracting birds and other wildlife to the garden by providing dense cover and colorful fruits in the winter. Plants such as autumn olive,

dogwood, bush honeysuckle, crabapple, juniper, holly, and bittersweet produce edible fruits for birds.

9. Increasing the value of your home and property by good landscaping. Real estate brokers say that it is hard to put a monetary value on good trees and shrubs in the home landscape (though insurance companies may for large trees), but that a well-maintained landscape certainly makes a difference in impressing prospective buyers.
10. Achieving peace of mind by gardening as a hobby, especially as it stimulates interest in people and provides a rewarding outlet for retired persons and others who benefit by escaping from the work-a-day world.
11. Achieving a sense of control over rising food costs by incorporating edible plants in the landscape plan.
12. Saving energy by careful analysis and planning of the landscape design. This can take many detailed forms, for example:
    a. Plant large deciduous shade trees on the south, southwest, and western side of a house (Figure 7–2). The effect of one large tree in blocking the sun and evaporating moisture can be equivalent to five 10,000 BTU air conditioners in the house. Use this deciduous

*FIGURE 7–2*
**Shade tree.** A large deciduous tree (silver maple) shades a picture window from hot afternoon sun. Note also the brick retaining wall along the front sidewalk with juniper ground cover forming interesting overhang. (Charlotte, N.C.)

tree to shade large picture windows and outdoor living areas such as porch or patio. In winter, the sun's rays stream through the naked branches to provide an important warming effect.

*b.* Plant deciduous vines on southern and western walls as insulation.

*c.* Plant trees or shrubs to shade and deflect hot air exhaust from air-conditioning units.

*d.* Plant one or two rows of evergreens on the north side of a home to block winter winds. A dense hedge or fence can reduce the chill factor by creating a dead air space behind it. Up to a 50 percent savings in heat loss can be achieved.

*e.* Plant trees and shrubs to provide a wind tunnel to channel summer breezes through the house, as a convective cooling effect.

*f.* Arrange plants around your house to allow the downhill flow of cold air in winter.

*g.* Use dark color on building exterior for winter heat absorption, then provide shade in the summer with deciduous trees.

## How to begin landscaping

In establishing the home landscape environment we are mainly concerned with proper use of those long-lasting, more or less permanent and conspicuous elements such as trees and shrubs; and non-living structures such as fences, walls, patios, driveways, and sidewalks. Annual and perennial plants, which are short-lived or easily moved, are not considered in the landscape plan, though appropriate beds or areas for such, including vegetable gardens, must be taken into consideration when designing the layout.

To execute a landscape project one would be wise to consider the following important elements, the five P's of landscaping, as part of an organized and systematic framework: *Plan, Procure, Plant, Prune, Protect.* Planning can be the most important step in the whole procedure because it is at this stage that considerations of specific problems and goals must be identified.

## Planning

### SITE ANALYSIS AND FAMILY INVENTORY

In order to best determine the nature of your property, what its strengths and weaknesses are, how it should be physically modified, and how your activities and attitudes towards the public, private, and service areas should be regarded, it is a good idea at the planning stage to first do a site analysis and family inventory (Table 7–1).

By systematically identifying specific needs and problems early, trouble can be avoided later. You certainly don't need any surprises after expensive work has been done to landscape your property. Drainage, slope, exposure, and views are among the more important factors to consider.

*TABLE 7–1*
*Family inventory and site analysis questionnaire*

A. Family Needs
1. Total number in family ________
2. Number of children ________
   Ages ________
3. Family activities (State *yes* or *no* and add comments where appropriate):
   Outdoor cooking or dining ________
   Particular activities or areas for children ________
   ________
   Swimming or wading ________
   Private sitting area ________
   Work areas ________
   Laundry drying area ________
   Special area for a particular sport ________
4. Entertaining:
   General type ________
   Usual number of guests ________
   Parking capacity ________
5. Gardening:
   Flower beds or shrub groupings ________
   Vegetable garden ________
   Cold frames ________
   Compost bin ________
   Greenhouse ________
   Herb garden ________
   Fruit trees or vines ________
   Water garden, pond or fountain ________
   ________
6. Animals:
   Pets (number and kinds) ________
   Any housing needed for pets (or special fencing) ________
   ________
   Interest in birds ________
   ________
7. Maintenance:
   Is the care of your garden of interest to you ________
   Would you prefer to minimize maintenance ________
   Are you interested in an automatic watering system ________
8. Storage:
   Any problems storing:
   Trailers ________
   Boats ________
   Gardening equipment ________
   Toys ________
   Cooking equipment ________
   Outdoor furniture ________

B. Site Analysis

1. Do you feel that any areas have too steep a slope ______

2. Are there any sunken areas that need to be filled ______

3. Does water drain onto walks and drives ______

4. Does water stand on paved areas ______

5. Is there bad drainage from the house or other directions ______

6. Are there drainage problems from or onto neighbors' property ______

7. Is septic tank drainage a consideration ______

8. Does drainage from roof affect ground drainage or plantings adversely ______

9. Are there eroded areas on the site ______

10. Are there any areas that cause dust problems ______

11. Do any trees or shrubs need to be removed (diseased, etc.) ______

12. Does lawn need improvement ______

13. Are there any trees or shrubs which you feel are particularly outstanding or of interest ______

14. Are there bodies of water or rock outcroppings that you may want to feature ______

15. Do you feel a general need for more protection from sun or wind outdoors ______

16. Do unshaded paved areas greatly increase the temperature of your yard in the summer ______

17. Are there any areas that are a problem in the winter ______

18. Could the "climate" within your home be improved (by shade to decrease heat or wind protection to protect from cold) ______

19. Are there specific good or bad views that should be enhanced or blocked ______

20. Are there any bright lights or signs which you find disturbing ______

21. Are there any noises from a nearby road, etc., that are a problem ______

*(continued)*

*TABLE 7–1* (continued)

22. Are there any views that your neighbors have of your property that you might want to change ____________________
23. Do people cut through your yard ____________________
24. Do any walks or drives need to be widened ____________________
25. Do you have adequate space for parking ____________________
26. Do you feel that too much of your yard is paved ____________________
27. Are any walks or drives inconvenient (do not lie where everyone tends to walk, dangerous intersection between drive and road, etc.) ____________________
28. Is the material used for the paved or graveled areas unsatisfactory (too much glare, slick in the rain, not in harmony with the materials used in the house, etc.) ____________________
29. Is any paved area cracked or uneven enough to cause possible injury ____________________

After a site analysis, the next step is to put down ideas on paper. Draw a simple, but accurate-to-scale outline of your property and home (Figure 7–3). Begin to outline important features that must be in the final plan, such as a large garden, pool, wooded area, driveway, patio, fences, and so forth. These structures or areas are "fixed" variables and cannot be changed later; it is better to identify them early and work around them. Do not be afraid to remove unnecessary trees or shrubs from an existing landscape if they do not fit in with your plan. Later we will consider how to determine the value of a plant; sometimes an existing plant can be more of a liability than an asset.

## SPACE DESIGN

The typical home environment should be viewed as the outdoor extension of the indoor living space and as such consists of three main areas: *public, private,* and *service.* One should deal with designing the outdoor space just as one would an interior room, with fences being the walls of the property and the sky the ceiling. The use of space for various activities should be carefully considered (Figure 7–4).

The public area generally consists of the front yard, that part of your property that is readily visible to others, and as such should show off your home and property at its best (just as the living room in the house has the finest furniture). The public area should present the best view of the house, with appropriate foundation planting and larger plants of proper scale to

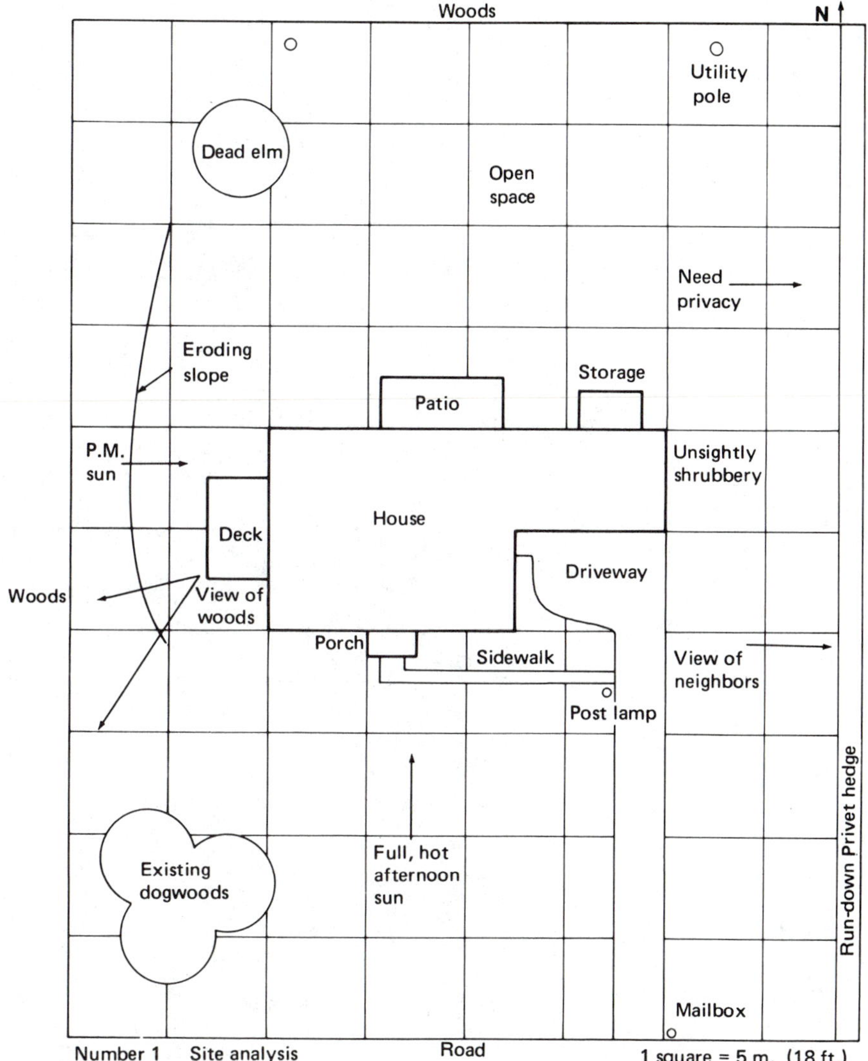


*FIGURE 7–3*

**Step 1—site analysis.** Identify environmental features, utilities, existing plants, landforms, views, and other important aspects.

frame the house. Good taste dictates that lawn ornaments should not be used as they detract from the natural setting.

The private area is the back yard living space, where your personal activities take place: relaxing, gardening, swimming, barbecuing, and children playing. It should be simple and tailored to fit your needs without undue maintenance. Choice of plants and landscape structures will depend on individual needs.

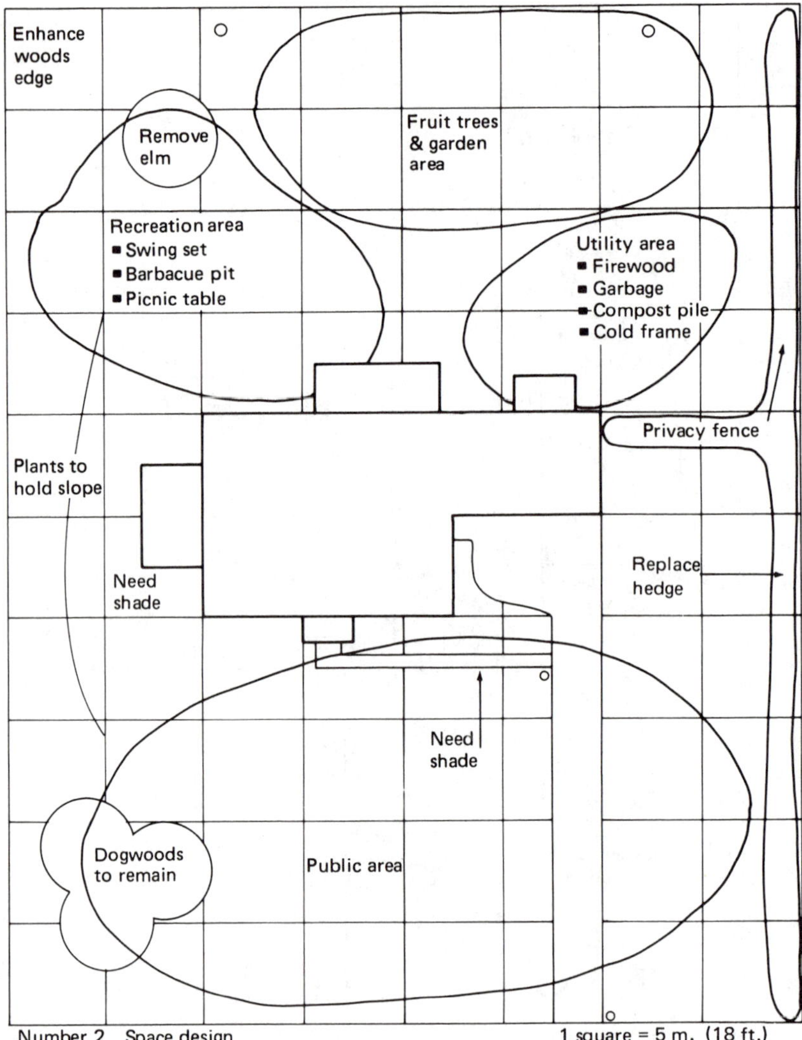


*FIGURE 7–4*

**Step 2—space design.** Identify public, private (recreation), and utility areas. Also identify problem areas and special use areas.

The service area consists of those areas where tools are stored, where the oil tank is situated, where the garbage cans are kept, the clothesline is strung, the compost pile is built, or the wood is chopped. Generally, the service areas should have convenient accesses but be rendered inconspicuous by the judicious use of screening plants or landscape structures.

When you have learned plant characteristics, maintenance requirements, how you want to use plants, and the principles of design, you will be ready to draw up your final landscape plan (Figure 7–5).

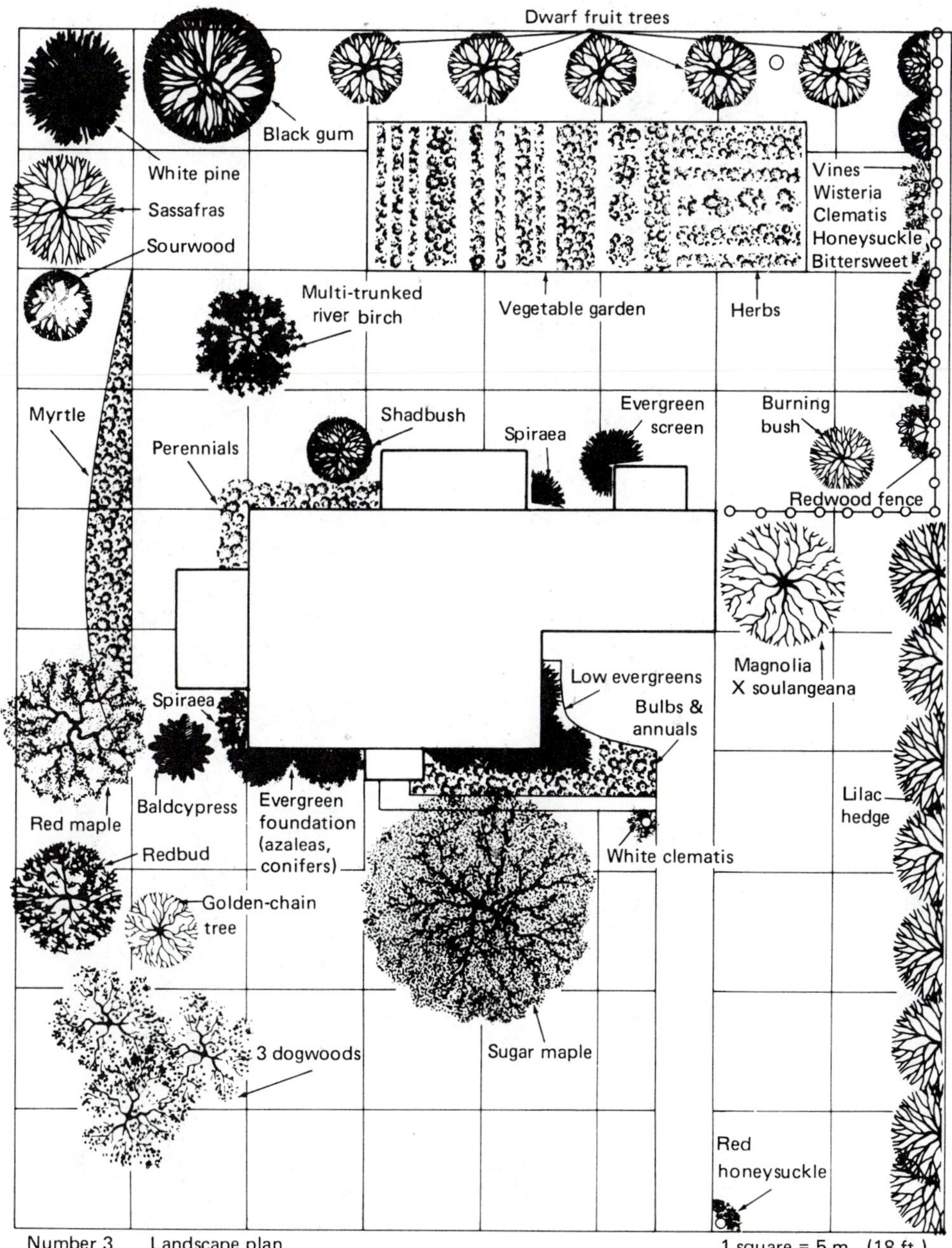


*FIGURE 7–5*

**Step 3—landscape plan.** Indicate name and position of actual plants (or types of plants available) to be used. Consider mature size, shape, growth rate, seasonal colors, ecological needs, and maintenance requirements of the species and cultivars you choose.

LANDSCAPE PLAN

*PLANT CHARACTERISTICS.* As you progress with your rough drawing, you begin to get ideas of the general types of plants you would like to incorporate into the landscape. At first you should not be concerned with a specific plant but more with general characteristics such as form and shape; size and growth rate; color and seasonal changes; texture and density; ecological and maintenance requirements.

1. *Form and shape* Trees and shrubs come in a variety of characteristic shapes and forms from the formal conifers and stately oaks to the informal weeping willow and free-growing crabapples (Figures 7–6, 7–7 and Table 7–2).
2. *Size and growth rate* Some people apparently do not realize that plants grow and get larger with age. This must be considered when planting trees and shrubs so that a large-growing plant is not put in a small space. Similarly, plants placed together should grow at the same rate so that one does not overcome the other as they mature.
3. *Color* Most plants are green, but there are variegate plants and fancy cultivars with leaves other than green (red Japanese maple, for example). Many plants are chosen for their seasonal colors: spring flowers, autumn foliage, winter fruits or twig colors. Such interesting charac-

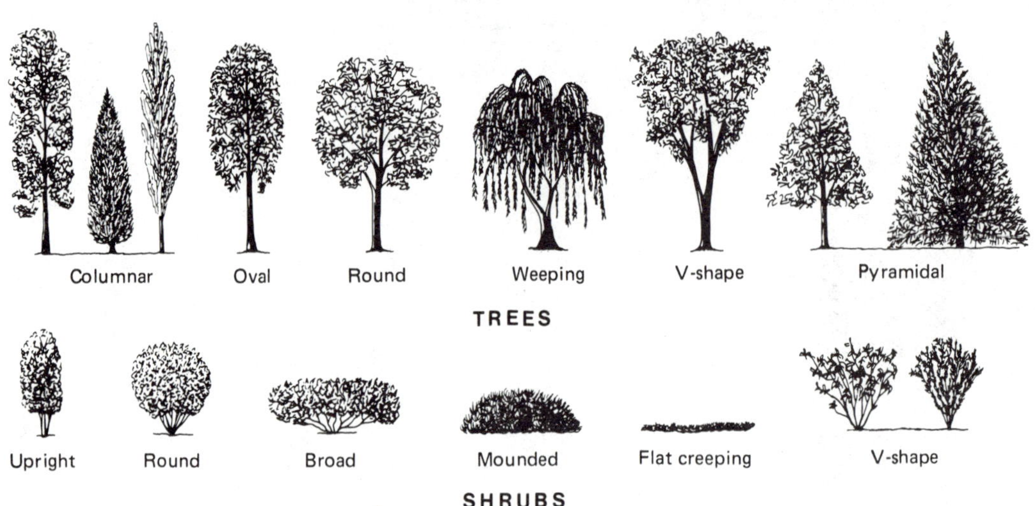


*FIGURE 7–6*
The most common forms of trees and shrubs. Species and cultivars may come in a variety of forms; consider mature size and position in the landscape when planning for desired effect.

*FIGURE 7–7*
Diverse planting of large landscape trees showing differences in shape, form, texture, and winter appearance. Compare the bold shapes of the pyramidal conifers (spruces) to the delicate branching patterns of the deciduous trees (maples, hickories); and note the distinctive form of the weeping elm (Camperdown elm) in the left foreground. (University of Michigan's North Campus, Ann Arbor)

## *TABLE 7–2*
## *Trees and shrubs of particular interest*

*Weeping Trees*

- Threadleaf Japanese maple (*Acer palmatum* 'Dissectum')
- Harry Lauder's walkingstick (*Corylus avellana* 'Contorta')
- Weeping beech (*Fagus sylvatica* 'Pendula')
- Weeping mulberry (*Morus alba* 'Pendula')
- Weeping cherry (*Prunus subhirtella* 'Pendula')
- Weeping willow (*Salix babylonica*)
- Sargent weeping hemlock (*Tsuga canadensis* 'Pendula')
- Camperdown elm (*Ulmus glabra* 'Camperdownii')

*Narrow, Erect-Growing Trees*

- Columnar Norway maple (*Acer platanoides* 'Columnare')
- Columnar red maple (*Acer rubrum* 'Scanlon')
- Sentry maple (*Acer saccharum* 'Monumentale')
- Hawthorn (*Crataegus monogyna* 'Stricta')
- Columnar ginkgo (*Ginkgo biloba* 'Fastigiata')
- Ascending American elm (*Ulmus americana* 'Ascendens')

*(continued)*

*Attractive Bark*

Japanese maple *(Acer palmatum)*
Paperbark maple *(Acer griseum)*
Striped maple *(Acer pensylvanicum)*
River birch *(Betula nigra)*
Paper birch *(Betula papyrifera)*
Red-twig dogwood *(Cornus stolonifera)*
Persimmon *(Diospyros virginiana)*
Crepe myrtle *(Lagerstroemia indica)*
Persian parrotia *(Parrotia persica)*
Amur corktree *(Phellodendron amurense)*
Lacebark pine *(Pinus bungeana)*
Plane tree *(Platanus X acerifolia)*
Mahogany-bark cherry *(Prunus serrula)*
Sassafras *(Sassafras albidum)*
Silk camellia *(Stewartia* species*)*

*Good Fall Color*

Japanese maple *(Acer palmatum)*—red
Red maple *(Acer rubrum)*—red
Sugar maple *(Acer saccharum)*—yellow, orange, red
Serviceberry *(Amelanchier* species*)*—orange to red
Birch *(Betula* species*)*—yellow
Hickory *(Carya* species*)*—yellow
Katsura tree *(Cercidiphyllum japonicum)*—yellow to scarlet
Redbud *(Cercis canadensis)*—yellow
Yellowwood *(Cladrastis lutea)*—yellow
Flowering dogwood *(Cornus florida)*—red
Beech *(Fagus grandifolia)*—bronze yellow
Fothergilla *(Fothergilla major)*—orange, scarlet
Franklinia *(Franklinia alatamaha)*—red
White ash *(Fraxinus americana)*—purple, maroon, yellow, red
Ginkgo *(Ginkgo biloba)*—golden yellow
Crepe myrtle *(Lagerstroemia indica)*—orange, red
Sweet gum *(Liquidambar styraciflua)*—purple, yellow, red
Black gum *(Nyssa sylvatica)*—scarlet
Sourwood *(Oxydendrum arboreum)*—scarlet
Parrotia *(Parrotia persica)*—yellow, orange, scarlet
Sargent cherry *(Prunus sargentii)*—bronze or red
Bradford pear *(Pyrus calleryana* 'Bradford'*)*—scarlet and purple
Scarlet oak *(Quercus coccinea)*—scarlet
Staghorn sumac *(Rhus typhina)*—red
Sassafras *(Sassafras albidum)*—yellow, orange, red
Blueberry *(Vaccinium corymbosum)*—yellow, bronze, orange, red

---

*Source:* See Countryside Books 1978; Dirr 1977; Nelson 1975; and Wyman 1965.

teristics can add interest to a landscape if used properly. Consider color combinations, backgrounds, and the number of nongreen plants. As a general rule you should use nine "normal green" foliage plants for each "colored" foliage plant in the landscape to avoid a cluttered, "busy" appearance. (Figures 7–8 and 7–9.)

4. *Texture* Variations in leaf and twig texture, both in summer and win-

ter, can add variety to the landscape just as color, but the same considerations should apply to avoid too much variation. The ultimate goal is to achieve an interesting harmony, not a hodgepodge of randomly placed variables. Carefully consider plant texture as it affects the appearance and use of the landscape. Plants may be characterized as coarse, medium, and fine textured depending on factors of leaf size, number, and arrangement (Figures 7–8 and 7–9).

5. *Ecological requirements* Matching a plant's growth requirements to your site is necessary for long-term success. Considerations of shade vs. sun, soil pH, soil texture (sandy, loamy, clay), drainage, and hardiness are most important. Some plants are very adaptable to a wide range of conditions (such as honeysuckles, privets, and junipers) while others are more particular (rhododendrons need a well-drained acid soil). Slight differences in sites can have significant effects on plant characteristics; some shrubs need full sun to flower well but will grow satisfactorily in the shade.

   Hardiness is a major factor since it pertains to the winter cold a plant can tolerate. The United States has been divided into hardiness zones (see USDA map on page 186) based on average minimum winter temperatures. You should know the hardiness rating of a plant you intend to use in your area. Some plants have a broad tolerance, some very narrow. Summer heat can affect plants just as much as winter cold, but this is usually less critical and more difficult to document. Remember it is the interaction of many ecological factors that truly determines the successful growth of a plant. To be safe, it is better to buy plants from a reputable local nursery that can certify that the plants are tolerant of local conditions rather than buying from an unknown (but often cheap) source through the mail (Tables 7–3 and 7–4).

6. *Maintenance requirements* Maintenance requirements of the various landscape plants should be considered along with ecological requirements. For example, how much pruning will be necessary to keep the plant in shape, how often will you have to fertilize, spray for pests, and water during droughts? If you are careful to choose plants which require little pruning and are relatively pest-free, and if you place them in the best ecological setting, then you should be able to minimize your efforts to keep the plant healthly and performing its landscape function.

   Shrubs with little or no pruning requirements include: gold-dust tree *(Aucuba)*, azalea, Carolina allspice *(Calycanthus)*, *Camellia*, some hollies *(Ilex)*, mountain laurel *(Kalmia)*, some junipers, *Nandina*, andromeda *(Pieris)*, *Rhododendron*, and *Yucca*. Other low maintenance shrubs may not require pruning if they are used in situations where they can grow unhindered such as quince *(Chaenomeles)*, *Exochorda*, *Forsythia*, fragrant sumac *(Rhus aromatica)*, *Spiraea*, some *Viburnum*, and ninebark *(Physocarpus)*. Of course, all plants may need occasional pruning to remove dead limbs or an odd branch.

*FIGURE 7–8*
A very effective and appealing foundation planting. All of the colors are various shades of green. The fine textures of the junipers and the hollies are similar. The conical shape of the juniper at the corner is distinctive (as an accent plant) but not overpowering. The ground cover junipers and the medium-textured bark mulch help to form a unified design. The concrete-formed edging helps to define space and make the overall aspect a little more formal. (Sterling Heights, Michigan)

*TABLE 7–3*
*Relative salt tolerance of some city trees*

| *Intolerant* | *Moderately Tolerant* | *Very Tolerant* |
|---|---|---|
| European alder | White ash | Black cherry |
| Pin oak | White birch | Red cedar |
| London plane tree | Catalpa | Jack pine |
| (Sycamore) | Crabapple | White oak |
| Hemlock | Douglas fir | Yellow birch |
| Sugar maple | Siberian elm | Black locust |
| White pine | American elm | |
| Shagbark hickory | Silver maple | |
| Beech | Red maple | |

*Source:* See Grey and Deneke 1978.

*FIGURE 7–9*
An unappealing, confusing foundation planting. There are too many different shapes, sizes, colors, and textures. There are three different variegated-leaved plants. There is not enough repetition of any elements in the design. The coarse mulch (lump coal) does not help to bring unity to this arrangement. (Sterling Heights, Michigan)

*TABLE 7–4*
*Tree sensitivity to city air pollution ($SO_2$ and Ozone)*

| *Sensitive* | *Tolerant* |
|---|---|
| Ash | Dogwood |
| Sweet gum | Maple |
| White oak | Red oak |
| Jack pine | Spruce |
| White pine | Juniper |
| Poplar | Black walnut |
| Linden | Pin oak |
| American elm | London plane tree |
| | Ginkgo |

*Source:* See Grey and Deneke 1978; Pirone 1978.

# Plant Hardiness Zone Map

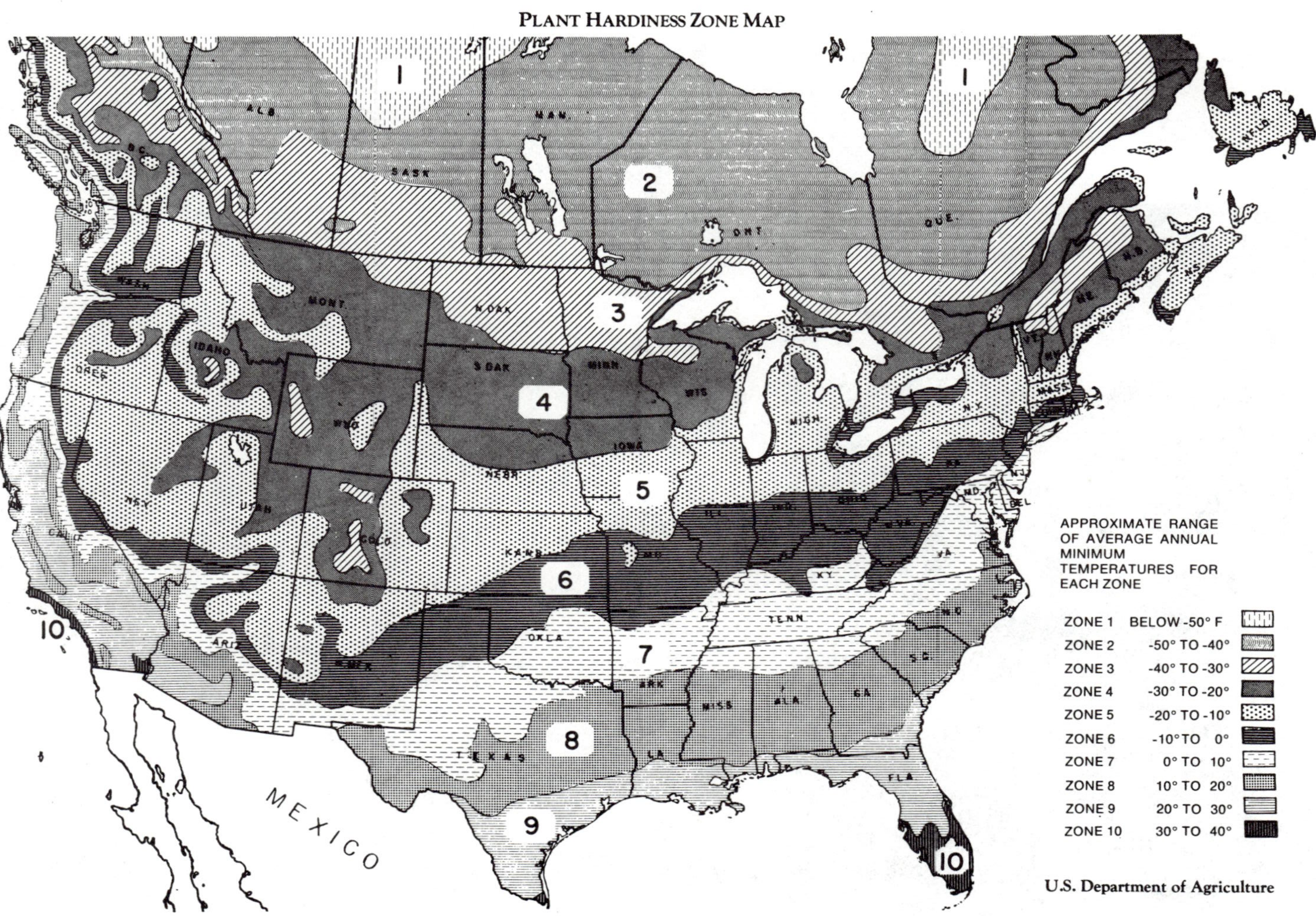


U.S. Department of Agriculture

*USES OF PLANTS.* Trees are usually used as large plants in the landscape to frame a house and to provide a background or shaded areas (Table 7–5). Small trees and shrubs have a wider range of uses which are listed as follows:

1. *Specimen plants* A single outstanding specimen that adds interest to a planting design. Should be placed so that it receives attention as an attraction inherent in itself. Examples: red Japanese maple (*Acer palmatum* 'Atropurpureum'), weeping flowering cherry (*Prunus subhirtella* 'Pendula'), corkscrew hazel or Harry Lauder's walkingstick (*Corylus avellana* 'Contorta'), cut-leaf sumac (*Rhus typhina* 'Laciniata') (Table 7–2).

2. *Accent plant* A plant that adds interest to the landscape when mixed in with other plants; not meant to be as spectacular as a specimen plant. A plant which, although basically grown for its foliage, may produce flowers or fruits at some time in the year. Examples include camellia *(Camellia japonica)*, pieris *(Pieris japonica)*, and Burford holly (*Ilex cornuta* 'Burfordii').

3. *Group planting* A small cluster of similar plants, like a group of three flowering dogwoods, several bush honeysuckles, or a group of junipers, which is used to provide a bold shape or to fill a small space. Mass plantings of azaleas around trees or in wooded borders would be another example.

4. *Border planting* A spread-out grouping of similar plants used to delineate space or provide a group of plants in a certain area, such as a shrub border along one edge of a yard. Usually low and informal.

5. *Hedge planting* A living wall created by fast-growing, utilitarian plants. Usually pruned into the desired shape; often neglected and left to grow wild and unattractive. Privet *(Ligustrum)*, barberry *(Berberis)*, holly *(Ilex)*, juniper *(Juniperus)*, and photinia *(Photinia)* are common hedge plants, but there are many possibilities. Hedges may be pruned to any desired height and can be used to block traffic flow, screen views, or reduce noise pollution.

6. *Foundation planting* This is the most basic type of planting. Plants are used to cover the foundation of the house, soften its architectural lines and corners, accent the house, and otherwise provide a pleasing setting for the house. Often, if neglected, foundation plants can actually detract from the appearance of the house. Homes without foundation plantings appear barren, cold, and uninviting. Foundation planting should be interesting without being too conspicuous. It should follow the rules of design to achieve a feeling of harmony (Compare Figure 7–8 with Figure 7–9.)

### *TABLE 7–5*
### *Some desirable trees for cities as street trees or lawn specimens*

Norway maple *(Acer platanoides)*
Sugar maple *(Acer saccharum)*
Pink horse-chestnut *(Aesculus* X *carnea)*
Yellowwood *(Cladrastis lutea)*
Flowering dogwood *(Cornus florida)*
Japanese dogwood *(Cornus kousa)*
Flowering ash *(Fraxinus ornus)*
Ginkgo *(Ginkgo biloba)*
Moraine locust *(Gleditsia triacanthos inermis* 'Moraine'*)*
Crepe myrtle *(Lagerstroemia indica)*
Sweet gum *(Liquidambar styraciflua)*
Arnold crabapple *(Malus arnoldiana)*
Kobus magnolia *(Magnolia kobus)*
London plane tree *(Platanus* X *acerifolia)*
Kwanzan cherry *(Prunus serrulata)*
Bradford pear *(Pyrus calleryana* 'Bradford'*)*
Scarlet oak *(Quercus coccinea)*
Pin oak *(Quercus palustris)*
Willow oak *(Quercus phellos)*
Red oak *(Quercus rubra)*
Littleleaf linden *(Tilia cordata)*
Zelkova *(Zelkova serrata)*

*Source:* See Pirone 1978; Dirr 1977.

*PRINCIPLES OF DESIGN.* To achieve good results in placing plants in the landscape it is worth the effort to apply a few simple principles of design when considering plant characteristics. Briefly, these include the following:

1. *Simplicity* Keep the design simple. Use the fewest variables. Have a central focus of color, texture, and form.
2. *Variety* Use several different types of plants, mix colors, textures, and forms appropriately. Too little variety yields monotony; too much yields confusion.
3. *Balance* Try to achieve a balance between the parts of a landscape design. For example, keep the two sides of a central axis in balance, use equal numbers of plants on each side of a center line, or balance textures and forms and mass of plants on both sides. If the plants are exactly the same on each side, you have a symmetrical, or formal design; if the plants are different, but balanced, you have an asymmetrical, or informal design.
4. *Repetition* Have a repeated pattern rather than a continuously changing variety to reduce confusion.

5. *Emphasis* Add interest to the design by having a focal point, something that stands out slightly to direct the eye to one portion of the composition—a single specimen plant, a group of shrubs, or a statue in a formal garden.
6. *Scale* Keep sizes in perspective. Do not use a three-story tall tree with a small house (Figure 7–10). Use the right size of shrub for foundation planting and do not let them become overgrown. Carefully choose the right size plant for the size of the yard.
7. *Harmony* Above all, ensure that the diverse elements in your final design have a purpose for being there and that the overall design is pleasing and fits together.

*FIGURE 7–10*
Trees out of scale to house. A major problem with new home developments is that forest trees that are left with the intention of adding to the value and aesthetics of the property are so badly disturbed during construction that they die within a year or two. This creates higher costs for careful removal and generates ill-will on the part of the homeowner. Sadly, home builders rarely realize that construction and grading can do such extensive damage to valuable trees. (Charlotte, N.C.)

*VALUE OF A PLANT.* Trees and good landscaping in general add value to a home depending on their desirable features. Important characteristics of trees include the following:

1. *Location* If the site chosen for trees and plants is appropriate to their use, the value of the home is enhanced. Thus, if a tree is to provide shade, its location must be carefully considered.
2. *Type* Some trees are more long-lived and sturdier than others. For example, an oak is more valuable than a poplar or mulberry.
3. *Size* A large old tree is more valuable if it provides shade where needed and enhances the house and property by fitting in the scale properly.
4. *Condition* A healthy tree is more valuable than a diseased one. For example, American elms may be a liability due to their susceptibility to the Dutch elm disease.

Some trees will naturally fulfill more of the valuable categories than others by having more desirable features. For example, oaks, hickories, hard maples, ashes, lindens, and hemlocks are generally more valuable than elms, soft maples, poplars *(Populus)*, hackberries, cherry, or horse chestnut. Each tree must be rated independently in its own setting for the use it was intended to fulfill. Most trees have at least some good characteristics, though poplars *(Populus)*, mulberries, Siberian elm, and box elder maple have almost no redeeming values and should be avoided. Even the widely planted silver maple is less than desirable because it rapidly grows into a large, unshapely tree (Table 7–6). Some trees are "messy." That is, they drop large leaves, twigs, fruits, and so forth, or they are brittle and their limbs break easily.

Some trees have much better fall foliage than others, such as maples, sassafras, ashes, hickories, sour gum, ginkgo, and dogwood; while still other trees may be desirable for their showy flowers, for example, horse chestnut, tulip poplar *(Liriodendron)*, cherries, yellow wood, redbud, and dogwood.

Visit parks, botanical gardens, and arboreta to learn about trees and see their characteristics as they mature. The best way is to make comparisons and carefully choose the best trees for you. Avoid buying bargain trees or those for which claims of rapid growth are made. These trees are usually short-lived and have other undesirable features.

## Landscape planning summary

1. Begin with the *site analysis and home inventory*—locate buildings and other structures, walks, drives, and so forth.
2. Then rough draw *plan of space* and incorporate *public, private,* and *service* areas, considering the following:

a. *Characteristics of plants* (form and shape, size, color, texture, ecology, and maintenance).

b. *Types and uses of plantings* (specimen, accent, groupings, borders, hedges, and foundation).

c. *Principles of design* (simplicity, variety, balance, repetition, emphasis, scale, and harmony).

d. *Values of a plant* (location, type, size or age, and condition).

3. *Finally*, draw up the *Landscape plan* and include the type and position of each plant to achieve unity in the design.

TABLE 7–6
Trees to avoid planting and reasons why

*Brittle wood*, rapid growth, may become unsightly; relatively short-lived

Box elder *(Acer negundo)*
Silver maple *(Acer saccharinum)*
Aspen (*Populus* spp.)
Poplar (*Populus* spp.)
Lombardy poplar (*Populus nigra* cv. 'Italica')
Siberian elm *(Ulmus pumila)*

*Objectional fruits*

Catalpa (*Catalpa* spp.)
Persimmon *(Diospyros virginiana)*
Female ginkgo *(Ginkgo biloba)*
Honey locust *(Gleditsia triacanthos)*
Black walnut *(Juglans nigra)*
Osage orange *(Maclura pomifera)*
Crabapple (*Malus*, certain spp.)
Mulberry (*Morus* spp.)
Black cherry *(Prunus serotina)*

*May become unsightly* due to diseases or cultural problems

Buckeye *(Aesculus octandra)* and horse chestnut *(A. hippocastanum)*
White birch *(Betula pendula)*
White ash *(Fraxinus americana)*
Crabapple (*Malus*, certain spp.)
Sycamore *(Platanus occidentalis)*
Black locust *(Robinia pseudoacacia)*
American elm *(Ulmus americana)*

*Source:* See Pirone 1978; Dirr 1977; and Wyman 1965.

## Procurement of plants for the landscape

During the planning stages there is no need to be concerned with specific plant names because there are literally hundreds of plants to choose from, many of which have very similar landscape features. Once you have finalized your landscape plan on paper, making some choices as to what types of plants you want and where to put them, then it is time to purchase the specific plants from a nursery. It is always better to buy a specific plant for a specific site rather than to buy plants first and then try and find places for them. The specific plants you obtain will depend on availability, cost, and personal preference. For the permanent plants, you will be choosing from three broad categories of woody plants: (1) trees, (2) shrubs, and (3) vines, ground covers, and hedges (Table 7–7).

### TYPES OF PLANTS

Woody plants come in two general types: deciduous and evergreen. The deciduous type lose their leaves in winter and offer the advantage of fall colors and stark textural contrasts as seasonal changes occur. Evergreens on the other hand provide year-round protection and covering as hedges or foundation plantings. Each of the two types have their particular landscape use. Woody plants may be further classified as follows:

1. *Coniferous* (or needle-leaved) *evergreens* such as pines, junipers, yews, arborvitae, cypress, hemlock, cedars, firs, and spruces. This is the largest category of nursery plants. Conifers may be particularly useful because of their often distinctive, bold forms which are evident year-round.
2. *Broad-leaved evergreens* such as boxwood, holly, azalea, rhododendron, winter-creeper *(Euonymus)*, privet *(Ligustrum)*, camellia, southern magnolia, some viburnums, osmanthus, red-tip photinia, and many ground covers like ivy and periwinkle.
3. *Deciduous plants* of trees or shrubs such as oak, ginkgo, maple, ash, dogwood, redbud, willow, linden, forsythia, some viburnums, cotoneaster, barberry, burning bush *(Euonymus alatus)*. Our most beautiful flowering plants, fruit trees, and shade trees with their attractive fall colors are deciduous. Roses are apparently America's most popular deciduous shrubs because so many are sold by nurseries.

While the primary basis for choosing specific plant material is appearance and function, certain ecological factors such as hardiness, moisture requirements, and sun/shade tolerances must be considered. Woody plants may be purchased in one of three ways: *bare root, balled and burlapped* (B&B), or *container grown.*

Bare root plants are moved from the nursery where they are grown to the planting site without any soil on the roots. This invariably means that the plants are deciduous trees or shrubs and transplanting must be done in

*TABLE 7–7*
*Plants worth knowing for home landscape use*

| Common Name | Scientific Name[1] | Hardiness[2] Zone | Mature Height | Mature Spread | Landscape Use |
|---|---|---|---|---|---|
| *Evergreen Trees* | | | | | |
| 1. White fir | *Abies concolor* | 4 | 9–15m (30–50′) | 4–9m (15–30′) | Broadly conical; gray green foliage |
| 2. Deodar cedar | *Cedrus deodara* | 7 | 12–21m (40–70′) | 4–15m (15–50′) | Extremely graceful habit; characteristic of southern gardens; cultivars hardy in zone 4 |
| 3. Port Orford cedar | *Chamaecyparis lawsoniana* | 5 | 12–18m (40–60′) | 4–9m (15–30′) | Beautiful foliage and graceful habit; many color forms |
| 4. Leyland hybrid cypress | X *Cupressocyparis leylandii* | 7 | 12–24m (40–80′) | 2–4m (7–15′) | Extremely fast growing; tall and columnar |
| 5. Eastern red cedar | *Juniperus virginiana* | 2 | 12–15m (40–50′) | 2–6m (7–20′) | Specimen; use as windbreak |
| 6. Southern magnolia | *Magnolia grandiflora* | 6 | 18–24m (60–80′) | 9–15m (30–50′) | Broadleaf evergreen; coarse texture; large, fragrant, summer flowers |
| 7. White spruce | *Picea glauca* | 2 | 12–18m (40–60′) | 3–6m (10–20′) | Specimen, mass, or windbreak |
| 8. Lacebark pine | *Pinus bungeana* | 4 | 9–15m (30–50′) | 6–11m (20–35′) | Good specimen tree because of attractive bark |
| 9. Swiss mountain pine | *Pinus mugo* | 2 | 4–6m (15–20′) | 8–9m (25–30′) | Small tree; dwarf cultivars particularly useful |
| 10. Austrian black pine | *Pinus nigra* | 4 | 15–18m (50–60′) | 6–12m (20–40′) | Dark green needles; specimen or windbreak |
| 11. Eastern white pine | *Pinus strobus* | 5 | 15–24m (50–80′) | 6–12m (20–40′) | Beautiful native pine; delicate, light green needles |
| 12. Umbrella pine | *Sciadopitys verticillata* | 5 | 6–9m (20–30′) | 4–6m (15–20′) | Needles borne in umbrellalike clusters; slow grower; few pests; attractive |
| 13. Eastern arborvitae or white cedar | *Thuja occidentalis* | 2 | 12–18m (40–60′) | 3–4m (10–15′) | Pyramidal form with good dark green foliage in winter |

[1]Species are arranged alphabetically by scientific name.
[2]See p. 186 for hardiness zone map. Most plants will grow in or below their listed zone, although many plants listed will not do well in zone 10 or lower parts of zone 9.

*(continued)*

*TABLE 7–7* (continued)

| | | | | | |
|---|---|---|---|---|---|
| 14. Canadian hemlock | *Tsuga canadensis* | 3 | 14–21m (45–70′) | 8–11m (25–35′) | Fine texture; good for graceful hedge and screening |
| 15. Carolina hemlock | *Tsuga caroliniana* | 4 | 14–18m (45–60′) | 6–8m (20–25′) | Excellent as specimen; pyramidal form; compact, dense, and dark green |

*Evergreen Shrubs*

| *Common Name* | *Scientific Name* | *Hardiness Zone* | *Mature Height* | *Landscape Use* |
|---|---|---|---|---|
| 1. Gold-dust tree | *Aucuba japonica* 'Variegata' | 7 | 4m (15′) | Excellent, bold broadleaf evergreen; bright red fruits produced if both sexes present |
| 2. Wintergreen barberry | *Berberis julianae* | 5 | 2m (7′) | A dependable dense shrub for hedges |
| 3. Boxwood | *Buxus sempervirens* | 5–6 | 4–6m (15–20′) | Fine textured; many cultivars; hedges, foundations, massing, topiary |
| 4. Camellia | *Camellia japonica* | 7 | 8m (25′) | Many cultivars; nice foliage; showy winter/spring flowers. Most cultivars smaller—to 4m (12′) |
| 5. Sasanqua | *Camellia sasanqua* | 7 | 6m (20′) | Smaller leaves; fall flowering |
| 6. Rock cotoneaster | *Cotoneaster horizontalis* | 4 | 1m (2–3′) | Semievergreen spreading ground cover; bright red fruit in winter |
| 7. Thorny elaeagnus | *Elaeagnus pungens* | 7 | 4m (14′) | Vigorous grower with long arching branches; silver leaves; fragrant flowers late fall |
| 8. Evergreen euonymus | *Euonymus japonicus* | 8 | 4m (15′) | Dark green leaves; variegated cultivars; susceptible to scale insects |
| 9. Gardenia | *Gardenia jasminoides* | 8 | 2m (7′) | Extremely fragrant white flowers in summer, subject to iron chlorosis |
| 10. Japanese holly | *Ilex crenata* 'Helleri' | 5 | 1m (4′) | Dwarf, fine-textured shrub, excellent hedge |
| 11. Burford holly | *Ilex cornuta* 'Burfordii' | 7 | 3m (10′) | Dense, dark green, shiny leaves with single terminal spine; heavy fruiter; excellent plant for southern gardens |

| | | | | |
|---|---|---|---|---|
| 12. Rotunda holly | *Ilex cornuta* 'Rotunda' | 7 | 1m (3′) | Shiny leaves with many sharp spines; forms a rounded plant without pruning |
| 13. Chinese juniper | *Juniperus chinensis* | 4 | 1–3m (3–10′) | Many good cultivars of various size, shape, and color |
| 14. Creeping juniper | *Juniperus horizontalis* | 3 | ⅓m (1′) | Good ground cover; many color forms |
| 15. Mountain laurel | *Kalmia latifolia* | 4 | 2–4m (7–15′) | Beautiful spring flowers; good foliage for shady shrub borders |
| 16. Amur privet | *Ligustrum amurense* | 3 | 3–4m (10–15′) | Hedge plant; will stand severe pruning; over-used |
| 17. Nandina | *Nandina domestica* | 7 | 2m (7′) | Clump forming; needs little pruning; attractive winter foliage and bright red berries for Christmas |
| 18. Fortune tea-olive | *Osmanthus* X *fortunei* | 7 | 4m (14′) | Handsome dark green foliage; very fragrant flowers in late summer |
| 19. Red-tip photinia | *Photinia* X *fraseri* | 7 | 4m (15′) | New growth is bright red; fast-growing hedge or screen plant |
| 20. Japanese pieris | *Pieris japonica* | 5 | 3–4m (10–14′) | Excellent as specimen or border plant; white flowers in earliest spring |
| 21. Podocarpus | *Podocarpus macrophyllus* | 8 | 8m (25′) | Good as a specimen plant; narrow, dark green leaves and informal growth habit |
| 22. Firethorn | *Pyracantha coccinea* | 6 | 2–5m (7–17′) | Informal barrier or hedge; very colorful fruit; susceptible to lace-wing bug |
| 23. Azalea | *Rhododendron* spp. and many cultivars | 4–8 | 1–3m (3–10′) | Low maintenance; outstanding as flowering specimens for foundations, hedges, and borders |
| 24. Rhododendron | *Rhododendron* spp. and many cultivars | 4–8 | 1–3m (3–10′) | Excellent specimen and foundation plants because of texture and colorful spring flowers; need well-drained acid soil and light shade |
| 25. Yew | *Taxus* X *media* | 4 | 9m (30′) | Attractive and popular foundation plants; females have red "seeds"; can be pruned |

*(continued)*

*TABLE 7–7* (continued)

| | | | | |
|---|---|---|---|---|
| 26. Tea | *Thea sinensis* | 8 | 2m (7′) | Interesting small shrub with showy white camellia-like flowers in October |
| 27. Yucca | *Yucca filamentosa* | 4 | 1m (3′) | Sharp, stiff swordlike leaves up to 2′ long; white flowers on tall spikes; for dry sites |

*Deciduous Trees*

| *Common Name* | *Scientific Name* | *Hardiness Zone* | *Mature Height* | *Mature Spread* | *Landscape use* |
|---|---|---|---|---|---|
| 1. Japanese maple | *Acer palmatum* | 6 | 4–8m (15–25′) | 2–8m (8–25′) | Many cultivars, some with dissected leaves; small tree for artistic touch |
| 2. Norway maple | *Acer platanoides* cv. 'Schwedleri' | 3 | 12–15m (40–50′) | 6–12m (20–40′) | Cultivar with purple spring leaves; fragrant flowers; bold leaf form |
| 3. Red maple | *Acer rubrum* | 3 | 12–18m (40–60′) | 9–14m (30–45′) | Good street or lawn tree; good fall color |
| 4. Sugar maple | *Acer saccharum* | 3 | 18–23m (60–75′) | 12–15m (40–50′) | Rounded shape; one of the best shade trees; does poorly in polluted conditions; outstanding fall color |
| 5. Common horse chestnut | *Aesculus hippocastanum* | 3 | 15–23m (50–75′) | 12–21m (40–70′) | Large tree; coarse texture; showy late-spring flowers; characteristic leaf scorch in late summer |
| 6. Mimosa, silk tree | *Albizia julibrissin* cv. 'Charlotte' | 7 | 8–11m (25–35′) | 9–12m (30–40′) | Wide-spreading, flat-topped tree; pink, brushlike flowers in summer; delicate foliage; brittle wood; species susceptible to blight |
| 7. Seviceberry | *Amelanchier arborea* | 4 | 4–8m (15–25′) | 2–3m (5–10′) | Informal shape; early spring flowers; blends well into shrub border |

| | | | | | |
|---|---|---|---|---|---|
| 8. River birch | *Betula nigra* | 4 | 12–21m (40–70′) | 12–18m (40–60′) | Handsome specimen tree, especially for its exfoliating bark; tolerates wet soil |
| 9. Canoe birch | *Betula papyrifera* | 2 | 12–21m (40–70′) | 9–18m (30–60′) | One of our best native trees because of its beautiful white bark and yellow fall color; not as susceptible to birch borer as other species |
| 10. Eastern redbud | *Cercis canadensis* | 4 | 6–9m (20–30′) | 8–11m (25–35′) | Specimen, grouping, or naturalized setting; deep pink spring flowers |
| 11. Flowering dogwood | *Cornus florida* | 4 | 6–9m (20–30′) | 9–12m (30–40′) | Outstanding as specimen plant in every season; good flowers, showy fruits, red fall color, interesting winter habit |
| 12. Cornelian-cherry dogwood | *Cornus mas* | 4 | 6–8m (20–25′) | 4–6m (15–20′) | Very early yellow flowers; attractive, edible red fruits |
| 13. European beech | *Fagus sylvatica* | 4 | 15–23m (50–75′) | 11–15m (35–50′) | Outstandingly beautiful large specimen tree for parks |
| 14. Ginkgo | *Ginkgo biloba* | 4 | 15–24m (50–80′) | 9–15m (30–50′) | Excellent city tree; not for small yard; striking golden fall color and interesting winter habit; few pests |
| 15. White ash | *Fraxinus americana* | 3 | 15–24m (50–80′) | 12–18m (40–60′) | Handsome large tree for parks; colorful multi-toned fall foliage |
| 16. Thornless honey locust | *Gleditsia triacanthos* var. *inermis* | 4 | 9–21m (30–70′) | 6–15m (20–50′) | Once a popular street tree; many serious disease problems |
| 17. Kentucky coffee tree | *Gymnocladus dioicus* | 4 | 18–23m (60–75′) | 12–15m (40–50′) | Bold winter form and interesting bark; messy tree—drops pods and leaves at different times |

*(continued)*

*TABLE 7–7* (continued)

| | | | | | |
|---|---|---|---|---|---|
| 18. European larch | *Larix decidua* | 2 | 21–23m (70–75′) | 8–9m (25–30′) | Specimen tree with interesting bark, fine foliage, and yellow fall color |
| 19. Sweet gum | *Liquidambar styraciflua* | 5 | 18–23m (60–75′) | 3–15m (10–50′) | Excellent specimen tree due to its variably colored fall foliage |
| 20. Saucer magnolia | *Magnolia* X *soulangeana* | 5 | 6–9m (20–30′) | 6–9m (20–30′) | Small tree with exquisite spring flowers; buds and flowers often affected by late frosts if unprotected |
| 21. Flowering crabapple | *Malus* X *atrosanguinea* | 5 | 4–6m (15–20′) | 3–4m (10–15′) | Many beautiful cultivars; some are resistant to troublesome scab and fireblight; grown for flowers and fruits |
| 22. Dawn redwood | *Metasequoia glyptostroboides* | 5 | 21–31m (70–100′) | 8m (25′) | Fast-growing pyramidal tree with feathery foliage; good fall color |
| 23. Black gum | *Nyssa sylvatica* | 4 | 9–15m (30–50′) | 6–9m (20–30′) | A most beautiful native tree as a specimen; brilliant red orange fall color; good plant habit |
| 24. Sourwood | *Oxydendrum arboreum* | 5 | 8–9m (25–30′) | 6m (20′) | Outstanding specimen tree at all seasons; fabulous red fall foliage |
| 25. London plane tree | *Platanus* X *acerifolia* | 4 | 21–31m (70–100′) | 20–24m (65–80′) | Will tolerate almost any growing condition; interesting bark; over-used and problematical |
| 26. Weeping higan cherry | *Prunus subhirtella* cv. 'Pendula' | 5 | 6–12m (20–40′) | 4–9m (15–30′) | Good pink flowers on graceful tree |
| 27. Bradford pear | *Pyrus calleryana* cv. 'Bradford' | 4 | 9–15m (30–50′) | 6–11m (20–35′) | Formal conical shape at maturity; good spring flowers and fall foliage |
| 28. Sawtooth oak | *Quercus acutissima* | 5 | 11–15m (35–50′) | 9–15m (30–50′) | Good lawn tree; unusual leaves |

| | | | | | |
|---|---|---|---|---|---|
| 29. Pin oak | *Quercus palustris* | 4 | 12–23m (40–75′) | 12–15m (40–50′) | Graceful tree; lower limbs spread to the ground |
| 30. Willow oak | *Quercus phellos* | 5 | 12–18m (40–60′) | 9–12m (30–40′) | Narrow leaves; handsome oak for more southern gardens and cities |
| 31. Scarlet oak | *Quercus coccinea* | 4 | 15–23m (50–75′) | 9–15m (30–50′) | Good fall color; hard to transplant |
| 32. Weeping willow | *Salix babylonica* | 6 | 9–12m (30–40′) | 9–12m (30–40′) | Graceful, delicate tree; good spring leaf color; susceptible to many diseases |
| 33. Sassafras | *Sassafras albidum* | 4 | 6–18m (20–60′) | 3–12m (10–40′) | Outstanding native tree; green twigs, yellow spring flowers; brilliant red orange fall color |
| 34. European mountain ash | *Sorbus aucuparia* | 3 | 6–12m (20–40′) | 4–8m (15–25′) | Showy fruits; but many disease problems |
| 35. Baldcypress | *Taxodium distichum* | 4 | 12–21m (40–70′) | 4–9m (15–30′) | Stately tree; formal conical shape; distinctive texture; good fall color |
| 36. Littleleaf linden | *Tilia cordata* | 3 | 18–21m (60–70′) | 9–18m (30–60′) | Excellent shade tree; attractive broad pyramidal shape; fragrant early summer flowers |
| 37. Zelkova | *Zelkova serrata* | 5 | 15–24m (50–80′) | 15–24m (50–80′) | Handsome tree with wide-spreading crown; good fall color |

*Deciduous Shrubs*

| *Common Name* | *Scientific Name* | *Hardiness Zone* | *Mature Height* | *Landscape Use* |
|---|---|---|---|---|
| 1. Japanese barberry | *Berberis thunbergii* | 4 | 2m (7′) | Excellent for hedges; red berries in winter; spiny stems |
| 2. Sweet shrub, Carolina allspice | *Calycanthus floridus* | 4 | 3m (9′) | Slowly spreading shrub; fragrant maroon flowers in late spring |
| 3. Flowering quince | *Chaenomeles speciosa* | 4 | 2m (6′) | Early spring flowers (red) and glossy green leaves; thorny stems |
| 4. Russian olive | *Elaeagnus angustifolia* | 2 | 4m (14′) | Unsurpassable silver foliage; interesting bark; prune to keep vigorous |

*(continued)*

*TABLE 7–7* (continued)

| | | | | |
|---|---|---|---|---|
| 5. Burning bush, winged euonymus | *Euonymus alatus* | 3 | 3m (9′) | Good hedge or specimen plant; striking scarlet fall foliage |
| 6. Pearlbush | *Exochorda racemosa* | 5 | 4m (15′) | Floriferous, large, vigorous shrub |
| 7. Forsythia | *Forsythia* X *intermedia* | 5 | 3m (9′) | Graceful, erect arching habit; large yellow flowers in early spring |
| 8. Fothergilla | *Fothergilla major* | 5 | 3m (9′) | Outstanding native shrub; peculiar spring flowers and brilliantly striking scarlet fall foliage |
| 9. Hydrangea | *Hydrangea macrophylla* | 5 | 4m (14′) | Large green leaves and huge rounded heads of pink or blue (depending on pH) sterile flowers |
| 10. Winterberry, Michigan holly | *Ilex verticillata* | 3 | 3m (9′) | Beautiful red fruits in bleakness of winter |
| 11. Crepe myrtle | *Lagerstroemia indica* | 7 | 6m (21′) | Outstanding profuse summer flowers; attractive peeling bark; characteristic southern plant amenable to heavy spring pruning to keep compact |
| 12. Winter honeysuckle, first-breath-of-spring | *Lonicera fragrantissima* | 5 | 2m (8′) | One of the very first shrubs to bloom in the spring, with very fragrant flowers |
| 13. Bush or Tartarian honeysuckle | *Lonicera tartarica* | 3 | 3m (10′) | Brightly colored summer flowers and fall fruits in profusion; can grow anywhere; many cultivars; may become a pest |
| 14. Star magnolia | *Magnolia stellata* | 5 | 6m (20′) | Small, early flowering magnolia |
| 15. Bush cinquefoil | *Potentilla fruticosa* | 2 | 1m (4′) | Compact, disease-free heavy summer flowering in full sun |
| 16. Pomegranate | *Punica granatum* | 8 | 4m (15′) | Bright red orange summer flowers; interesting edible fruits in fall |
| 17. Deciduous azaleas:<br>Exbury hybrids<br>Native azaleas | *Rhododendron* spp. and cultivars | 5–6 | 1–3m (3–10′) | Huge colorful flowers provide outstanding spring bloom on Exbury hybrids; many have good fall color; native azaleas add a natural effect to shady shrub borders and woodlands |

| | | | | |
|---|---|---|---|---|
| 18. Cut-leaf staghorn sumac | *Rhus typhina* 'Laciniata' | 3 | 9m (30′) | Delicate texture; slowly spreading; excellent red fall color and winter habit |
| 19. Fragrant sumac | *Rhus aromatica* | 3 | 1m (4′) | Pubescent foliage; colorful fruits and scarlet fall foliage; spreading |
| 20. Alpine currant | *Ribes alpinum* | 2 | 2m (8′) | Attractive hedge plant |
| 21. Clove currant | *Ribes odoratum* | 4 | 2m (7′) | Slowly spreading; clove-scented showy yellow flowers in spring |
| 22. Vanhoutte spiraea | *Spiraea* X *vanhouttei* | 4 | 2m (7′) | Superior spiraea for prolific white flowers in late spring; foundation, border, or mass planting |
| 23. Common lilac | *Syringa vulgaris* | 3 | 6m (20′) | Many colorful cultivars of this popular northern shrub; needs careful spraying and pruning maintenance; prefers slightly alkaline soil |
| 24. Fragrant snowball | *Viburnum* X *carlcephalum* and *Viburnum* X *juddii* | 5 | 3m (9′) | Large heads of fragrant white flowers; good foliage; specimen or border shrub |
| 25. Japanese snowball | *Viburnum plicatum* cv. 'Mariesii' | 4 | 3m (10′) | Large flower clusters with outer flowers sterile; attractive plant habit |
| 26. Highbush cranberry | *Viburnum trilobum* *Viburnum opulus* | 2 | 4m (14′) | Showy flowers; medium-textured foliage; especially attractive persistent red fruits which are edible |
| 27. Chaste tree | *Vitex agnus-castus* | 6–7 | 3–4m (10–15′) | Coarse, aromatic foliage; prominent blue flowers in late summer; needs pruning to maintain neat, rounded habit |

(continued)

*TABLE 7–7* (continued)

*Climbing Vines*

| *Common Name* | *Scientific Name* | *Hardiness Zone* | *Mature Height* | *Landscape Use* |
|---|---|---|---|---|
| 1. Fiveleaf akebia | *Akebia quinata* | 4 | 9–12m (30–40′) | Twining semi-evergreen vine; excellent foliage; rapid, aggressive grower |
| 2. Trumpet vine | *Campsis radicans* | 4 | 9m (30′) | Clinging, deciduous vine; attractive red orange summer flowers |
| 3. American bittersweet | *Celastrus scandens* | 2 | 6m (20′) | Twining deciduous vine; very persistent ornamental with yellow and red fruits; other species may become pests |
| 4. Creeping fig | *Ficus pumila* | 8 | 4m (15′) | Delicate clinging vine for wall covering in protected areas |
| 5. Yellow jessamine | *Gelsimium sempervirens* | 8 | 6m (20′) | Twining evergreen vine with showy yellow flowers in early spring |
| 6. English ivy | *Hedera helix* | 5 | 27m (90′) | Vigorous, clinging, evergreen vine for walls or groundcover. Difficult to remove from mortar once attached; may become a pest; climbs trees |
| 7. Coral honeysuckle | *Lonicera sempervirens* | 3 | 15m (50′) | Twining deciduous vine with long tubular scarlet flowers in late spring |
| 8. Boston ivy | *Parthenocissus tricuspidata* | 4 | 18m (60′) | Tenacious clinging deciduous vine with lustrous green leaves and good fall color. Best vine for covering mortar walls |
| 9. Chinese wisteria | *Wisteria sinensis* | 5 | 8–15m (25–50′) | High twining vine with abundant, pendulous bunches of deliciously fragrant flowers in spring. May strangle and outcompete trees used for support |

*Ground Covers*

| *Common Name* | *Scientific Name* | *Hardiness Zone* | *Mature Height* | *Landscape Use* |
|---|---|---|---|---|
| 1. Ajuga, bugler | *Ajuga reptans* | 4 | ⅓m (1′) | Rapid spreader; shiny green or bronze leaves; blue spring flowers. Sun or shade |
| 2. Bishop's weed | *Aegopodium podagraria* | 4 | ⅓m (1′) | Deciduous, rhizomatous herb with compound leaves, often variegated; very vigorous and often troublesome grower; sun or shade |
| 3. English ivy | *Hedera helix* | 5 | ⅓m (1′) | Vigorous woody vine; glossy green leaves; will climb trees and shrubs; sun or shade |
| 4. Day lily | *Hemerocallis fulva* | 4 | 1m (3′) | Spreading perennial with beautiful orange flowers in summer; sun |
| 5. Lilyturf, monkey grass | *Liriope muscari* | 6 | ⅔m (2′) | Long, narrow leaves; evergreen perennial; clump forming; sun or shade |
| 6. Spreading lilyturf | *Liriope spicata* | 6 | ⅓m (1′) | Spreads by underground rhizomes; excellent soil binder for shade or sun; purple flowers in summer |
| 7. Pachysandra | *Pachysandra terminalis* | 5 | ⅓m (1′) | Excellent evergreen perennial for shade; has terminal spike of white flowers in late spring; spreads relatively slowly |
| 8. Periwinkle, myrtle | *Vinca minor* | 5 | ⅓m (1′) | An excellent evergreen vine; grows rapidly but does not climb trees; roots at the nodes of runners; attractive blue flowers in early spring. Shade or sun |

late fall or early spring while the plants are dormant and the ground is not frozen. Bare root plants can survive the shock of transplanting only because they are dormant, and if planted and watered properly, new feeder roots can form before new growth commences in the spring. Transplanting bare root plants means that you can eliminate the need to move heavy masses of soil with the plant. It is very important that bare root plant roots not be allowed to dry. The roots should be soaked for a few hours in water prior to planting to make sure they have adequate moisture in them. At the nursery, bare root plants will be "heeled in" by covering the roots temporarily with moist sawdust or soil mixes. Examples include fruit trees, most shade trees if not more than six feet high, and many shrubs if planted small (about one foot high).

Balled and burlapped plants will have been grown in the ground in nursery fields, and prior to selling they will be dug with a good ball of soil around the roots. The root ball is snuggly wrapped with burlap or heavy plastic to hold it together (Figure 7–11, 7–13 A and B). Plants thus transplanted suffer less shock than bare root plants, but it is still better to move

*FIGURE 7–11*
**Balled and burlapped trees.** Note that these rather large specimens of maples and oaks are dormant (leafless) and the branches are tied up to protect them during moving to a nursery where they will be planted. (Newell-House Nursery, Charlotte, N.C.)

B&B plants during the dormant season to minimize water loss and reduce the work load on the roots (having to supply water to the transpiring leaves) until a period of recovery has passed (time for damaged root hairs to be replaced). Any form of transplanting produces some degree of disturbance and shock to a plant.

With bare root and B&B plants, if transplanting necessitates removal or severance of a large proportion of the roots, it is advisable to prune back part of the top portion of the plant to minimize shock. Examples include large shade trees, evergreens, and shrubs.

Container-grown plants are grown in plastic pots, metal containers, or more recently in heavy plastic bags simulating pots (Figure 7–12). These plants experience the least transplant shock since their root systems have been confined to a small space for some growing time already and it is not necessary to cut through large roots in transplanting. They may be safely transplanted at any season. It is still best to transplant at a time which produces the least strain on the newly moved plant, and thus the dormant period is preferred. Examples are small trees and any type evergreen or deciduous shrubs.

*FIGURE 7–12*
**Container-grown plants.** A nursery with an extensive selection of both deciduous and evergreen types. Container-grown specimens may be safely transplanted at any time if care is taken to provide adequate moisture. While container-grown plants may not need to be pruned back after transplanting, if the root mass has become tightly pot-bound, they may need some disturbance in order to stimulate new root formation in the planting hole. (Turtle Creek Nursery, Mooresville, N.C.)

## SELECTING THE BEST PLANTS

When obtaining specific plants from a reputable nursery, you may be confronted with choosing from among many specimens (Figure 7–12). Attention paid to the following features will allow you to choose the best specimens.

1. Well-formed shape (shape can be adjusted by pruning).
2. Good color of foliage: no yellowing (chlorosis) or brown leaves from winter burn or sunburn.
3. Absence of dead branches and insect damage. Insects can cause discolored leaves, blotches, or missing tissue.
4. Good root ball on B&B plants (root ball should be almost as wide as the top part of the plant); several large roots present on bare root plants; well-rooted specimens in containers with the top part of plant only slightly wider than the container.
5. Well-formed buds and most recent growth in proportion (texture, shape, size) with older growth, indicating uniformly good growing conditions of the past few growing seasons.
6. Good color of foliage and flowers. If you are selecting a plant with a specific color characteristic in mind, such as red fall foliage or pink spring flowers, try to visit the nursery at that time of year to pick out the most appropriate specimen, since there can be significant variations in characteristics from plant to plant.

Selecting the best plants also involves choosing plants by named varieties and, sometimes, cultivars. The importance of scientific names seems to be misunderstood by most people, who want to avoid those unpronounceable Latin names at all costs. Actually rhododendron, magnolia, ginkgo, viburnum, forsythia, and cotoneaster are all accepted Latin scientific generic names of landscape plants; they just happen to be the common names also. Of course, landscape plants can certainly be appreciated without having name tags on them or your having to remember their names. However, the real significance of scientific names comes in improving communication about the plants. (See Foreword, "Naming Plants.") Not all plants have widely accepted standard common names; for example, forsythia may be called "yellow bells" by some. But each plant has an accepted Latin name which is understood around the world. If you have to ask a question about a plant, look up the information in a book and visit a nursery to find the very specific cultivar you are looking for—these are the times when the scientific name is useful.

If you purchase a named cultivar (*cultiv*ated *var*iety), you have a greater assurance of getting a specimen true to its characteristics that you want to emphasize in your landscape. For example, *Acer rubrum* is the common red maple. It is widely available as inexpensive specimens grown from unselected seeds. *Acer rubrum* cultivar 'October Glory' is outstanding for its compact shape and fine fall color. It would be better than an unnamed seed-grown specimen, though it will cost a little more since cultivars must be carefully grafted or propagated from cuttings.

### COST AND GUARANTEES

The cost of a plant will be determined by how difficult it has been for the nursery owner to produce in large numbers (ease of propagation), how long he or she has had to grow it to salable size (rate of growth and size), and the final labor involved in preparing it for market (ease of transport). Thus, large trees will cost more than small shrubs, but B&B or container-grown trees will cost more than bare root trees. Rare or unusual plants will cost more because they are harder to obtain and propagate or are slower growing. They may be no more difficult to grow than common plants.

One final factor to consider when purchasing plants from a nursery: inquire about a guarantee covering survival and quality of the plant. Some nurseries guarantee only that the plant is healthy and true to name and will replace the plant only if it is proven to be unhealthy or mislabeled at the time of purchase. If a plant dies in your yard, it's your fault. Other nurseries guarantee a plant's survival up to one year after purchase. Keep a record of purchase until you are sure the plant is satisfactorily established in your landscape and that it is performing as you had anticipated. A reputable nursery should be able to answer all of your questions about their plants to help you choose the best for your particular situation and to care for them to ensure their survival. A nursery that cannot help you with problems should be avoided unless you know exactly what you are doing.

## Planting

The next step in landscaping is to plant your specimens. The two most important things to remember in transplanting trees and shrubs are to consider the time of year and, especially, to prepare the soil properly.

### TIME OF YEAR

The time of year may not be critical unless you are transplanting bare root plants. They must be planted when they are dormant. If they have a healthy root system, and you soak the roots before planting, you should have no problems. If you *must* transplant bare root, or even balled and burlapped, plants during the growing season, cut back the top part of the plant *severely.* This reduces strain on the disturbed root system to supply water to a mass of stems and leaves. Otherwise, you will likely lose the entire plant. Container-grown plants may be successfully transplanted anytime because there is minimal disturbance to the root system. Still, in all cases, be careful to water the plants after transplanting and make sure they do not dry out until they become established and are out of shock. The first year is certainly the most critical for establishment of a new plant. Consider watering it *weekly* unless there is a thorough soaking rain (at least one-half inch). After the first year, most plants can withstand a dry spell of several weeks, although it may stunt potential growth and weaken the plant.

In general, the best time of year to transplant all types of plants is late autumn after growth has stopped but before the onset of cold weather and permanently frozen ground; or better yet, early spring. If you transplant dormant plants of any type in early spring as soon as the ground has thawed but before new growth commences, you will give the plant a chance to form a few new roots in order to prepare for the coming growing season (Figure 7–13). Some new roots may be formed in autumn if you plant early enough, but most will be formed in the spring anyway. In the south, you can transplant anytime during the winter when the ground is workable. Above all, avoid transplanting while new leaves are flushing out or during the heat and drought of the summer.

## SOIL PREPARATION

You may have heard the old saying, "Don't put a five dollar plant in a fifty-cent hole." This is very true. Site preparation should be done right the first time because it is expensive and time consuming to redo it after plants have been set in. Spare no effort in properly preparing the hole for a choice specimen. Poor drainage, soil too heavy or too sandy or infertile, and improper pH are some of the more important problems to be considered and corrected.

Knowing the plant's ecological requirements will determine what must be done to improve site conditions. Most plants prefer a loose, friable, well-drained soil with plenty of organic matter to retain water and supply nutrients. Some plants, however, such as blueberry, alder, baldcypress, pin oak, sycamore, river birch, and black gum are quite tolerant of water-logged soil (in nature they are characteristic of swampy habitats). It has been said that a strategically placed alder or pin oak can actually transpire enough water to help keep a wet area drier (sort of like a living sump pump). Some plants can even grow in clay, especially those native to such soils in the southeastern United States.

On the other hand, rhododendrons and azaleas are easily killed by planting too deeply in heavy, wet soil. They must have perfect drainage in an organic soil with an acid pH (around 5.5 to 6). In one case, for example, azaleas used as foundation plantings around a house where mortar and brick rubble had been buried constantly appeared yellow (chlorotic) and sickly until the owner took a soil test and corrected the alkaline pH by adding ferrous sulphate or powdered sulfur. He had not properly analyzed the site and prepared it for planting. Never add horticultural lime (alkaline) to soils for gardenias, azaleas, rhododendrons, or other members of the heath family (Ericaceae).

If your soil is red clay, as in the Piedmont area of the southeast, or

(A)

*FIGURE 7–13*
**Transplanting** (A) First step in transplanting a large tree. Dig around the drip line to form a root ball as large as the crown of the tree. Moving such a large specimen must be done while it is dormant. (B) The root ball is then tightly wrapped with burlap and rope to prevent any disturbance of the delicate root mass during transport. After placing in a properly prepared hole, the rope and burlap may be loosened and left in the hole to decompose. (Colonial Williamsburg, Virginia)

(B)

basic (alkaline) glacial sands, as in the north, you may have to resort to building raised beds in order to bring in soil of the proper texture and pH (Figures 7–14 and 7–15). Raised beds are easily constructed from a variety of materials (logs, rocks, railroad crossties) and can add interest and variety to the landscape (Figure 7–16). They may also act to terrace a slope to provide a site for additional plantings (Figure 7–17).

*FIGURE 7–14*
**Raised beds.** A raised bed will allow the use of soil of the proper pH and will improve drainage. Raised beds may also be constructed to add interest to a landscape design or to provide a terracing effect on a sloping grade. (Charlotte, N.C.)

*FIGURE 7–15*
**Berm bed.** In this situation on the University of North Carolina at Charlotte campus the soil was too rocky in places, and too poorly drained due to high clay content in other places, to allow satisfactory growth of trees. The problem was solved by building up a linear mound of soil above grade in which the trees were planted, along with soil-holding ground covers. The mound, called a *berm,* is both effective and aesthetically pleasing in such a setting. A berm is thus a type of raised bed without restraining walls.

*FIGURE 7–16*
**Raised beds in urban area.** In downtown Ann Arbor, Michigan, raised beds or planters have been constructed to allow the planting of shade trees like Moraine locust and littleleaf linden.

*FIGURE 7–17*
**Walled bed.** In this case, a bed has been formed by the construction of a dry wall from railroad crossties. The wall acts to terrace a steep slope and provide a waist-level planting bed for rock garden perennials. (Ann Arbor, Michigan)

In general, dig a hole twice as large as the root ball of the specimen you are planting. Discard the soil altogether if it is pure clay or sand; or mix sand, peat moss, loam, leaf mold, composted organic matter, decomposed bark or well-rotted sawdust in with the soil to make it suitable. This is called *amending* the soil, and these materials are called *soil amendments.* Fertilizing newly planted trees is not recommended during the first year. Wait until an established root system is capable of absorbing the added fertilizer.

Once you have prepared the soil in the hole, dig out a large enough volume to accept the plant's root system (Figure 7–18). Place the plant in the hole, fill in the hole around the plant, firming the soil and watering as you go layer by layer. Then mound the last bit of soil around the plant and apply a thick layer of mulch (four to six inches). Plant the specimen a little higher than the surrounding soil level (mound it up) as it will sink a little as the soil settles. Never plant a specimen deeper than it was growing before. Rhododendrons actually like to be planted high (mounded up) so there is no chance of too much water on the roots in a heavy soil. The mulch acts to make the planting look neat, retain moisture, prevent weeds, and supply nutrients slowly as it decays. Use shredded bark, dead leaves, pine needles, wood chips, gravel, coco hulls, or rotten sawdust to blend in the plantings with your overall landscape design (Table 7–8).

If a container-grown plant has become pot-bound with tight, dense masses of thick roots, cut some of these roots and loosen the ball before planting to enable new roots to form. Otherwise, the root ball will remain dense and the plant will not grow after transplanting.

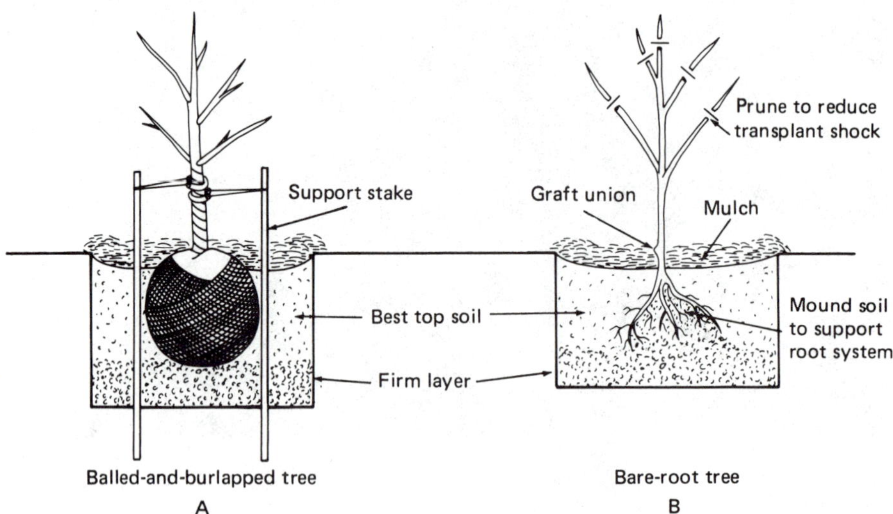


*FIGURE 7–18*
Proper planting of a (A) balled and burlapped and (B) bare root plant.

*TABLE 7–8*
*Proper planting of a balled and burlapped or bare root plant*

1. Soak bare root plants in tub of water one to two hours before planting. Carefully cut off any broken or dead roots back to healthy tissue.
2. Make the hole about twice as big as the root ball. Remove topsoil and subsoil; keep separate. Mix soil with organic matter or loam to improve drainage and quality if necessary. Test pH if you are planting acid-loving plants such as rhododendrons and azaleas; add peat moss to make pH 5 to 6.
3. Measure hole depth to root depth so root-shoot union zone will be just above ground level. Fill in first layer of soil to bottom of root depth and pack down to remove air pockets.
4. Place plant in hole. Fill in with good soil mix, firming each layer of soil around roots and watering to help settle soil. Leave shallow "well" around plant at ground level to hold water until it soaks in.
5. For balled and burlapped plants, remove tie-rope from around stem but do not remove burlap—it will decompose. For bare root plant, mound soil up to support center of plant, spreading roots around mound.
6. After firming soil, drive supporting stakes into hole six to twelve inches from trunk without damaging roots. Fasten wires to stakes and run ends through short sections of old garden hose around trunk so as not to girdle or injure the tree. Leave tree staked for one to two years until established. Small trees may not need staking.
7. Wrap stem with tree-wrap tape to first branch to prevent suburn, reduce freezing damage, ward off insects and rodents, and prevent sprouts from forming.
8. Prune plants if they were not prepruned. Remove branches so that branch system is about as large as the root system on both kinds of plants. This reduces transplant shock.
9. Mulch well (two to four inches), mounding mulch slightly around outer edge to help hold in water. Soak area well.
10. Prune properly after the first season of growth to help establish desired shape for the tree or shrub.

The common foundation plants are quite tolerant of a variety of soil conditions. Such over-used plants as junipers, boxwoods, yews, hollies, and privet grow rapidly and are easily propagated by nurseries. You might consider some of the more interesting plants available. They may require more careful site preparation but will pay off by requiring less pruning and providing a more satisfying enhancement to your home.

## Pruning

There are several reasons to prune woody plants and certain best times to do it. Pruning is best done with sharp tools: hand-pruning shears, long-handled lopping shears for thick twigs, pole-type pruners, and curve-bladed pruning saws for large branches. If there is difficulty with large trees or ones that overhang your house, it is best to call on a reliable tree service company to remove them safely. It is not necessary to paint over wounds with tree paint. Healthy plants are perfectly capable of sealing their wounds with sap.

Do not allow a recently pruned plant to be under stress, such as lack of water.

There are two broad categories of pruning: (1) *thinning out,* where you remove whole branches and (2) *heading back* or cutting back, where you cut a branch partly off, back to a bud, to cause it to grow another branch or several branches (Figure 7–19).

## *Why prune?*

1. *To shape a plant*
   a. After transplanting, to remove damaged or broken branches.
   b. During growth, to remove an odd branch that grows too long or out of character with the rest of the plant.
   c. To produce a dense hedge of desired dimensions.
2. *To reduce stress;* after transplanting bare root specimens, *cut back* the crown to reduce shock of water loss to the roots and reduce the volume of leaves and twigs the roots must support. New growth will come back strong.
3. *To control the quantity of branches* or flowers produced on new growth, *head back* to stimulate many new branches to grow. This applies to annual and perennial flowering plants, hedges, and foundation plants.
4. *To thin out* by removing dead wood, diseased branches, old flowering branches, crossed branches, crowded branches, or weak crotches, especially in fruit bushes and trees.
5. *To rejuvenate an old plant.* By removing older stems of dense shrubs, even cutting the entire plant back to the ground, stimulates new, strong growth to come forth. You may do this by removing old flowering or fruiting branches.
6. *To maintain foundation plants.* Overgrown plants are unsightly and may cover windows, doorways, and driveways (Figure 7–20).
7. *To increase fruit size* on grape vines.
8. *To create interesting forms,* the artistic shaping of plants by strict control of growth. The best known types of plant training are:
   a. *Topiary* Pruning plants into geometrical shapes, either fanciful figures or simply more formal shapes (Figures 7–21 and 7–22). Frequent pruning to achieve a carefully designed shape and removing new growth is necessary. Some plants useful for topiary training are boxwood, yew, juniper, dwarf fig *(Ficus pumila),* holly, privet, hemlock, and cherry laurel *(Prunus laurocerasus).*
   b. *Espalier* The technique of selective pruning to produce a plant which grows vertically in one plane such as flat against a wall (Figures 7–23 and 7–24).

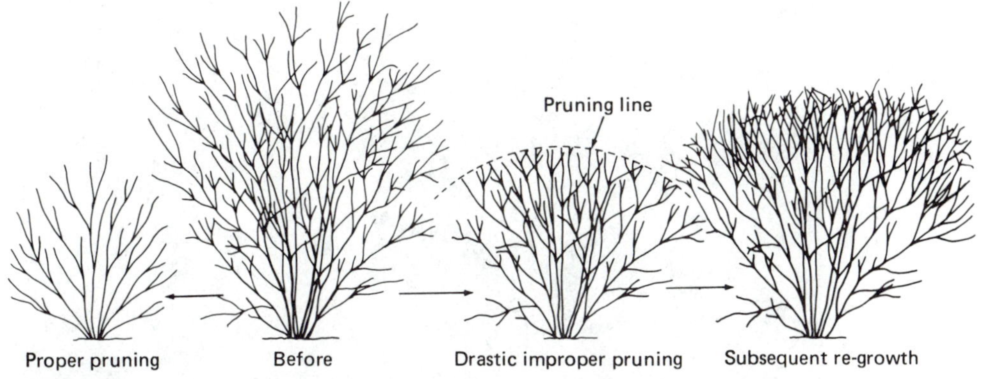


*FIGURE 7–19*
**Proper pruning of shrubs.** Prune to retain the natural shape as far as possible, unless you are making a clipped hedge which requires frequent pruning. Pruning involves *heading back* (cutting branches back to a bud to induce branching) and *thinning out* (removing whole branches). Prune to rejuvenate, thin out an overgrown shrub, remove dead wood, or reduce in size. Shrubs which are properly thinned out will receive light down to all branches and bear leaves below. Drastically pruned shrubs—with their tops cut off flat or rounded—will regrow a dense network of small stems that will shade out lower branches causing a "naked trunk" appearance (Fig. 7–28).

*FIGURE 7–20*
**Neglected foundation plants.** Fast-growing plants such as privet or photinia may soon overtake the house. Regular pruning throughout the growing season is the correct method, cutting back a little bit at a time and stimulating new growth, which provides a full, dense hedge effect. Avoid drastic pruning in the fall, as the newly stimulated growth will be susceptible to frosts. (Charlotte, N.C.)

*FIGURE 7–21*
Topiary pruning. Plants may be carefully pruned to form interesting geometrical shapes. It has been used historically in very formal garden settings. Here, Canada hemlocks *(Tsuga canadensis)* have been pruned meticulously into tear-drop shapes. They certainly produce a striking feature in the landscape. *(Elizabethton, Tennessee)*

*FIGURE 7–22*
**Topiary pruning.** A formal landscape with boxwood *(Buxus sempervirens).* (Colonial Williamsburg, Virginia)

*FIGURE 7–23*
**Espalier pruning.** A pear tree on the side of a house with lateral branches pruned and trained to grow horizontally in one plane, flat against the wall of a house. Branches are tied with rope through metal screw eyes affixed to the wall. Winter pruning must be done annually to maintain the desired form.

*FIGURE 7–24*
**Espalier pruning.** A fig tree is trained against the brick wall of a house with branches tied to a vertical wooden trellis. The branches are pruned and trained to grow in a fan pattern on a vertical plane.

c. *Cordon* A fruit tree or vine that has been pruned like an espalier, except that it is not flat against anything but is pruned to grow horizontally or along a trellis or fence (Figure 7–25).

d. *Pollarding* The technique of drastic pruning where a large tree has its largest branches cut way back to save space. These massive limbs sprout out and give a full appearance without the massive volume of the normal plant (Figure 7–26). Pollarding destroys the natural shape of the tree, but it may be necessary when a large species of tree has been planted in a small space or other unsuitable area such as under power lines.

*FIGURE 7–25*
**Cordon pruning.** A dwarf apple tree on a wooden fence. Four lateral trunks rise from the short main trunk. Each of these lateral trunks is called a *cordon*.

*FIGURE 7–26*
**Pollarding or topping.** Drastic pruning of overgrown shade trees. This reduces tree size and avoids interference with power lines. (North Carolina)

*When to prune*

1. *Foundation plants and hedges* Maintenance pruning for such fast-growing plants should be done during the growing season so the dense foliage cover is maintained. Drastic rejuvenation pruning should be done in late spring so the plants can quickly replace themselves (Figures 7–27 and 7–28). Examples include holly, boxwood, privet, azalea, nandina, juniper, red-tip photinia.
2. *Spring-flowering trees and shrubs* Plants flower in the spring on *old* wood; prune immediately after flowering so that new growth can be produced in the summer, bearing buds for next spring's flowers. Never prune in summer or fall unless you are aware that by doing so, you will diminish the number of next spring's flowers. Examples include forsythia, lilac, flowering quince, spring-flowering trees, sweet shrub (Carolina allspice), serviceberry, azalea, rhododendron, camellia, barberry, mahonia, and magnolia.
3. *Summer-flowering trees and shrubs* Plants flower in the summer on *new* wood; prune in fall or spring to induce abundant new growth and to maximize flowers on new growth. Examples include potentilla, crepe myrtle, hydrangea, butterfly bush, and honeysuckle.

*FIGURE 7–27*
**Rejuvenation pruning.** Foundation shrubs (hollies) have been drastically cut back to stimulate new growth with the potential for completely reshaping the specimens. The problem is that the pruning was done in late fall, and now the homeowner must wait all winter until new growth appears in the spring before the "naked" appearance can be alleviated. (Charlotte, N.C.)

*FIGURE 7–28*

**Improper pruning.** The effect of overpruning or improper pruning is thin foliage and naked stems on the foundation planting. They may become more of a distraction than an asset to the home landscape. In this case where large trees shade the front of the house, the foundation plants are not under optimal growing conditions to start with, and allowing the tops of the shrubs to be wider than the bottoms causes more shading of the lower leaves of the shrubs and hence the thin appearance. One possible solution would be to cut the plants back to within a foot of the ground in the spring, and as new growth appears, begin a periodic, gradual pruning campaign to reshape the new, dense growth into suitable foundation plantings. In settings such as this where dense shade is created by the large forest trees in the yard, all aspects of shrub maintenance must be reconsidered. (Charlotte, N.C.)

4. *Fruit trees and berry bushes* Prune in late winter after severely cold weather when the plants are dormant but before bud-break. Prune to remove excess and crossed branches of fruit trees (Figure 7–29). Remove sucker sprouts which occur along sides and tops of limbs and frequently at the bases of trees (Figure 7–30). Cut back old branches of grapes and berry canes which have borne fruit. Remember, however, that berry bushes bear flowers and fruits on last year's unflowered stems or canes only; it takes two years for a given branch to grow and bear.
5. *Summer-flowering perennials and annuals* To stimulate more flowering and branching, remove flower heads before seeds are formed.

(A)

(B)

(C)

*FIGURE 7–29*
**Proper pruning** of an apple tree, preferably just before spring commences. (A) Before pruning. Note the large number of small branches in the interior of the tree, many of which cross each other and produce a crowded situation. (B) During pruning. Sharp pruning shears of the scissors type should be used to remove undesired twigs by cutting off flush at the main branches. (C) After pruning. Most inwardly growing branches have been thinned out or cut back to allow even space for all branches to develop into an uncrowded crown. When certain branches are headed-back, an outward facing bud is left just below the cut so that the new branch growing from it will be directed into the most desirable space.

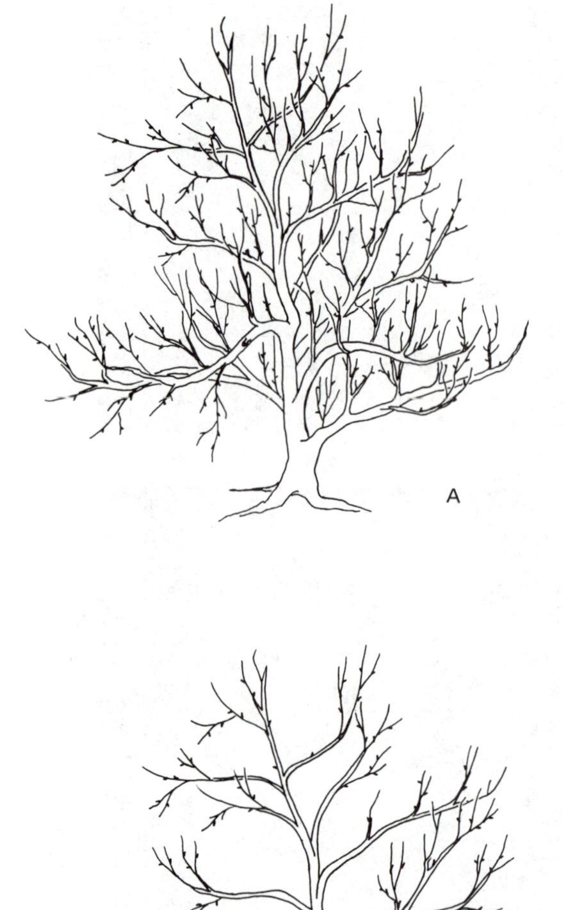


*FIGURE 7–30*
Pruning a crabapple tree that is filled with sucker shoots on its main branches: (A) unpruned tree; (B) tree after pruning.

So, as you can see, fall is not necessarily the ideal time to do one's yearly pruning as so many people believe. You need to understand the growing cycles of your plants and prune them accordingly.

## Protecting plants

There are various techniques that you can employ to help protect your plants from various problems. Following a few simple guidelines and recognizing the potential weaknesses of your plants will enable you to make positive decisions, without guesswork, that will help your plants survive and thrive.

### *Transplanted plants*

1. After transplanting, prune back if necessary to reduce shock.
2. Water thoroughly and mulch.
3. Keep a close watch on watering during a drought, especially during the first year when you will need to water every week if necessary.
4. If you have planted a large or top-heavy tree or shrub, you may need to stake the plant to hold it erect until its roots become established and can anchor the plant. Use one or two strong stakes near the plant and attach the plant by a rope or wire, pulling tightly to take up any slack in the wire or the direction the plant is leaning. Be careful to protect the trunk where the wire touches so as not to rub a wound. Similarly, a large tree may be staked to the ground like a tent, with rope or wire extending from the trunk (about six feet up) to the ground. In this case, it is best to use three guy-wires to ensure stability during a windstorm.
5. Some young trees require that their trunks be wrapped with tree tape to protect from sunburn, cold winds, rodents, or to retard sucker sprouts from emerging.

### *Established plants*

1. For plants that have been growing on the property for some years, fertilize regularly to ensure a strong tree with good flowers and foliage. Use liquid or dry fertilizer applied in holes dug around the drip line of a tree or solid fertilizer spikes driven into the soil. The *drip line* of a tree is an imaginary line on the ground below the plant's outermost branch tips. In the soil below this line fertilizer can reach the plant's feeder roots.
2. Water if drought is prolonged. Do not water just a little as this promotes shallow root formation. Shallow roots will die if they are not kept watered. Whenever you do water—water well and let it soak, as if there had been an inch of rain.
3. Check for insects, pests, and diseases and spray if necessary. Keep a constant eye on your plants to catch any problems at the onset. Scale insects are common garden pests on euonymus and camellia. Lacewing bugs on full-sun grown pyracantha, azalea, and other evergreen shrubs, wax scale on many plants, and aphids on maples and the soft new wood of many shrubs are easy to control. Japanese beetles, fall web worms, gypsy moths, weevils, and many other insects are more difficult to permanently eradicate.

4. Watch for signs of nutrient deficiency. Yellowing (chlorosis) in azalea, rhododendron, and gardenia due to nonacid conditions or lack of iron; or yellowing leaves due to lack of nitrogen. (See Chapter 4.)
5. Protect tender plants in winter from drying winds that can cause leaf burn due to drying out. Screening with evergreens, wrapping with burlap, planting in a protected corner, or spraying with an antidessicant can help reduce damage. Make sure the soil is moist. When the soil is frozen, however, moisture is not available, and there is nothing you can do about that.
6. Watch for potential injury to plants:
   a. From pedestrian or vehicular traffic bumping into plants.
   b. "Cut-through" traffic in hedges.
   c. Dogs urinating on shrubs (too many concentrated nitrogen, potassium, and sodium salts can burn roots and leaves and kill plants much like overfertilizing).
   d. Salt from sidewalk de-icing can build up in the soil and kill plants; use wood ashes if available for de-icing.
   e. Pollution from city smog can harm some trees.
   f. Wires and ropes wrapped around trees can choke them as they grow by *girdling* which cuts through the vascular system.

## Landscaping with woody plants for fruit

Growing fruits in your garden can be a most satisfying experience. Fruit trees as well as shrubs and hedges which bear edible fruits can easily be substituted for strictly ornamental counterparts. Many of the small fruits, including grapes, currants, gooseberries, elderberries, highbush cranberries, the cane berries (raspberries, blackberries, youngberries, boysenberries), and blueberries, can be grown in the garden as "living fences" in place of, or together with, wooden or wire fences along the border of your property. In the spring, they can provide a magnificent display of flowers. And, they can furnish you with fresh fruits that are becoming increasingly more expensive to buy. In this section, we discuss how to grow small fruits and fruit trees in the garden, whether you have a small lot in the city or several acres in the country.

### GRAPE VINES

Grow the grape cultivars hardy for your area. These cultivars are generally grafted onto vigorous, hardy rootstocks of the American grape *(Vitis labrusca)*. Grape vines can be grown on several kinds of supports. Variations are single stakes (as in many vineyards in California, Michigan, and New York); wires supported by widely spaced stakes; espalier treatment on trellises or sides of buildings; and elaborate frame structures such as arbors and pergolas.

To train the branches (canes) of the grape plant, support them on one or two horizontal wires. Allow the two strongest canes to develop to the left and right of the plant. Two canes on one wire refers to the "two-cane kniffen system"; four canes, two each on two wires one-half meter above one another, refers to the "four-cane kniffen system." Either system assures that there is maximum exposure of the canes' leaves to sunlight, and at the same time, provides for maximized, vigorous canes for bearing flowers and fruit.

Here, we shall simply consider grapes grown on two-wire supports, using the four-cane kniffen system. This system is most easily adaptable to home garden living fences. A five-year sequence of pruning steps is begun with usually a two- or three-year-old transplanted grape vine (Figure 7–31). The first several years are spent in selecting two sets of canes at two levels to serve as the primary branch system, then training them on the two wires. Old or dead canes must be pruned out annually, preferably soon after dormancy in the fall or early in the spring before bud-break. In pruning, each cane that is left to grow on a given wire should have four to five buds on it to produce new canes the next growing season. Sucker canes are removed from the bases of the plants, especially those below the graft line which are from the wild rootstock. In addition, weak lateral canes are removed and four primary arms are established as new fruiting canes for the next year. The primary purpose of all this "heavy pruning" is to ensure development of large grape clusters and large grape berries. The pruning principles for espaliered grapes are the same as for grapes grown on wires.

You may prefer to grow your grapes on a supporting structure such as a trellis, arbor, or pergola. Prune out the dead canes in the fall before the leaves fall off or in early spring when the buds begin to swell on live canes. New canes that start to develop from the buds are very fragile and are easily broken, so pruning is done before cane development. Grape vines grown on frame structures produce many more clusters, but the berries and the clusters are generally smaller than with grapes grown on wires and pruned in the four-cane kniffen system. But growing grapes on trellises, arbors, or pergolas requires much less work, and the copious yields are often sufficient for both table grapes and a sizeable wine-making operation for the home gardener.

Grape plants are "heavy feeders" so must be supplied with fertilizer, especially nitrogen, annually. Horse or cow manure is one of the best organic sources of nitrogen for these plants. Dried, sterile manure can be purchased at garden centers.

## BERRY BUSHES

Cane berries have special needs for proper growth and are trained in various ways, depending on their habit of growth. Upright cane berries such as red raspberries *(Rubus idaeus)* are usually grown in beds about one or two meters wide without wire support. Black raspberries *(Rubus occidentalis)* are

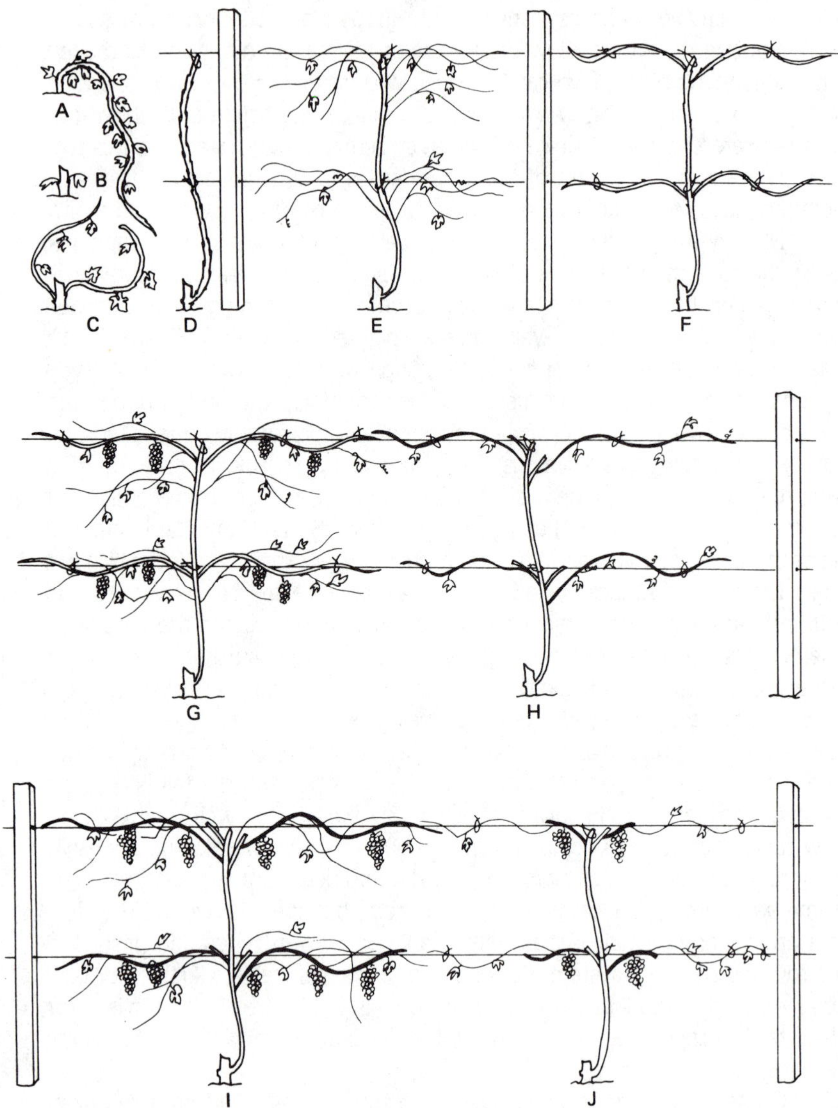


*FIGURE 7–31*
**Four-cane kniffen system** for pruning and training a grape vine from the time you start with young plants (Year I) (A–D) through and including three more years (Years II–IV). Steps are as follows: **Year I:** (A) obtain grape plant; (B) cut back to leave two basal buds; (C) two buds form two new shoots; (D) remove lower shoot, tie topped upper shoot to wires. **Year II:** (E) laterals formed. **Year III:** (F) spring—tie four best laterals to wires, remove all other shoots; (G) fall—fruits formed on four main laterals, new canes formed during summer; (H) fall—after fruits removed, tie four new shoots to wires. **Year IV:** (I) fall—fruits formed, new canes formed; (J) fall—select four good fruiting canes for next year's growth, prune others.

best supported on stakes because of the long canes they produce (Figure 7–32). Boysenberries *(Rubus loganobaccus)* and youngberries *(Rubus ursinus)* produce still longer canes (some up to ten meters long) and must be supported in wires as mentioned above for grapes. Elderberries *(Sambucus canadensis)* must be covered with loosely woven fabric (cheesecloth or netting) or harvested as soon as ripe, or the birds will harvest them.

All of the cane fruits require copious amounts of water and heavy applications of fertilizer, especially of a nitrogen type. In fact, the berries often fail to ripen properly when soil moisture is low. To prevent this, the plants should be watered well once a week during periods of drought. Soil moisture is retained by mulching the plants with compost and leaves. The best source of nitrogen fertilizer for these plants is animal manure.

Blueberries *(Vaccinium corymbosum)*, huckleberries *(Gaylussacia baccata)* and cranberries *(Vaccinium oxycoccus)* must be grown in an acid soil (pH 4 to 5). This is achieved by replacing the garden soil with peat to create a miniature bog. Maintain a low pH by adding more peat or annually treating the soil with aluminum sulfate or with elemental sulfur dust. The peaty soil must also be kept moist throughout the growing season. Other shrub fruits such as elderberries, gooseberries *(Ribes grossularia)*, currants *(Ribes sativum)* highbush cranberries *(Viburnum trilobum)*, and the cane berries do not have this requirement for low pH soils and are fairly tolerant of a wide acid to alkaline pH range.

*FIGURE 7–32*
Training and support system for black raspberries in an inexpensive wood-post frame.

Each type of cane fruit is pruned differently. Red and purple raspberries are pruned in the autumn by cutting out the old canes that fruited in the spring. Boysenberries are pruned in the fall by cutting out the old canes and winding the current year's new cane around one or two wires. In both these cases, new canes are produced annually during and after fruiting of the previous year's canes. Red raspberries often fruit twice in a season, spring and fall; canes from the fall fruiting are not pruned but saved for the next year. Blackberries *(Rubus allegheniensis)* and black raspberries *(Rubus occidentalis)* produce long canes that root at the tips (tip layering) and produce fruit the second season. That's how they form large patches or thickets in nature. After fruiting, old canes are pruned out and the new, current year's growth of canes are pruned to desired height the following spring. If you leave them unpruned, you will have a thicket. This is fine if you have ample space. However, if trained, as on a fence, they must be rigorously pruned (Figure 7–33). All prunings, which are generally woody, can be burned on the garden soil to provide potash (potassium) and other elements for the soil. They do not break down well in a compost pile unless they are well shredded.

## FRUIT TREES

The first consideration in growing fruit trees is the amount of space available and the ultimate size of the mature trees. If space is at a premium, consider growing dwarf fruit trees. Dwarf trees result from upper-portion grafts onto rootstocks which cause dwarfing. Fruit trees with dwarfing rootstocks are available for apple (*Malus* sp.), pear *(Pyrus communis)*, peach *(Prunus persica)*, cherry *(Prunus avium)*, and some citrus. Dwarf fruit trees may be grown in the open with no special support or unusual type of pruning; they may be espaliered against a wall or side of a building; or they may be cordon trained to grow in one plane on a fence. Dwarf trees can take up as little as one-fourth the space of a standard fruit tree, thereby allowing a greater variety to be grown.

Fruit trees in the garden need care throughout the year. In the spring, before the buds break dormancy, the trees can be sprayed with dormant oil for insect pest control. Pruning out dead wood, diseased parts, root sprouts, and crowded limbs is also done at this time of year. Pruning is best done annually to avoid a buildup of problems that are not so easily solved later. After flowers and leaves appear, begin a consistent program of pest control, either with repeated applications of fungicides and insecticides or with biological control methods. (See Chapter 8 on pest control.) In addition, fertilizer applications are needed when active shoot growth is under way. Manure, a balanced inorganic fertilizer, and/or compost fortified with nutrients serve this purpose well. Fertilizer preparations, liquid or dry, can be applied directly on the soil, in the soil to feed the roots, or as a spray applied to the foliage. Fertilizer should not be applied in the later part of the growing season, as this encourages weak, late shoot growth which is prone to winterkilling.

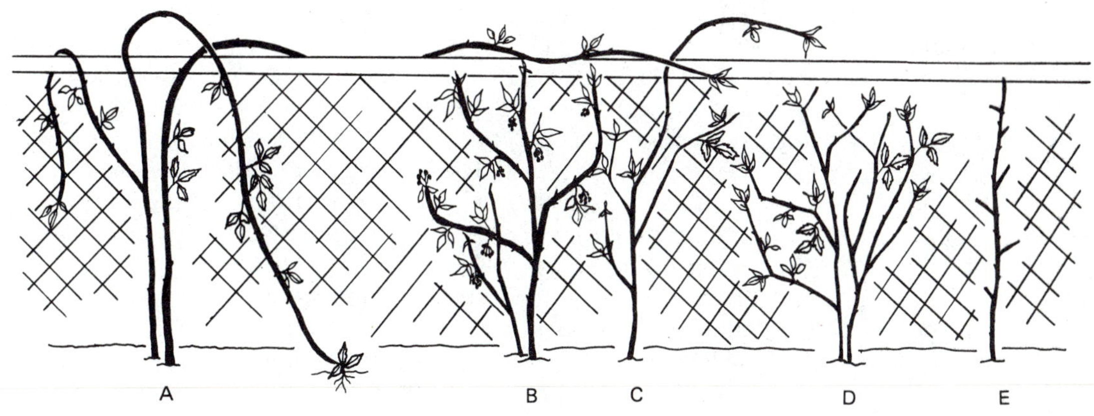


*FIGURE 7–33*
Steps in pruning black raspberries: (A) unpruned, last year's new growth; (B) pruned last year, last year's new growth; (C) unpruned, this year's growth; (D) pruned, this year's new growth; (E) next spring, prune laterals to six to eight inches after fruiting. Note that these raspberries are trained and supported on a cyclone-type wire fence.

Watering must be done throughout the growing season, especially in regions where there are no summer rains, such as desert regions. In other areas, summer drought periods dictate supplemental watering of fruit trees if they are to attain good shoot growth and produce reasonable size fruits. And remember, next year's flower buds are produced during this year's growing season, so keep growing conditions optimal. Sprinkle irrigation is used for fruit trees in arid regions of the western United States. In these regions many orchards are irrigated by allowing water to flood the orchard periodically from irrigation ditches. To conserve water, use trickle irrigation; allow the water to trickle slowly from plastic pipe over long periods directly into the soil root zone near the tree trunks. A modification of this method is to irrigate the trees well by allowing water to flow into circular depressions around each tree base. With all this kind attention paid to fruit trees through pest control, pruning, fertilizing, and watering, one can obtain bountiful harvests of fruit that can be eaten fresh, frozen, or stored for later eating. Also fruit can be made into jams and jellies or used for producing the nicest of wines.

To prolong the storage life of fleshy fruits, place in a cold (not freezing) dry place. Apples should not be stored near other fruit because they give off so much of the fruit-ripening hormone, ethylene. Cold storage of apples in plastic bags or tightly sealed plastic containers obviates this problem. Storage life is prolonged if the fruits are stored dry and if any bruised or soft fruits are eliminated before and during storage.

## NUT TREES

Some of the nut trees frequently grown in the home garden are filberts or hazelnuts *(Corylus americana)*, Persian, English, or Carpathian walnuts *(Juglans regia)*, black walnuts *(Juglans nigra)*, shagbark hickory *(Carya ovata)*, pignut hickory *(Carya glabra)*, butternut or white walnut *(Juglans cinerea)*, Chinese chestnuts *(Castanea mollissima)*, pecans *(Carya illinoensis)*, and almonds *(Prunus dulcis)*. All of them require considerable space. Since most of these plants are good size trees (with the exception of shrub size filberts) with no common dwarf forms, they are best used in the home landscape as shade trees that can also provide food. Generally, the nut trees have little ornamental value and poor fall color. Most are large and coarse. Some, such as black walnut and black cherry *(Prunus serotina)*, even poison the soil, stunting other plants that may grow near them.

Many nut trees require cross-pollination, either by wind or by insects, so provision must be made for this in the planting scheme. For example, with filberts or hazelnuts, preferably two different varieties should be planted several meters apart. With chestnut trees, plant at least two trees 10 to 15 meters apart; several trees planted in a "grove" are still better. Hickory trees and Persian or English walnut trees may bear poor crops unless other varieties are nearby. Butternut and black walnut trees have no pollination problems.

Walnut trees are typically budded or grafted in order to produce nuts of better quality. This is especially true of the black walnut. The cultivar, 'Thomas', produces a large nut that can be easily hand picked. In contrast, nuts of nongrafted black walnut trees are very hard to crack, and extraction of the "meats" is equally difficult.

Most of the native nut trees are very hardy in northern or southern temperature regions. These include the shagbark hickory, butternut hickory, and beech *(Fagus grandifolia)*, which are native to northern temperate regions. The nuts of these trees are considered to be excellent sources of "snack foods" for those who enjoy wild edible foods. (See Chapter 11 on edible wild plants.) Cultivated nut trees, such as those of Chinese chestnut, some Persian walnuts, and some filbert cultivars, will withstand temperatures as low as −30° C (−20° F) and thus can be grown in northern temperate regions. However, with these trees, the growing season should be at least 150 days long (the period between last-spring and first-fall killing frosts).

When the nuts of a given species are mature and start falling to the ground, gather them before squirrels and other rodents collect them! Removal of the husks from black walnut fruits is difficult and their tannins stain the hands. One of the best ways to remove husks is to drive car wheels over the nuts. The husks may be saved as a natural plant dye source. (See Chapter 13, on natural dyes.)

Generally, nuts are placed on newspapers, in a cool, dry, well-ventilated place to dry. After drying for several weeks, the nut "meats" can be extracted. An exception is with the nuts of Chinese chestnut which do not need to be dried when mature. They can be placed in perforated polyethylene bags when mature and stored at 0° C (32° F). So that other nut fruits can be used as seed sources or for eating during the winter months after fall harvest, it is best to keep them in loosely tied plastic bags or in glass or plastic containers with loosely sealed lids. Store in a cool, dry place at 5° to 8° C (40° to 45° F). Airflow into the containers prevents moisture accumulation and attendant problems with mold-causing fungus. Low temperature results in lowered respiration rate in the fruits, thus reducing the rate at which food reserves are used (metabolized).

## Summary of landscaping errors

Some common errors encountered in home landscapes may be caused by one or more of the following:

1. Failure to preplan the landscape design on paper. This leads to impulsive placement of plants, inappropriate spacing, interference with permanent structures, and other problems.
2. Lack of knowledge about soil conditions and terrain, resulting in plants having poor drainage, improper soil pH, unsuitable soil texture (clay or sand), and nutrient deficiencies.
3. Lack of knowledge about the ecological requirements of plants, especially hardiness, soil preference, soil moisture, and sun or shade requirements.
4. Lack of knowledge of maintenance requirements and the importance of this factor when considering a plant's space allowance at maturity.
5. Lack of knowledge about specific plant species or cultivars, resulting in plant choices of poor color, wrong shape, inappropriate texture, disease susceptibility, and unattractiveness.
6. Using plants that are too large or too small for the situation.
7. Planting a poor specimen, in a poor hole, with poor care which leads to a poor sight for some poor soul.
8. Purchasing a plant first, then looking for a place to put it.
9. Improper pruning, such as pruning too much, without consideration for the form of the plant, or at the wrong season.
10. Allowing shrubs to become neglected, choked with old stems and dead branches, or overgrown by vines and weedy shrubs.

## REFERENCES

Bloom, Adrian. 1972. *Conifers for your garden.* France: Sachets Floraisse. (Excellent photos of conifers; good introduction on choosing, planting, and caring for trees.)

______. 1979. *A year round garden.* Nottingham, England: Floraprint. (Step-by-step guide to garden design and planting.)

Bonnie, Fred. 1976. *Flowering trees, shrubs, and vines: a guide for home gardeners.* Birmingham: Oxmoor House. (Good lists of various plants for various uses, and discussion of techniques.)

Brickell, Christopher. 1979. *Pruning.* New York: Simon and Schuster.

Brooklyn Botanic Garden. 1952. *Handbook on pruning.* Plants and Gardens 8(2): 115–76.

______. 1971. *Handbook on fruit trees and shrubs.* Plants and Gardens 27(3): 1–84.

Carpenter, Philip L.; Walker, T. D.; and Lanphear, F. O. 1975. *Plants in the landscape.* San Francisco: W. H. Freeman and Co. (Complete book covering all aspects of planning and planting, operating a landscape business, and understanding the natural landscape.)

Countryside Books. 1974. *Home landscaping.* Barrington, Ill.: A. B. Morse Co. (Color photos of many important landscape plants of all types.)

Dirr, Michael A. 1983. *Manual of woody landscape plants.* Champaign, Ill.: Stipes. (An exhaustive reference containing many details on landscape plants for midwestern and northern regions.)

Foster, Ruth S. 1978. *Landscaping that saves energy dollars.* New York: David McKay Co. (A compilation of ways to utilize landscape planting to reduce home energy consumption.)

Grey, Gene W., and Deneke, F. J. 1978. *Urban forestry.* New York: John Wiley and Sons. (Standard textbook covering all aspects of using woody plants in the city.)

Janick, J. 1979. *Horticultural science.* 3d ed. San Francisco: W. H. Freeman and Co.

Johnson, W. T., and Lyon, H. H. 1976. *Insects that feed on trees and shrubs.* Ithaca, N.Y.: Cornell University Press. (Beautifully illustrated reference on the many pests of woody plants, their life cycles, and how to control them.)

Lancaster, Roy. 1974. *Trees for your garden.* New York: Charles Scribner's Sons. (Excellent photos of unusual ornamental trees, with lists of specific uses.)

MacDaniels, L. H., and Lieberman, A. S. 1979. Tree crops: a neglected source of food and forage from marginal lands. *Bio Science* 29:173–75.

Menninger, D. 1977. *Edible nuts of the world.* Stuart, Florida: Horticultural Books.

Nelson, William R. 1975. *Landscaping your home.* Rev. ed. Urbana, Ill.: University of Illinois Press. (Complete and detailed account of all aspects of planning and planting the residential landscape; includes information on specific trees and shrubs.)

Pirone, P. P. 1978. *Tree maintenance.* 5th ed. New York: Oxford University Press. (The standard reference on all aspects of tree use and maintenance.)

Reader's Digest Association. 1978. *Illustrated guide to gardening.* Pleasantville, N.Y.: Reader's Digest Association. (A beautifully illustrated book containing a wealth of information on all aspects of gardening: lists of plants and their characteristics and uses; planting, pruning, and maintenance techniques; chapters on the major groups of plants.)

Sunset Books. 1975. *Sunset ideas for landscaping.* Menlo Park, Calif.: Lane Books.

Time-Life Editors. 1972–1980. *Time-Life encyclopedia of gardening.* New York: Time-Life Books. (Volumes on evergreens, flowering shrubs, garden construction, Japanese gardens, lawns and ground covers, pruning and grafting, trees, wildflower gardening, and winter gardens. An excellent series; a must for serious gardeners.)

USDA. 1972. *Yearbook of agriculture: landscape for living.* Washington, D.C. (An excellent idea book.)

Wyman, Donald. 1965. *Trees for American gardens.* New York: Macmillan Co. (Discusses the best choices of trees; recommends specific cultivars; good introductory information, including lists of trees for various uses.)

———.1969. *Shrubs and vines for American gardens.* New York: Macmillan Co. (Companion volume to the above.)

Zion, Robert L. 1968. *Trees for architecture and the landscape.* Condensed ed. New York: Van Nostrand Reinhold Co. (Offers a different perspective on tree use; gives good and bad qualities; consideration of form; and lists of recommended trees for each state.)

# Plant pests and their control 8

A pest is usually defined as a thing or organism that causes trouble, annoyance, or discomfort. Although it need not be destructive, it often is. The variety of plant pests is practically limitless, exhibiting great diversity in form, size, and habitat. The range of organisms extends from the microorganisms (such as viruses, mycoplasmas, and bacteria), to invertebrates (nematodes, mites, and insects) and vertebrates (birds, mammals), to the plants themselves (fungi and higher plants including weeds and nongreen parasites).

Traditionally, *plant diseases* are distinguished from *plant pests* in that the former are caused by microorganisms or environmental conditions affecting the *internal* environment and functioning of the plant, while pests cause only *external* or localized damage. There are many situations of overlap between these designations, however. Examples of organism-induced diseases are blotches, blights, wilts, damping-off, rots, rusts, galls, and witches' brooms. Environmentally caused diseases include mechanical injury, sunburn, excess water, insufficient water, air pollution, nutrient deficiency with symptoms such as chlorosis (yellowing), soluble salts damage, insufficient light, and damage from improper temperatures. Common houseplant pests are arthropods (animals with jointed feet and hard outer skeletons), including mites, mealybugs, scale insects, aphids, and whiteflies, and vertebrates (animals which have an internal spinal column) that chew or suck on plant tissues (Table 8–1). There are many situations where the household cat, who likes to chew on plants, can be your biggest problem. In that case, you need to protect the animal from such potentially poisonous plants as *Dieffenbachia* or Jerusalem cherry.

Pests and diseases can occur on economically important plants including food and feed grains, legumes, vegetable crops, ornamentals, trees of the forest, lawn grasses, and houseplants. No matter what the stage of plant development or plant structure, pests are in the environment and can attack the plant or plant part; and the seed, seedling, and the mature plant with all its organs—roots, stems, leaves, flowers, and fruits—are susceptible to them.

Plants, then, upon which mankind is so dependent for food, shelter, and clothing, must contend not only with nutrient deficiencies and the unfavorable climatic or environmental conditions of both summer and winter, but with ever-present pests and the disease-causing organisms.

## Major types of plant pests and diseases

### VIRUSES

These are electron microscopic-sized living particles consisting of a core of nucleic acid enclosed by a sheath of protein. Viral diseases of crop plants include tobacco mosaic; maize (corn) dwarf mosaic, which also produces infection in sorghum and sugarcane; tomato and sugar beet curly top; and bud blight in soybeans. Flower beds containing dahlias, narcissus, nasturtium, tulip, sweet pea, and petunia, among others, are susceptible to virus infection. Dozens of virus diseases of stone fruits (plums, prunes, peaches, and cherries) are known.

Symptoms on host plants are not always readily observable, although alternate light green (chlorosis) and normal green areas, giving the plant a mottled or mosaic appearance, is the most common symptom (Figure 8–1). Other effects include the development of ring spots, which can become chlorotic and necrotic (dead areas), curling of leaves, black blotches, abnormal growths on host organs, stunting, reduced yield, and sometimes, death of the host. In addition, viruses are responsible for some cases of highly prized variegations in the leaves and flowers of such ornamentals as *Camellia*.

Means of transmission of viruses are numerous and varied, but many of the animal pests of plants are vectors. These are insects such as aphids, leaf and tree hoppers, mealybugs, spittle insects, thrips, earwigs, grasshoppers, and beetles; spider mites (arachnids); nematodes; fungi; and dodder *(Cuscuta)*, a parasitic seed plant in the morning glory family. Other modes of transmission are by mechanical means, vegetative propagation of infected tissues, and use of infected seed.

Recovery of a virus-ridden plant is rare, although some control of viruses is possible even though "viruscides" do not exist. A few viruses can be inactivated after exposure to heat. The actively growing shoot tip of a virus-infected plant can be excised (cut out) under sterile conditions and used to develop a new plant by mericlone culture, as with virus-infected orchids or potatoes (see Chapter 6). Such virus-free plants should be isolated from

*FIGURE 8–1*
Illustration of chlorosis, a symptom of nutritional deficiency, on red oak *(Quercus rubra).* Chlorosis is a mottled, yellowing of the leaves and can be caused by deficiency of iron, nitrogen, manganese, sulfur, or magnesium.

virus-infected plants, if, in fact, the latter are not destroyed. Other options include utilization of pesticides or biological means of pest control (e.g., predators) to get rid of the vector and plant breeding to produce virus-resistant plants.

## MYCOPLASMAS

Mycoplasmas are little understood disease-causing organisms, but they are actually somewhere in between viruses and bacteria in size and complexity. The best known effects of mycoplasmas are the variegated, mottled, doubled, and otherwise deformed flowers of such plants as white trillium *(Trillium grandiflorum)* and parrot tulips. They can spread throughout a population by unknown means and can infect healthy plants. Variegated trilliums should not be planted near healthy trilliums. Mycoplasma can also be responsible for aster yellows and witches' brooms of some woody ornamentals. Some workers have claimed to have obtained mycoplasma-free tissue from infected plants by treating them with streptomycin in the proper concentration, causing the mycoplasma to retreat from the shoot apex, then excising this apical meristem and culturing it by mericloning.

## BACTERIA

Generally, the bacteria which cause disease in plants are rod-shaped (bacillus) forms. These bacterial organisms (also called *bacterial pathogens*) can be disseminated by means of air, water, insects, and various agricultural practices. The bacterial organisms enter the plant (called the *host*) through natural openings in the epidermis (stomata or hydathodes, used for water vapor and gas exchange) or bark (lenticels, pores also adapted for water and gas exchange). On occasion, bacterial entry may be through nectar-producing tissue, fragile root hairs, or wounds.

Bacterial diseases, some of which are very destructive, occur on many economically important plants, such as vegetables, fruit trees, and ornamentals. Some important diseases are bacterial soft rot of vegetables caused by *Erwinia carotovora* and fireblight of pear and apple caused by *Erwinia amylovora.* A disease familiar to many, and a worldwide problem affecting a number of plant families, including tree fruits and other woody plants, is crown gall, produced by the pathogen, *Agrobacterium tumefaciens.* Recognizable symptoms are rots, galls, wilts, leaf spots, leaf distortion, imperfect edible plant parts, and tissue death. Host plants are susceptible at any stage of development. Bacterial black spot on the petals of orchids and other plants is promoted by a very humid environment with little ventilation, as can occur in the greenhouse.

Control of bacterial pathogens can be achieved in a variety of ways, including sanitation, use of insecticides and antibiotics, crop rotation, and planting of resistant cultivars. The sterilization of cutting tools (by heat or chemical liquid dip) is an important means of preventing the spread of viral and bacterial pathogens. This practice, along with good general sanitation, is strictly adhered to by orchid growers who do a great deal of pruning, dividing, and propagating of very valuable, normally long-lived orchid plants.

## ANIMAL PESTS

These pests range from roundworms and some mollusks to the arthropods and chordates (Table 8–1).

Just as the nematodes cause many serious and sometimes fatal diseases of animals, including man, so do these unsegmented roundworms parasitize economically important plants. These common, almost microscopic organisms are soil borne and produce plant diseases that can be very destructive to a number of crop and ornamental plants. Because of their habitat, the worm often gains entry to the host through the roots, resulting in a condition called root-knot, characterized by the presence of root nodules or galls. Other plant parts attacked include stems, leaves, and flowers, and ultimately symptoms appear such as swollen stems, wilted and deformed leaves, and reduced floral development. In nurseries, nematodes can move through the soil from infected plants to healthy plants growing in containers. It is always best to sterilize soil with heat (e.g., steam or in an oven) before use to kill

*TABLE 8–1*
*Classification of some common animal pests that attack plants*

| Phylum | Class | Order | Examples |
|---|---|---|---|
| Nematoda | | | Nematodes |
| Mollusca | Gastropoda | | Slugs, snails |
| Arthropoda | Arachnida | | Mites |
| | Crustacea | | Sowbugs, pillbugs |
| | Insecta | Orthoptera | Grasshoppers |
| | | Isoptera | Termites |
| | | Thysanoptera | Thrips |
| | | Hemiptera | True bugs |
| | | Homoptera[1] | Leafhoppers, whiteflies, aphids, scale insects, mealybugs, treehoppers |
| | | Coleoptera | Beetles |
| | | Lepidoptera | Butterflies, moths |
| | | Diptera | Flies, midges |
| | | Hymenoptera | Bees, wasps, ants |
| | Chilopoda | | Centipedes |
| | Diplopoda | | Millipedes |
| Chordata | Aves | | Birds |
| | Mammalia | | Moles, mice, rats, squirrels, gophers, rabbits, raccoons |

[1]This order includes our most serious houseplant pests.

nematodes and never to reuse greenhouse or nursery soil which has the potential of harboring nematodes. Nematodes are difficult to detect until their damage is done, and they are just about impossible to eradicate. Sanitation is the best prevention.

Larger pests, more prevalent in the spring, are garden slugs, such as the spotted garden slug *(Limax maximus)* and land snails. These are annoying pests which feed on garden plants at night; evidence of their presence the night before is eaten vegetation and the secretion of slime trails.

Spider mites, which are not insects, because they lack the antennae and wings found in insects and possess four pairs of jointed legs rather than the three pairs of an insect, will more often appear when the weather is hot and dry. Host organisms include vegetable and field crops, ornamentals, greenhouse plants, and houseplants. A curse of fruit trees, from the northwest through the midwest to the northeast United States, is the European red mite *(Panonychus ulmi)*. The pest, which damages fruit and reduces yield, has developed some resistance to pesticides, which kill off many natural parasites.

The most common house and greenhouse plant mite pest is the two-spotted mite *(Tetranychus urticae)* (Figure 8–2). Its populations seemingly come from nowhere (though bringing infected plants into your collection is usually the source), may explode almost overnight, and can be very destructive in a short time. The pests live on the undersides of the leaves, particularly on plants in a hot, dry, stagnant environment. They become evident only

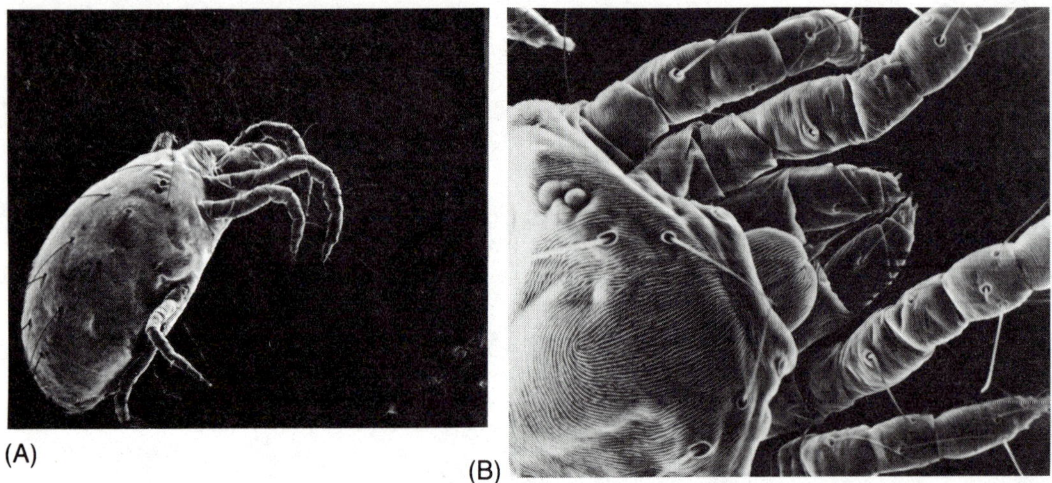


*FIGURE 8–2*
Scanning electron microscope photographs of the two-spotted spider mite, *Tetranychus urticae*, a very common pest of house and greenhouse plants. (A) is magnified ×36 and (B) ×198! *(Photographs courtesy of P. Dayanandan)*

when they have formed an extensive fine web along the stems and leaves (Figure 8–3). By then, the mites have chewed out the green chlorophyll layers (mesophyll) of the leaf, producing a massive area of characteristic small white specks on the leaves (Figure 8–4). This area of destruction grows larger as the population of mites increases. There are two good ways to control mites: (1) keep the leaves of your plants washed off and in a humid, well-ventilated, uncrowded environment; at the first sign of mites, wash off the leaves periodically with a strong stream of water (e.g., a kitchen sink dish-washing spray) to literally knock the mites off; (2) spray with a chemical containing the miticide, kelthane (either kelthane itself, mixed according to bottle directions, or Isotox, a mixture of various chemicals including kelthane).

Although insects can be extremely beneficial in many ways, they are still the major pests of plants. They are ubiquitous, occupying a variety of habitats, and are never too far from a delectable plant or plant part.

We are all too familiar with insect pests that plague our house and greenhouse plants, garden vegetables, lawns, and fruit and shade trees. Houseplant pests are usually less common than those pests found on greenhouse plants and on garden plants. They are not immune, however, to whiteflies, mealybugs, scales, aphids, mites, and thrips.

*Whiteflies* (Figure 8–5), which were relatively uncommon pests before 1970, have become resistant to normal controls and have become serious pests outdoors (e.g., on tomato and gardenia) and indoors (e.g., on fuchsia, *Solanum, Nicotiana,* gardenia, hibiscus, and many other broad-leaved, nonsuc-

*FIGURE 8–3*
Scale and red spider mite on avocado *(Persea).* The rounded objects on the stem are the adult scale insects; the delicate webbing is evidence of chronic spider mite infection.

*FIGURE 8–4*
Red spider on clover *(Trifolium).*

*FIGURE 8–5*
Scanning electron micrograph of a whitefly on a lower leaf surface, ×31. *(Photograph by P. Bagavandos)*

culent plants). Normal chemical sprays such as Isotox, Malathion, and Diazinon easily kill adults, but the larvae continue to hatch, and control is thus very temporary. Three alternative ways to kill whitefly include: (1) use of Resmethrin, which is a hormonelike chemical which disrupts the normal development of the larvae (read the label on whitefly sprays and make sure they contain Resmethrin); (2) hang square masonite boards that are painted bright yellow and covered with heavy grease or oil near the plants (the whiteflies are attracted to yellow and get immobilized in the grease or oil); and (3) release one of the relatively harmless, small (about one millimeter in diameter) predator wasps that will selectively feed on whiteflies and keep the population under control (this method is best for plants grown in a greenhouse).

Likewise *mealybugs* can become a real problem, especially root mealybugs which remain unseen below the soil level until it is too late. Mealybugs look like masses of cottony material on the undersides of the leaves (Figure 8–6), in the leaf axils, and in the growing tips of cacti and succulents. If only a few mealybugs are present, remove them with a cotton swab (Q-tip) dipped in rubbing alcohol. If heavy infestations are found, you may try repeated spraying of a systemic spray which is absorbed by the plant and kills such sucking pests as trichogramma wasps (a predator on mealybugs), or simply throw the plant away to prevent more widespread infestations throughout your collection.

*Scale insects* (Figure 8–7) are even more difficult to control because the adults produce a hard shell (or scale) which protects them from chemical

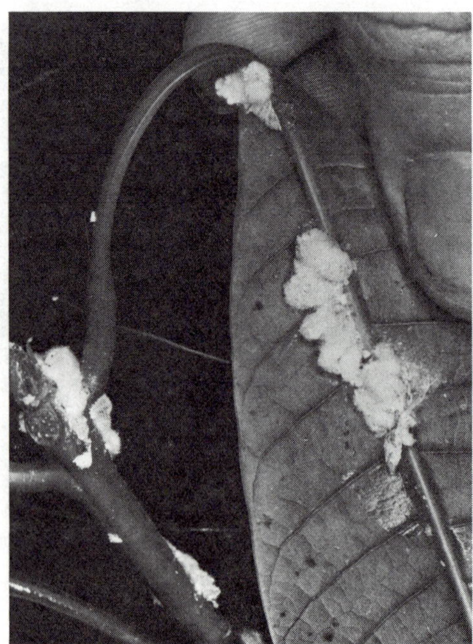

*FIGURE 8–6*
Mealybugs on mango *(Mangifera indica).* The cottony masses are groupings of adults. Juvenile crawlers move along the plant parts to suitable new sites.

control. The young, or crawlers, which are produced frequently and abundantly from each scale-covered adult, are vulnerable to common sprays (such as Isotox) when they leave the protection of the adult "scale" to crawl along the plant stems in search of a suitable place to settle down. The young, upon finding such a place, insert their mouth parts into the tender stem, begin sucking juices, enlarge into mature scale insects which shed their legs, and become nonmotile, covered by their protective scale. The sole purpose of these adults is to suck plant juices and produce hundreds of new young crawlers. Understanding the life cycle of such pests—with a motile crawler stage and a nonmotile, resistant adult stage—can help to determine just how they move about on your plants and what the best time would be to spray to control them. Outdoors, we use dormant oil spray applied just before budbreak to scale-infested fruit trees or wintercreeper *(Euonymus)*. Such sprays are routinely used by gardeners on fruit trees (such as apple, pear, and peach) as a preventive measure to curtail the multiplication of scale insects. The oil is especially effective, since it suffocates the pest it covers. However, it can only be used on still dormant woody plants; it will kill emerging buds or young shoots on such trees if applied too late and is *never* used on house or greenhouse plants, most of which are herbaceous, or if woody, actively growing.

It is probably significant that our worst houseplant pests (indeed, our worst greenhouse and agricultural insect pests) are all members of the order Homoptera, class Insecta (see Table 8–1). While aphids (Figure 8–8) are relatively easy to remove with a stream of water or ordinary chemicals, whiteflies, mealybugs, and scale insects are not so easily removed and have led to

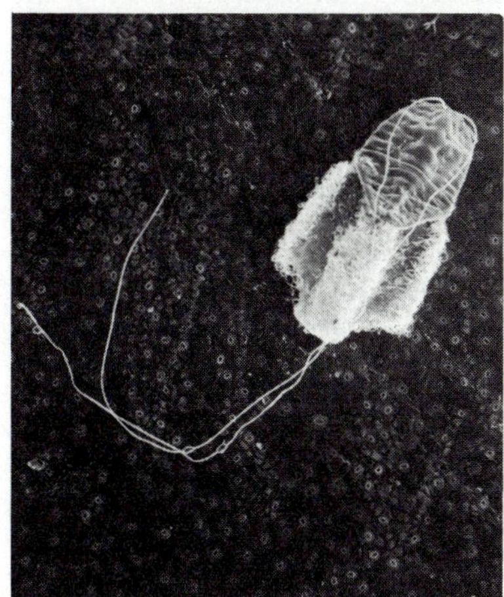

*FIGURE 8–7*
Scanning electron micrograph of a scale insect on a leaf surface. The two threads are the insect's antennae, ×48. *(Photographed by P. Bagavandos)*

*FIGURE 8–8*
Scanning electron micrograph of an aphid, a common pest of house and garden plants, ×30. *(Photographed by P. Bagavandos)*

untold damage in the California agriculture industry. In such situations, biological control (using predators such as ladybugs and small wasps) has begun to become an effective means of control.

In a world in which people are dying of starvation, those insects which annually damage or destroy millions of acres of economically important plants, including farm crops, are of special concern. The world's most important fiber plant, cotton, is subject to two major insect pests: boll weevil *(Anthonomus grandis)* and pink bollworm *(Pectinophora gossypiella).* The cotton boll weevil, which produces many generations in one season, destroys cotton over a wide area from Texas to the East Coast in spite of the use of insecticides. Unlike the weevils that flourish in rainy periods, the pink bollworm thrives on cotton in the southwest, from Texas to California. One of the dozens of insect pests of corn is the corn earworm *(Heliotis zea),* which is worldwide in distribution.

Insect pests make no exception in the case of garden vegetables. Cutworms, the larvae of moths, feed at night, cutting off the plants at or near the soil level, and thus destroy many plants including tomato and cabbage. The larval stage of click beetles *(Agriotes),* called wireworms, attack the roots of many vegetable crops.

Everyone who grows tomatoes is familiar with the rather large, green tomato hornworm *(Manduca quinquemaculata)* which can eat most of the leaves off a medium-sized plant overnight. These pests are difficult to see, but their damage is evident, and upon such discovery, you should search the plants carefully until you find the caterpillars. Biological control may be obtained by using Dipel or Thuricide, which is dust made up of a bacterium, *Bacillus thuringiensis.* Figure 8–9 illustrates biological control of the green tomato hornworm caterpillars, which are covered with the small white cocoons of the parasitic Brachonid wasps. These tiny wasps lay their eggs in the body of the living caterpillar. The eggs then hatch, and the young wasps feed on and kill the caterpillar; finally, they pupate and spin the characteristic external cocoons. You should allow these helpful parasitic types of insects to remain in the garden.

Although people seem most concerned about weed pests in the lawn, there are also many surface and soil insects that attack lawns, including sod webworms and grubs. Symptoms of webworm infestation are chewed blades of grass and, eventually, irregular brown patches. Webworms, the larvae of moths, are gray with black spots, less than 25 millimeters in length, and they produce several generations in a single season. Grubs, which are the larvae of several kinds of beetles, chew off young grass roots, and, as a result, patches of dead grass can be lifted.

Trees, too, whether they are beautiful shade trees or the edible fruit variety, are vulnerable to injurious insects. One such insect is the caterpillar of the gypsy moth *(Porthetria dispar).* This pest has defoliated forests of hardwood trees including oak, maple, birch, and aspen from Maine and other New England states to Pennsylvania and Michigan. When hardwoods run out,

*FIGURE 8–9*
Tomato worm with wasp cocoons. *(Photographed by K. S. Walter)*

the pest will feed on the leaves of softwoods such as apple and willow. In 1981, 5,200,000 hectares of trees in the United States were defoliated by gypsy moths. Although pesticides have not solved the problem in the heavily infested forests of the northeastern United States, authorities in Michigan have reported some success following aerial spraying and the release of sterile male moths. California has announced that they are fighting the moth invasion with aerial spraying of the pesticide, Sevin. Recently, researchers at the U.S. Department of Agriculture reported that parasitic wasps from India have proved effective in laboratory and limited field tests in controlling gypsy moths.

An insect that virtually dominated the news during the summer of 1980 and 1981 is the devastating Mediterranean fruit fly, *Ceratitis capitata,* which was found in several counties of California and imperiled the state's $14 billion-a-year agriculture industry. Debate over correct control measures, hesitancy, indecision, and mismanagement among those in authority pitted the insect against man. The decision to use natural controls, including fruit stripping, sterile males, and ground spraying, resulted in some control but not complete eradication. Compounding the problem was the accidental release of fertile flies that were supposedly sterile. Amid insistence from agricultural interests to utilize aerial spraying of Malathion and pressure from environmentalists to resort to other means of pest control, the governor, faced with the threat of a broader quarantine than had been imposed earlier, reluctantly ordered aerial spraying. At the present time, the results of the eradication program appear promising. Some scientists are of the opinion that a return to more normal, cooler winter temperatures in California may be the most effective means of control of the Mediterranean fruit fly.

Birds and rodents are the most serious vertebrate pests of plants. Fruit crops are especially vulnerable to starlings, crows, and blackbirds, and the latter cause heavy damage to grain crops. One of the most ravaging birds is the quelea, a sparrow-sized bird, that roves by the millions in parts of Africa, feeding and annually destroying thousands of metric tons of grain. We should hasten to mention that, just as in the case of many insects, where the majority are beneficial, so are many species of birds the "farmer's and gardener's friend." Many are insectivores, such as purple martins, which consume countless insects daily.

Moles, not to be confused with molds, are mammals that live underground and often burrow within 10 centimeters of the surface. As a result, ridges are produced, and when these pests tunnel in the lawn, the sod is broken and root systems are severed. Moles are controlled by means of metal traps, flooding with water, and smoke from a slow-burning fire.

Among the rodents, or gnawing animals, rats and mice are the most injurious to plants. These pests, which frequently outnumber the human population in a given area, thrive on field crops, vegetables, and fruit either before harvest or during storage. Although they are not as destructive as some pests, rabbits and raccoons are familiar to most gardeners. Rabbits have enormous appetites and devour garden vegetables such as cabbage, lettuce, peas, and beans, and also, grasses and clovers, tree bark, fruit trees, and young tulip leaves, one of their most favored foods. Raccoons are predators of insects, birds, and mice, but in turn, they will feast on sweet corn and berries. Wrapping the lower trunks of tender trees and shrubs will deter rodents from gnawing on them in winter. Likewise, the use of poultry wire barriers and chemical repellents have been successfully used to prevent serious damage. The poultry wire needs to be at least a meter high, especially in areas that receive heavy winter snowfall.

## FUNGI

Fungi rank not far below insects as serious competitors of useful plants. The typical structure of this plant is a threadlike filament, called a *hypha,* which forms a mass known as a *mycelium.* Since fungi lack chlorophyll and, therefore, cannot photosynthesize, many of them must obtain their nutrients parasitically from living organisms; others exist harmlessly off dead organic matter as saprophytes. A very common fungus, one that most people have seen, and probably, unknown to them, eaten, is *Rhizopus* or black bread mold. In addition to growing on bread, it is often present on fruits such as strawberries, bulbs of lily, corms, and sweet potatoes.

The rose gardener has also probably encountered serious fungal diseases. Common diseases are black spot *(Diplocarpon rosae)* of leaves, Botrytis blight *(Botrytis cinerea)* of some hybrid tea roses, and powdery mildew *(Sphaerotheca pannosa).* Numerous fungi attack various lawn grasses and produce diseased areas that are generally circular and vary in diameter from

2.5 centimeters to 1.3 meters or more depending on the disease. Snow mold usually occurs where the snow is slow to melt. Leaf spot, one of the most common diseases from spring to fall, is most likely to occur during cool, moist weather and is characterized by brown to blue black spots. The hot, humid days of July and August and more than average moisture level are ideal conditions for the development of brown patch. One of the most destructive lawn diseases, which can occur following drought stress and high temperature, is Fusarium blight; it is characterized by rings of dead grass.

Two fungi in particular, that had a major impact on millions of people, are *Plasmopara viticola,* the cause of downy mildew of grape, and *Phytophthora infestans,* the late blight of potato. The French depend heavily on the grape industry, and downy mildew was the cause of great concern, while the very destructive late blight of potato led to the Irish potato famine of 1845.

Plant disease epidemics have also had a major impact on our forests. White pine blister rust, caused by *Cronartium ribicola,* has meant the virtual elimination of large stands of white pine. *Endothia parasitica,* the causal organism of the blight of the American chestnut, has destroyed an extremely important source of timber and edible nuts. More recently, *Ceratocystis ulmi,* the organism responsible for the Dutch elm disease, has ravaged the American elm.

Not to be overlooked are the fungal diseases produced by the rust (Figure 8–10) and smut fungi. As parasites of fruit trees cereal grains, these pests have been responsible for gigantic crop losses, and therefore are of tremendous economic importance.

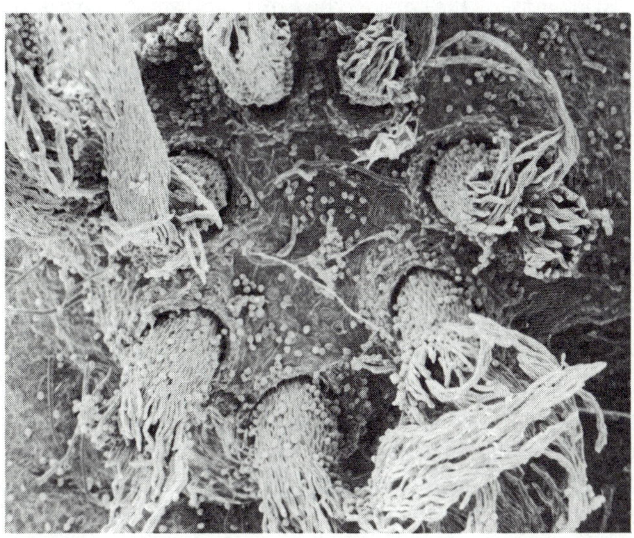

*FIGURE 8–10*
Scanning electron micrograph of the spore-bearing fruiting structure (called a *pustule*) of a rust fungus on apple. This fungus causes the cedar-apple rust disease, ×70. *(Photographed by P. Bagavandos)*

## WEEDS

Just as microorganisms, insects, fungi, and larger animals can be serious plant pests, so too can other green plants, namely weeds, compete with desirable garden and crop plants. Whether they are in the garden, lawn, or cultivated cropland, the definition of a weed as "a plant growing where it is not wanted" is an appropriate one, since they are competitors for space, nutrients, water, and light. Even indirectly, they do harm to desirable plants since they are a habitat for insect and fungal pests. On the other hand, they may be a haven for beneficial insects.

Weeds possess many attributes which enable them to be serious pests. These include (1) production of thousands of seeds per plant; (2) easily and widely disseminated seeds; (3) seed dormancy that may be long term; (4) rapid seedling growth and seed production at a very early stage; (5) vegetative reproduction; (6) biennial and perennial habit as well as annual; and (7) some resistance to drought, frost, and pests. To many homeowners, the prime pest in the lawn is the annual crabgrass *(Digitaria sanguinalis).* Unfortunately, there are many others, including annual bluegrass *(Poa annua),* chickweed, dandelion, henbit, knotweed, oxalis, plantain, purslane, creeping spurge, speedwell *(Veronica),* hawkweed, and quackgrass. Some people consider these weeds to be pests and therefore would prefer to rid their lawns of all of them. On the other hand, other people like them as a "polyculture" in their lawns and do not consider them to be pests! However, many weeds cause serious harm to crop and garden plants because of their competitive properties. These are truly pests that must be contended with if yields or landscape esthetics are impaired by their presence.

While on this topic, it is important to mention that there are definite benefits to be derived from some weeds in the garden or in the cropland. These include the following: (1) they may occupy fresh soil and prevent erosion; (2) their shoots and roots may add organic matter (and even nitrogen from leguminous weeds) to the soil when they die; (3) many weeds are edible at certain stages in their development (see Chapter 11); and (4) many weeds are very useful as potential energy sources (see Chapter 15), in crafts (see Chapter 13), in making household items such as mats and wall coverings, and as sources for natural plant dyes (see Chapter 13).

## PARASITIC HIGHER PLANTS

There are some, although relatively few, seed plants that are pests of other useful seed plants. These are parasitic species and include the true mistletoes *(Phoradendron),* which attack oaks, walnuts, and hickories, and dwarf mistletoes *(Arceuthobium),* which parasitize coniferous species, including pine, spruce, and fir. Another parasite is dodder or love-vine *(Cuscuta),* an orange, twining vine, which is destructive to some crops and ornamentals, especially clover and alfalfa. In some parts of the southern United States, the hemiparasite, witch-weed *(Striga lutea),* is a very serious pest of corn and has caused such damage that several counties in eastern North Carolina have crop quarantines in an effort to control this pest. In parts of Europe, the nongreen broomrapes *(Orobanche)* are destructive pests in crop fields.

## The impact of pests

In a world in which overpopulation and starvation are becoming more and more of a problem, the protection of our food sources is increasingly important. For centuries, the farmer and gardener have harvested their crops at the mercy of "Mother Nature." Very good years would be followed by years characterized by poor yields as hordes of pests of various kinds invaded the fields. The loss of potential food and the monetary losses have been enormous. Upwards of $18.2 billion have been lost as a result of bacteria, nematodes, insects, fungi, and weeds. More important, however, than the economic loss caused by these pests is the 35 percent of the worldwide production of food crops that will not be available to a hungry world because they were destroyed prior to harvest.

There are more species of insects than of all other plants and animals combined; and, of these, 10,000 different species of insects attack man, and his food and fiber crops. So detrimental to our livelihood are these that an immense agrichemical industry has evolved, devoted to eliminating them. The federal government, also, has expended many millions of dollars to try to eradicate specific insect pests, and in 1979, a $10 million program was initiated in an attempt to stem a grasshopper onslaught in 17 of our western and central states. In 1978, the gypsy moth caused a timber loss of $18 million, and a $3 million program of control was not completely successful. Expenditures of over $50 million per year have still not completely eradicated the cotton boll weevil. Important food crops, such as corn, are victimized by many different insect pests, and these pests can cause the loss of millions of dollars to the farmer and millions of liters of much needed food to the hungry of the world.

To a great extent, the diseases to which crop plants, ornamentals, fiber plants, and timber trees are susceptible have had an economic impact on humankind. Nations have suffered the loss of large numbers of their population because of the starvation brought on by plant disease. Other nations have lost many of their inhabitants through emigration; for example, 1.5 million Irish left their homeland during the mid-1800s because of famine brought about by the late blight of potato disease, caused by the fungus, *Phytophthora infestans*. In 1970, 15 percent of the corn crop in the United States was lost due to an outbreak of southern corn leaf blight, caused by another fungus, *Helminthosporium maydis*.

Weeds have always been a "thorn-in-the-side" of farmers and home gardeners, alike. Losses amounting to $16.1 million per year, both in the value of the harvestable crops and in the cost involved in weed control, can be attributed to the presence of these undesirable plants.

Nematode infestations have always caused many problems for the farmer, and it has been estimated that as much as 10 percent of the annual crop potential in the United States is not attained because of this pest.

All in all, as long as humankind has been depending on plants for so many of the essentials of life, there has been a never-ending war against the pests that would destroy these plants. Those battles are sometimes won, and sometimes lost, with resultant suffering, starvation, and economic loss. We must win more if we are to have enough food to feed the hungry world in the years to come.

## Control of pests

With a current world population of 4.5 billion that is constantly increasing, it should be obvious, in spite of the tremendous cost and risk, that we provide practical and effective means of protecting our crops from pests. Although other methods of control are available and are more desirable, controlling pests with chemicals has been the chief means, even though hazardous at times, for maintaining agricultural productivity.

### CHEMICAL CONTROL OF PESTS

A pesticide is a substance which controls or kills pests, whether they be bacteria (bacteriocides), nematodes (nematicides), mites (miticides or acaricides), insects (insecticides), rodents (rodenticides), fungi (fungicides), or green plants (herbicides). Of the various types of pesticides, the inorganic ones are not now used very widely, whereas the organic pesticides are. Table 8–2 lists some of the common pesticides in use today.

*TABLE 8–2*
*Some common pesticides and their uses*

| *Group* | *Example* | *Type* | *$LD_{50}$*[1] | *Uses* |
|---|---|---|---|---|
| Inorganic | Lime sulfur | Fungicide<br>Miticide | — | Powdery mildews<br>spider mites |
| | Lead arsenic | Insecticide | — | Chewing insects |
| | Copper sulfate | Fungicide | — | Bordeaux mixture for fungi |
| Organic | | | | |
| **Botanicals** | Cycloheximide (Actidione) | Fungicide | — | Powdery mildews, rusts, turf diseases |
| | Streptomycin | Bacteriocide | — | Blight of pear and apple, soft rot of vegetables |
| | Milky spore *(Bacillus popilliae)* | Insecticide | — | Japanese beetles |
| | Nicotine sulfate | Insecticide | 50–60 | Banned by EPA (once used to control sucking insects) |
| | Pyrethrins | Insecticide | 1,000 | Insects |
| | Rotenone | Insecticide | 300–1,500 | Insects |
| | Dipel or Thuricide *(Bacillus thuringiensis)* | Insecticide | Harmless | Caterpillars (lepidopterous larvae) |

*TABLE 8–2 (continued)*

| *Group* | *Example* | *Type* | $LD_{50}$[1] | *Uses* |
|---|---|---|---|---|
| **Synthetics** | | | | |
| Oil-based hydrocarbons | Dormant oil (Volck) | Insecticide | — | Controls overwintering pests (e.g., scale insects) |
| Organochlorines (Chlorinated hydrocarbons) | DDT (dichloro-diphenyl-trichlorethane) | Insecticide | — | Banned by EPA (once used to control beetles, mosquitoes, gypsy moths, earworms, bollworms, and budworms) |
| | Chlordane | Insecticide | 475 | Banned by EPA (once used to control soil insects, ants, termites) |
| | Kelthane | Miticide | 809 | Spider mites |
| Organo-phosphates | Malathion | Insecticide | 1,375 | One of most widely used insecticides for sucking and chewing insects |
| | Meta-Systox R | Insecticide | 56–65 | Systemic; controls many insect pests |
| | Diazinon | Insecticide | 76–108 | Controls many insect pests of ornamentals |
| Carbamates | Benlate or Benomyl | Fungicide | 10,000 | Fungus diseases of ornamentals |
| | Captan | Fungicide | 9,000 | Black spot of roses |
| | Phaltan | Fungicide | 10,000 | Powdery mildew |
| | Truban | Fungicide | 2,000 | Damping-off |
| | Sevin (Carbaryl) | Insecticide | 500 | Insects on ornamentals; fleas on pets |
| | Resmethrin | Insecticide | 4,240 | Synthetic pyrethroid; controls whiteflies |
| | Metaldehyde | Molluscicide | 630 | Kills slugs and snails |
| Phenoxy acids | 2,4-D (2,4 dichloro-phenoxyacetic acid) | Herbicide | — | Broadleaf weeds |
| | 2,4,5-T (2,4,5 trichloro-phenoxyacetic acid) | Herbicide | — | Banned by EPA[2] (once used to control woody vegetation) |

[1]The $LD_{50}$ number indicates the relative toxicity of various pest control chemicals. $LD_{50}$ stands for "lethal dose required to kill half (50 percent) of a group of test animals." The amount (dose) is expressed as the amount of pesticide, in milligrams, per 1,000 grams of body weight of the test animals (usually rats). The lower the $LD_{50}$ value, the more toxic the chemical, as it takes less to kill the test animals.
[2]Because of the presence of dioxin, which is very toxic to humans and thought to cause cancer.
Source: Values cited are from Pirone 1978.

Although the use of pesticides has been beneficial in many ways, pesticides can be hazardous, and they have had a major impact on the environment. After decades of worldwide use and misuse, thousands of harmful chemicals have been disseminated into the ecosystem, causing untold damage to plants (from herbicides) and all forms of animal life. Important cases are documented in the following discussion.

The chlorinated hydrocarbons, including such insecticides as DDT, aldrin, dieldrin, and chlordane, have effectively controlled insects, but because they are nonselective, other insects and wildlife, such as birds and fish, have been killed and some reduction in the reproductive rate of other animals has also occurred. Some of these nontarget organisms are predator insects and birds which act as natural controls of pests. With the eradication of the natural enemies, the target pests proliferate. As a result of spraying to control the cotton boll weevil, for example, the natural predators of such destructive pests as the cotton bollworm and the tobacco budworm were also destroyed, and the cotton boll weevil developed resistance to the spray. Chlorinated hydrocarbons are fat soluble, not very biodegradable, and, as persistent pesticides, tend to accumulate and increase in concentration as they move through the food chain. After repeated exposure to chlorinated hydrocarbons, such as DDT, many species of insects are capable of developing a genetic resistance to an insecticide. Hundreds of pests that were at one time susceptible to pesticides are now immune "superbugs." Repeated use of Sevin, for instance, has resulted in resistant populations of insects and an increase in numbers in spider mite populations.

Careless spraying of pesticides has also had a deleterious effect on the honeybee population. Honeybees, which are responsible for the pollination of fruits, vegetables, and other economically important plants, have been the victims of massive kills.

The deleterious effects of pesticides on humans are well documented. Evidence is accumulating that pesticides such as Kepone, Malathion, and others can cause nausea, headaches, altered brain activity, and chest pains. Cancer in animals has been found to be caused by high doses of some pesticides.

Control of pests is essential if sufficient food for a hungry world is to be produced. Although our heavy reliance on the use of chemicals has provided good pest control, they have had an adverse effect on the environment and on all types of organisms in different food chains. Thus, alternative means of control must be implemented.

## BIOLOGICAL CONTROL

The application of known biological principles can be an aid in the development of pest control methods that will produce the desired effect without harming the environment.

One of the most effective of these natural methods is the use of pheromones and hormones. Pheromones are chemical substances, secreted by insects, that serve to attract the male to the female. Extremely small amounts of this sex lure are given off by the female during the mating period and can be detected by the male as far away as one and a half kilometers. Herbert Brown of Purdue University has succeeded in producing synthetic sex attractant pheromones, and these, along with naturally occurring pheromones, can be used in a variety of ways to control insects. For example, they have been used to monitor pests, thus allowing the scientist to alert the farmer to a potential problem. The pheromones have also been used to bait traps containing poison that will kill the pest before it damages the crop plant. This method has been successful against the gypsy moth in the United States and bark beetles in some Scandinavian countries. Another way in which the pheromones are useful in controlling pests is by using a small amount of the pheromone in a field to confuse the male insect, so that he cannot locate the female and mating does not occur. Pheromones are either naturally occurring or synthetic reproductions of these natural products, and are not poisons that can harm the environment. A specific pheromone will only attract a specific species of pest, so that they can be used without the danger that they will kill off useful, as well as harmful, insects.

Hormones are internally produced, natural chemicals that will affect the biology of a certain pest in much the same way that hormones affect humans. They can be used to interrupt the normal life cycle of an insect. For example, the juvenile hormone, when used at the correct stage in the life cycle, will stop the insect from reaching sexual maturity. The hormone, ecdysone, will accelerate the maturation process to such an extent that the pest will die. Antijuvenile hormones, Precocene I and II, have been isolated recently from the common mistflower *(Ageratum grandiflorum)*, a blue flower often used as a border planting in home gardens. Precocenes I and II will prevent the pest from making the juvenile hormone that is necessary for it to reach normal adulthood, and insects so treated will either fail to develop normally or will produce sterile adults. The precocenes may prove to be very useful in controlling pests because they are effective, nontoxic to the environment, biodegradable, nonpersistent, and will act on more than one species.

Interrupting the normal process of reproduction will also serve to limit pest populations. It may not curtail a current infestation, but it will certainly go far to eliminate future problems. The screwworm of cattle has been brought under control in the southeast and southwest United States by utilizing this method. The males are subjected to enough gamma radiation to make them sterile. They are then released to mate with normal females, who mate only once in their lifetime. This method has also been successful in helping to control the pink bollworm, and the U.S. Department of Agriculture is currently conducting tests in which boll weevils and gypsy moths are subjected to synthetic growth regulators that will leave them sexually immature.

Plant geneticists stand in the front lines in the ongoing war on pests. The development of pest-resistant cultivars of some of our most important grain crops, wheat, oats, barley, and corn, have saved us millions of dollars. However, this method is slow; up to 20 years may be necessary to develop a plant that is resistant. Moreover, as the scientist is developing the disease-resistant cultivar, the pests are adapting themselves to this new source of food. The Hessian fly, a pest of wheat, is a good example of an insect that has adapted itself to new, supposedly resistant cultivars a number of times. The continuing work of the plant breeders has, however, brought this pest under control. Cultivars resistant to the spotted alfalfa aphid, the European corn borer, boll weevil, and the wheat stem sawfly have also been developed.

The introduction of natural predators, such as viruses, bacteria, and insects, has been effective in the control of insect pest populations. Probably a great deal of control is carried on in nature with the action of natural predators on undesirable insects (Table 8–3). Insects, regarded as pests, have evolved ways to prevent poisoning by natural plant toxins. So, by the use of these mechanisms, it is easy for them to resist poisons that are applied to plants by humans. But predatory insects, regarded as "friendly," have little experience with poisons and are destroyed when they feed on pesticide-infused pests. The parasitic wasp, *Trichogramma,* is effective in controlling the cotton bollworm, and the Japanese beetle has been controlled by milky spore bacteria (*Bacillus popilliae* and *B. lentimorbus*). Another bacterium, *Bacillus thuringiensis,* sold under the trade names of Dipel or Thuricide, is very effective in controlling many kinds of caterpillars (on cabbage and broccoli, for example) and is harmless to humans and pets. The tent caterpillar, Mediterranean fruit fly, and alligator weed have all been controlled by natural predators. Predatory mites have been used successfully in greenhouses to control spider mites that attack plants. Related to the cotton boll weevil, but a beneficial insect rather than a harmful one, is the small brown beetle *(Cryptolaemus montrouzieri).* Both indoors and outdoors, it is an effective predator on mealybugs. Just as some nematodes can be the scourge of the garden, parasitizing edible plants, so too are there nematodes that can parasitize harmful garden insects. One such helpful nematode is the caterpillar nematode *(Neoaplectana carpocapsae)* which, acting symbiotically with a bacterium *(Xenorhabdus nematophilus),* can kill most insects within 24 hours. Among those insects that are the prey of this relationship are caterpillars, root beetles, weevils, carpenter worms, cutworms, armyworms, and squash borers. Recent U.S. Department of Agriculture research indicates that the caterpillar nematode may be an effective means of biologically controlling gypsy moth larvae. In such cases of biological control, the pest cannot be totally eradicated but can be kept to an acceptable level of tolerance.

## INTEGRATED PEST MANAGEMENT (IPM)

At the present time, it is becoming more and more apparent that a system of integrated pest management is the most feasible method of dealing with the control of pests, while simultaneously, preserving our environment and allowing the farmer adequate economic return for expended labor. Unfortunately, IPM cannot be purchased at the local feed and grain or garden supply center. Rather, it is a plan of attack that utilizes many methods of control—biological and cultivation practices as well as the careful and limited application of chemicals as a last resort.

IPM is a new way of thinking and is interdisciplinary in scope. The sciences of genetics, entomology, plant pathology, and agriculture are all instruments to be used in controlling the pest population. The farmer and home gardener must utilize a combination of methods to offset the ravages of plant pests. Rather than reaching immediately for the insecticide or herbicide, one might do well to look at the methods of cultivation.

*TABLE 8–3*
*Common animal predators that attack pests*

| *Class* | *Order* | *Predator* | *Stage* | *Pest* |
|---|---|---|---|---|
| Insecta | Coleoptera | Ladybug, ladybird | Larvae<br>Adult | Aphids, mealybugs, moth eggs, spider mites, scale insects |
| | | Ground beetle | Larvae<br>Adult | Gypsy moth larvae, cankerworms, grasshoppers, snails, slugs, and many caterpillars |
| | | Tiger beetle | Larvae | Many insects |
| | | | Adult | Many insects |
| | Odonata | Dragonfly, skimmer | All stages | Mosquitoes, midges |
| | | Damselfly | All stages | Mosquitoes, midges |
| | Orthoptera | Praying mantis | All stages | Flies, grasshoppers, aphids, white grubs, beetles, chinch bugs, caterpillars |
| | Neuroptera | Lacewing | Larvae | Aphids, mealybugs, mites, leafhoppers, thrips, moth eggs, bollworm larvae, whiteflies, scale insects |
| | Diptera | Syrphid fly, hover fly | Larvae | Aphids, scale insects, ants |
| | | Tachinid fly | Larvae | Larvae of butterfly and moth, grasshoppers, armyworms, bollworms, wood lice, centipedes |
| | | Robber fly | Larvae | Larvae of insects |
| | | | Adult | Grasshoppers, flies, beetles |
| | Hemiptera | Damsel bug | | Aphids, leafhoppers, lygus bugs, spider mites, treehoppers, small caterpillars |
| | Hymenoptera | Trichogramma wasps | Larvae | Eggs of 200 species of insect pests (bollworms, leafworms, fruitworms, cutworms, most moths and butterflies) |
| | | Fly parasite wasps | Larvae | Maggots |
| | | Scale parasite wasps | Larvae | Scale insects |
| | | Brachonid wasps | Larvae | Tomato hornworms, corn borers, butterflies and moth caterpillars, birch leaf-mining sawflies, wood-boring beetles, bark beetles, ants |
| Arachnida | | Spiders | | Insects |
| | | Greenhouse mites | | Spider mites |
| | | Pacific mites | | Spider mites |
| | | Willamette mites | | Spider mites |
| Aves | | Birds | | Insects |
| Amphibia | | Toads | | Insects |
| Reptilia | | Snakes | | Insects |

Long before the use of chemicals, certain agricultural practices were followed. Most of them were developed as a result of the "tried and true" method—they had proven to be successful, and the farmer was able to harvest a fair return for labor. This approach may be termed a common-sense method of farming. Although more physical work was often required and the farmer had to tolerate more pests than with extensive chemical spraying, the benefits gained from a cleaner environment more than offset the disadvantages.

Presently, much of our farming is done under monoculture practices; that is, many adjacent fields are used to grow genetically similar strains of the same crop. Rather than this, utilizing a more natural ecosystem of polyculture (mixed crops on the same piece of land), practicing crop rotation, and the alternation of crops in adjoining areas will allow the checks and balances of nature to serve as barriers to pest infestations.

*Intercropping* is a term used for alternating types of crops rather than growing pure stands of one crop plant. Weeds which attract pest predators can also be interspersed in areas near crops. Examples are goldenrod (attracts 75 species of predators), pigweed, dandelion, wild carrot, lamb's-quarters, evening primrose, buckwheat, dill and mustard. For adult insects they provide nectar, rest, and shelter cover. These weeds also produce odors that, while attracting predatory insects, repel or confuse pest insects. Neutral insects on the weeds provide an alternate for predators. Herbs which reportedly repel pests are rue, savory, basil, sage, lavender, thyme, garlic, and tansy.

Along with such cultural methods, the farmer would be wise to choose plant strains that are genetically resistant to the pests that are common for the particular environment and to keep a watchful eye out for climatic conditions that might lead to an increase in pest populations. By becoming aware of a potential infestation before it becomes full blown, the pest problem will be much simpler to control. In some cases, the introduction of bacteria, parasites, or insects has served as a means of controlling undesirable plants and insects. The introduced species are natural predators of the pests and act as another means of cutting down the pest numbers. A knowledge of the life cycle of a pest can be useful to the farmer so as to determine at what stage in its development the pest will be most controllable. The cotton boll weevil has been controlled to some extent by timing the planting and harvesting of the cotton crop so that it does not coincide with the pest's life cycle requirements. And soil cultivation has done much to eradicate the corn earworm which winters-over in the soil.

In IPM all of these methods can be used in combination with pesticides only when epidemics occur. The natural methods of control will allow the farmer to greatly curtail the use of chemical pesticides, thus resulting in a harvest with a lower cost input. In our times of high energy costs, that can be a considerable saving for the average farmer. To reduce the need for extensive chemical spraying there are several options. By varying types of pesticides, resistance develops more slowly in pests. Selection of disease-resistant plants and plowing under crop remains as soon as possible after harvest prevents runaway pest infestation by preventing pest overwintering on old plant remains. By learning about the life cycle of pests, timing of the crop growth cycle can be changed by planting cold-resistant crop plants which mature and are harvested before pest egg-laying time. Plant breeding for an increase in a crop's natural defenses, such as poisons in parts that we do not eat, is another method.

Since the farmer has depended so heavily on the use of chemical sprays to keep crops pest-free for the last 30 years or so, it will take time and effort for the IPM system of pest control to be widely accepted. The universities and extension services must do their part by initiating curricula that will provide the broad background and expertise needed to implement IPM. Homeowners and farmers will then be able to combine nature's controls without the widespread use of pesticides. Farmers will find that a possible slight decrease in yield will be offset by the decreased costs for pest control (by not relying as heavily on costly pesticides). In some cases, the use of IPM techniques has actually resulted in higher yields. Not only is the IPM system economically feasible for the farmer, but equally important to all mankind, ensuring an environmentally safe world in which to live.

## Pest control in the home garden

As food prices continue to rise, more and more people are planting home gardens. In 1980, over 40 million households in the United States grew vegetables at home. We do not want to see our harvest ravaged by pest infestations. It would be wise for the home gardener to know which organisms are harmful as well as which are beneficial.

Just as indiscriminate spraying by the farmer has polluted our environment, the careless use of chemicals by millions of home gardeners has also contributed to the pesticide problem. The use of pesticides by the home gardener did decline in the late 1970s, but a majority still resort to chemical control.

Rather than depending on the sole use of well-known pesticides which were considered to be safe at one time, biological control is a viable alternative. Many hard-working organisms, actually friends, are naturally present where pesticides have not been used. These include predatory insects, flowering plants, worms, birds, toads, and microorganisms.

Many beneficial insects (natural predators of pests) such as lacewings, ladybugs, and praying mantises can be purchased in a garden supply center or from mail-order suppliers if they are lacking in your garden. The lacewing is active at night and after grasping an aphid will suck juice from the pest; other favorites of this predator are mealybugs, mites, and scale insects. Ladybugs also enjoy aphids, in addition to weevils, beetles, and other insects. The praying mantis feeds on insect pests including caterpillars and aphids. However, it is not very selective since it will also feed on natural predators and, in addition, is cannibalistic. It may be difficult to keep natural predators around unless you have plenty of pests for them to feed on and unless you have a natural area (old field or buffer zone around the garden) in which they can breed and produce new generations. A supply of drinking water for predatory insects is an attractant. Birds and wasps sharing water in a bird bath is a common sight in a natural garden.

Another method of biological control involves the use of companion plants—plants which are used in combination with other garden plants and which ward off many insect pests because of the volatile, often aromatic, compounds they produce. One of these companion plants is the garden marigold. It produces chemicals from the roots, called *root diffusates,* which are toxic to and control nematodes. Other companion plants, which give off natural insecticides that control many garden pests include garlic, chives, wormwood, radishes, tansy, and nasturtiums.

Many, though not all, birds are insect eaters. These include the thrushes, chickadees, mockingbirds, and wrens. The wrens will also chase away other berry-eating birds. If one does not object to having them around, salamanders, snakes, and toads feed on many pests. The toad has a liking for the very damaging cutworm.

The bacterial microorganism, *Bacillus thuringiensis,* attacks a variety of larval pests including corn earworms, tent caterpillars, and the peach tree borer. A word of caution is in order, however, if you have these predators in your garden: do not use pesticides since they are toxic to many of these beneficial organisms.

Also available on the market are various control devices for insect pests, birds, and other animals. These include such items as Japanese beetle traps with pheromone bait, electronic insect lanterns, whitefly spray bombs, antibird meshes, mole traps, and special traps for raccoons, groundhogs, and possums where the animal can be released elsewhere unharmed.

## REFERENCES

Agrios, G. N. 1978. *Plant Pathology.* 2nd ed. New York: Academic Press. (Deals with disease cycles, concepts of plant pathology, pathogens, and symptoms as well as practical aspects of diagnosis and control.)

Allen, G. E., and Bath, J. E. 1980. The conceptual and institutional aspects of integrated pest management. *BioScience* 30(10): 658–64.

Anagnostakis, S. L. 1982. Biological control of chestnut blight. *Science* 215: 466–71.

Baldwin, F. L., and Santelmann, P. W. 1980. Weed science in integrated pest management. *BioScience* 30(10): 675–78.

Barfield, C. S., and Stimac, J. L. 1980. Pest management: an entomological perspective. *BioScience* 30(10): 683–89.

Barrons, K. C. 1981. Contributions of pesticides to land and energy conservation. *Down to Earth* 37(2): 5–8.

Batra, S. W. T. 1982. Biological control in agroecosystems. *Science* 215: 134–39.

Bird, G. W., and Thomason, I. J. 1980. Integrated pest management: the role of nematology. *BioScience* 30(10): 670–74.

Boraiko, A. A. 1980. The pesticide dilemma. *National Geographic* 157(2): 145–83.

Carson, R. 1962. *Silent Spring.* Boston: Houghton Mifflin Co. (A landmark publication stating the hazardous effects of pesticides on the environment.)

Cravens, R. H. 1977. *Pests and diseases.* Alexandria, Va.: Time-Life.

DeBach, P. 1974. *Biological control by natural enemies.* New York: Cambridge University Press. (Treatment of biological pest control and the application of those methods to control specific pests.)

Edens, T. C., and Koenig, H. E. 1980. Agroecosystem management in a resource-limited world. *BioScience* 30(10): 697–701.

Foreman, K. 1982. Mealybug destroyers. *Organic Gardening* 29(3): 69.

Fry, W. E., and Thurston, H. D. 1980. The relationship of plant pathology to integrated pest management. *BioScience* 30(10): 665–69.

Haynes, D. L.; Lal Tummala, R.; and Ellis, T. L. 1980. Ecosystem management for pest control. *BioScience* 30(10): 690–96.

Hooper, G. R.; Case, F. W., Jr.; and Myers, R. 1971. Mycoplasma-like bodies associated with a flower greening disorder in *Trillium grandiflorum.* Mich. Agri. Exper. Sta. Article No. 5625. In *Plant Disease Reporter* 55 (12):1108–10.

Huffaker, C. B., ed. 1980. *New technology of pest control.* New York: Wiley-Interscience. (Comprehensive, ecologically based, summary report of the significant progress toward integrated pest management systems.)

Jepson, R. B., Jr., ed. 1977. *Organic plant protection.* Emmaus, Pa.: Rodale Press.

Johnson, W. T., and Lyon, H. H. 1976. *Insects that feed on trees and shrubs.* Ithaca, New York: Cornell University Press.

Jordan, W. 1982. A fruitless pursuit. *Science 82* 3(3): 62–68.

Lewert, H. V. 1976. *A closer look at the pesticide question.* Midland, Mich.: Dow Chemical Co. (Acquaints the public with little known facts about pesticides and their uses with quantitative data from respected impartial sources.)

National Academy of Sciences. 1976. *Pest control: an assessment of present and alternative technologies.* Washington, D.C.

Norris, J. R. 1969. Sporeformers as insecticides. In *The Bacterial Spore,* edited by G. W. Gould and A. Hurst, pp. 485–516. New York: Academic Press.

Pirone, P. P. 1978. *Diseases and pests of ornamental plants.* 5th ed. New York: John Wiley and Sons. (Discusses diseases and pests by symptoms and by causes and their control.)

Poinar, G. O. 1982. New light on nematodes. *Organic Gardening* 29(1): 90–93.

Silverstein, R. M. 1981. Pheromones: background and potential for use in insect pest control. *Science* 213: 1326–32.

Ware, G. W. 1975. *Pesticides.* San Francisco: W. H. Freeman and Co.

USDA. 1952 *Insects.* Yearbook of Agriculture. Washington, D.C.: U.S. Government Printing Office. (Summarizes knowledge about protecting crops from insects.)

USDA. 1953. *Plant diseases.* Yearbook of Agriculture. Washington, D.C.: U.S. Government Printing Office. (Presents information on the causes and control of many diseases of important crop plants.)

# Growing plants in controlled environments 9

Greenhouses, cold frames, terraria, bonsai containers, and growth chambers are controlled environments. They allow you to exert a remarkable degree of control in providing for growing plants. They also add a second dimension to plant cultivation. Terraria and bonsai plantings, for example, provide you with miniature indoor landscapes. Cold frames can be used year-round to grow herbs and salad-type vegetables. A greenhouse allows you to grow a diverse array of tropical houseplants and food plants, and at the same time, provide energy for your home or apartment as a passive solar collector. Growth chambers have controlled environments that are invaluable in the classroom, commercial nursery, or research laboratory. It is important to understand the function of these structures, how to set them up, and how to use them most effectively for growing plants.

## Cold frames

### WHAT IS A COLD FRAME?

Cold frames are actually mini-solar greenhouses. Because of their size they are much easier and less expensive to build than ordinary greenhouses. Sunlight is the external heat source, but they may contain an underground electric heating cable for minimal bottom heat. Basically, a cold frame is a box consisting of a south-facing, inclined top to increase the capture of solar energy; well-insulated sides; and planting beds inside that are usually submerged below ground level. Cold frames have many uses for growing plants: starting plants in the spring, propagating cuttings, storing hardy bulbs in containers for cold treatment prior to forcing, growing salad herbs and vegetables during the darker winter months, overwintering herbaceous perennials that have been started in late summer, and stratifying seeds that require after-ripening treatments.

## BUILDING AN ENERGY-EFFICIENT COLD FRAME

Cold frames may either be attached to a structure such as a house foundation, garage, or barn; or, they may be free-standing structures. To be energy-efficient, they must be well insulated, be able to retain through the night much of the heat they capture during the daytime, and maximize their ability to capture solar energy. Heat can also be generated by decomposing manure placed in berms as inclined banks against the side walls or inside the cold frame well below the plant root zone. In addition, a less energy-efficient source of heat can be provided by electric heating cables placed in the soil of the growing area (well below the root zone) or by heat provided from a house basement when the cold frame is mounted against a basement window (Figure 9–1).

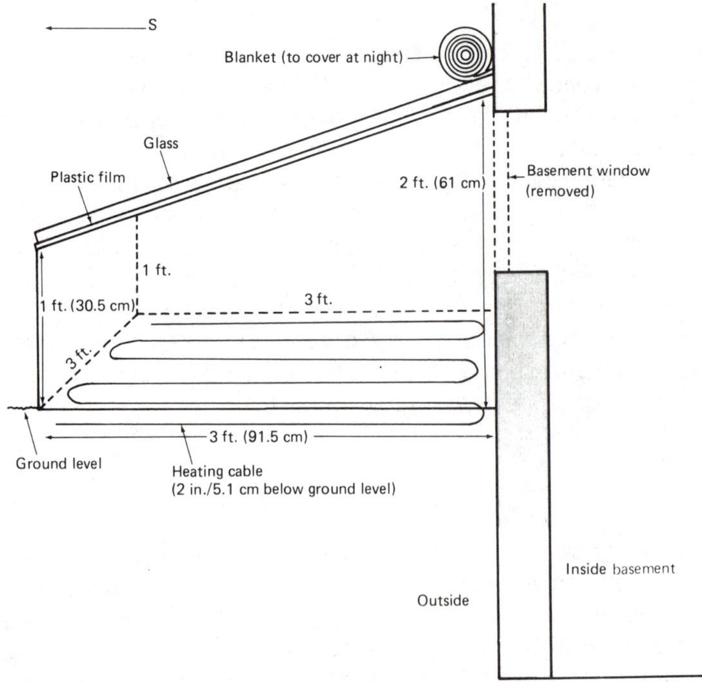


*FIGURE 9–1*
Solar-heated cold frame. Cold frames can be constructed against the basement window of a house, on the south side.

A solar-heated cold frame, on the other hand, utilizes only heat that is captured from the sun. The cold frame with its transparent top acts as a passive solar collector. In addition, some of the sun's heat may be stored in water-filled, black-painted, one gallon plastic milk jugs placed in tiers against the north vertical wall of the cold frame (Figure 9–2).

The cold frame itself may be constructed of wood, concrete, cinder blocks, brick, or stone. To provide the best insulation, add at least four inches of styrofoam or other appropriate insulation material to the side walls outside of the frame, both above and below ground. The top covering of the cold frame should consist of a clear cover, such as an old window frame fitted with glass or any kind of close-fitting frame with a clear covering (glass, fiberglass, plastic film). Further, it is recommended that this covering be double glazed; that is, the glass or plastic should be covered with a second layer such as clear plastic polyethylene film, creating a dead air space in between. This extra insulation helps to retard heat loss. At night, cover the cold frame with insulation material such as a heavy blanket, styrofoam, or plastic bags filled with straw. The window frame must fit snugly over the top of the cold frame. This can be achieved by constructing a frame with hinged windows that have strips of insulation felt along the under sides of the window frames.

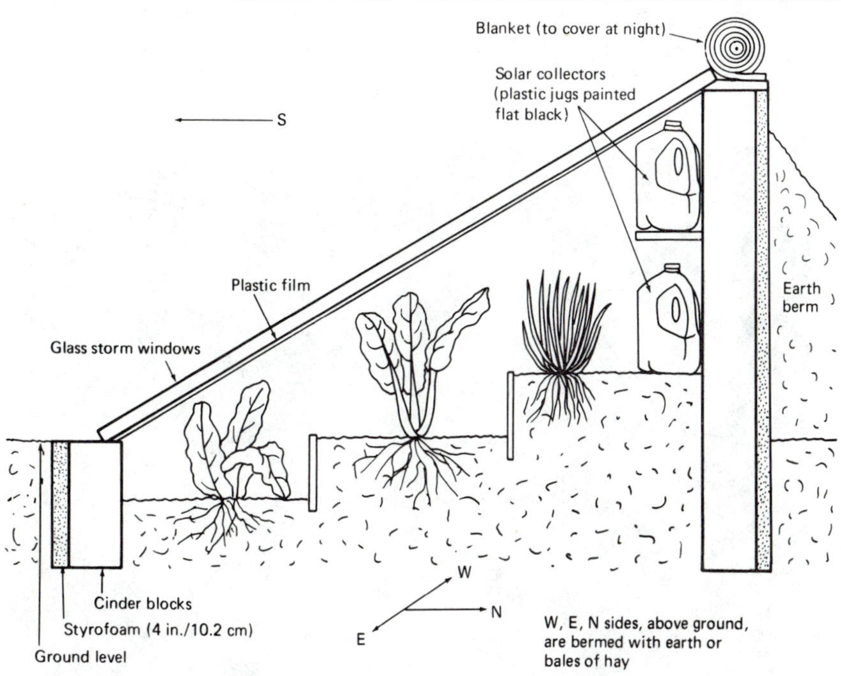


*FIGURE 9–2*
Solar-heated cold frame as detached unit. South-facing glazing and water-filled, black-painted milk jugs on two shelves against north wall absorb and store heat.

To increase the amount of light reaching the plants growing in the cold frame, one can simply construct terraces rising toward the north wall of the cold frame (Figure 9–2). This is especially effective in northern and southern temperate latitudes where the winter sun makes a low-angle arc over the horizon during the daytime.

## COLD FRAME USES

We have mentioned a few of the uses for a cold frame. The main point is that it should not be abandoned after one-time use during any particular time of year. It can be used year-round! In the autumn various perennial herbs and vegetables such as sage, rosemary, chives, parsley, basil, and Swiss chard can be transplanted into the cold frame. At the same time, you can start salad plants such as Chinese cabbage, endive, lettuce, radishes, and onion "sets" (bulbs). These can be harvested all winter long. In spring, the cold frame can be used to start annual flowers and vegetables such as peppers, tomatoes, melons, squash, cucumbers, eggplant, and leafy vegetables (Figure 9–3). In summer use the cold frame to start cuttings or store potted plants. At this time it is important to open the windows or simply remove them and to make sure plants receive sufficient water. By autumn you can begin the cycle over again. When you do, this time you might try adding some pots or ceramic containers of hardy bulbs such as hyacinths, crocuses, or daffodils to chill before you bring them inside to force in midwinter. Some bonsai fanciers use the cold frame to overwinter their hardy bonsai plants.

*FIGURE 9–3*
Free-standing cold frame with supports for glass-panel coverings.

## HOW ARE GREENHOUSES USED?

Greenhouses in times past were expensive and considered as luxury items that only the wealthy could afford. Today, that has changed. Greenhouses bought from commercial firms can be expensive but are much cheaper than before because of the increase in volume of sales and because smaller units are now available. However, they are cheaper still if you build them yourself. And, they take on greater relevance today because of their use not only for ornamental plants, but for year-round food production. At the same time they serve as home-heating passive solar collectors. Even when space is very limited, a greenhouse window attached to the house provides space for winter-grown tomatoes (determinate-growth plants), herbs, or houseplants.

## TYPES OF GREENHOUSES

Greenhouses can be constructed in various ways. The foundation is usually cinder block, poured concrete, brick, or stone. The upper framework is made of wood or metal. Their shapes are quite variable: lean-to, A-frame, geodesic dome, Quonset-type, or with straight sides and gabled roof (Figures 9–4, 9–5, and 9–6). Greenhouses are covered with glass panes, fiberglass, plastic film, or a combination of glass and plastic film. Plastic film alone is temporary at best. It tears easily, tends to get dirty, is easily ripped by strong winds, and tends to transmit less light as it ages, but it is by far the cheapest greenhouse covering (Figure 9–7). Fiberglass is good and strong, but its light-transmitting characteristics become poor as it ages, and this limits what

*FIGURE 9–4*
Lean-to-type greenhouse attached to a home.

*FIGURE 9–5*
Quonset-type greenhouse. Shows benches full of seedlings of annuals to be used for outdoor plantings. Note semi-circular curving tubular frame-work used to support the polyethylene film covering.

*FIGURE 9–6*
Detached greenhouse with straight sides and gabled roof at Longwood Gardens, Kennett Square, Pennsylvania.

*FIGURE 9–7*
Home-made wooden-frame greenhouse covered with plastic polyethylene film.

can be grown in the greenhouse, especially during times of the year when sunlight is limited. Glass has good light transmission qualities, but it is expensive and easily broken by falling objects such as hailstones. A combination of glass covered by a layer of clear plastic film combines relatively clear light transmission capability with a double-glazing system that is absolutely necessary for solar-heated structures. Glass and film should be separated by 2.5 to 5 centimeters (one or two inches) of dead air space. Many commercial greenhouses now use the Quonset-type structure with a double layer of polyethylene film for maximum heat retention. Of course, they have to put on new coverings every two years.

## HEATING A GREENHOUSE

With the advent of the energy crisis, heating a greenhouse has become extremely expensive. Fossil fuels such as oil, propane, butane, and natural gas, as well as electricity, are becoming so expensive that alternative heating systems become much more attractive. Many people have gone to solar heating as a backup supplementary heating system. Some have gone completely to solar heating. Fossil fuels are not only very expensive, but also may be the source of the air pollutant, ethylene, following combustion. Ethylene causes fading of flowers, leaf drop, and premature fruit ripening. So, solar heating is definitely a viable alternative. Other alternatives include use of steam heat from geothermal sources or from utility power plants for commercial production of crop plants. However, for the individual, a solar heating system is the best alternative.

Solar heating systems may be either passive or active. Active systems can utilize solar panels mounted on the ground outside the greenhouse. Such systems can utilize heated air that is stored in beds of rock underground the greenhouse floor and released by means of reversible fans. Water systems can involve heat stored in water tanks, but these take up a lot of space. Passive systems are much less expensive. They rely on the *greenhouse effect,* which is the trapping of heat energy inside the greenhouse. This occurs when visible light and high energy (short wavelength) ultraviolet rays enter the greenhouse through the covering, strike the ground or objects inside, and are converted to invisible, low energy infrared heat rays which cannot escape back through the covering. Heat buildup can occur even on overcast days because the UV rays can still penetrate the cloud layer and enter the greenhouse. The greenhouse effect is especially desirable in the winter to help heat the greenhouse, and double glazing aids in retaining the heat. The heat inside the greenhouse can be absorbed by water in large containers (tanks, plastic jugs) painted black or by the cement, brick, or stone walk and walls.

A further way of retaining trapped heat is to build the greenhouse below the frost line or where the soil is not frozen, a pit-type greenhouse, relying on the natural thermal insulation of the soil (average 8° to 12° C or 45° to 55° F) (Figure 9–8). Passive solar greenhouses need no backup fuel system in more southern climates, but in northern latitudes some supplemental system is necessary. This can be methane obtained from a methane generator (methane is like natural gas), a heat pump, or heat from the house if the greenhouse is attached and used as a passive solar collector. A combination of active plus passive solar heating combined with pit-type construction, double glazing, super insulation, and manure berms against the north side of the greenhouse can provide for all the heating needs of the greenhouse, even in northern latitudes. The success of such a system is conservation of heat trapped and stored in the greenhouse during the day by preventing, as much as possible, its loss at night. This means placing curtains under the glass on the inside or some kind of effective insulation on the outside of the glass such as sheets of styrofoam or an insulated shutter, or filling the dead air space between the layers of glazing with styrofoam beads.

## GREENHOUSE VENTILATION AND COOLING

A greenhouse must have provision for ventilation. During warm weather, especially in the summer, the greenhouse effect is more of a hindrance than a help to greenhouse operations because excess heat builds up and must be removed by convection, exhaust fans, or evaporative cooling. On a bright sunny day in a closed greenhouse, it is not uncommon that solar radiation will cause temperatures to soar to over 43.5° C (110° F). Automatically controlled vents (windows which can be opened) mounted at the sides and top of the greenhouse are standard ventilation aids in most greenhouses. The vents can be triggered to open automatically when the thermostat reaches

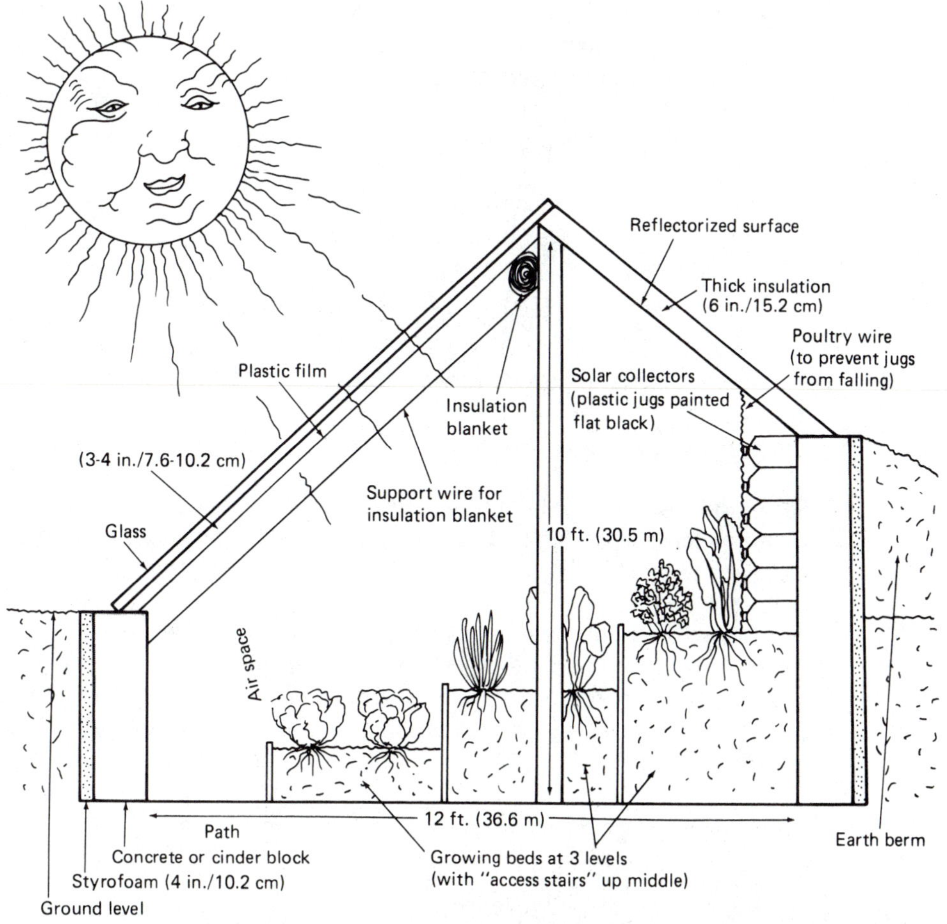


*FIGURE 9–8*

A pit-type greenhouse with lean-to configuration. Inside are three terraces for growing plants, established so that plants intercept more solar radiation from the south (especially in winter).

a preselected upper limit, and to close when the greenhouse temperature reaches a set lower limit. Alternatively, there are hand-operated vents controls. By opening these vents, some hot air can escape by convection.

In addition to vents and fans, many greenhouses have a water-use cooling system. The simplest one relies on cold water running over the glass to absorb infrared radiation, if a large, inexpensive water supply is available. A more complicated system is called evaporative cooling, where exhaust fans in one wall of the greenhouse pull outside air through a spongelike "pad" or "mesh" of water on the opposite wall. The water column, through which the air moves, absorbs heat from the air and adds humidity to the greenhouse.

These "cooling pads" may be made of cellular fiber, treated cardboard, or plastic. Water which trickles down over the "cells" or mesh is collected in a trough and recirculated through a pump (Figure 9–9).

A greenhouse must have some provision for shading, especially during summer months. This is achieved by one of several ways:

1. Use of shading compound (slaked lime) sprayed or whitewashed onto the glass, which then gradually washes off or which can be scrubbed off in autumn.
2. Lath shades which can be rolled up or down over the glass roof.
3. Installation of plastic mesh shade cloth over the plants using one or several layers to provide varying degrees of effective shading.

## SUPPLEMENTAL LIGHTING

Some greenhouses must have provision for supplemental illumination during winter months, depending on the types of plants being grown. Effective supplemental lighting in the greenhouse may be provided by pairs of four- or eight-foot fluorescent lamps mounted under reflectors. These are hung as close above the plants as necessary. A combination of cool white and warm white fluorescent tubes have proved to provide an adequate spectrum (light) quality for most plants' growth. One drawback to using several such fluorescent fixtures is that they may shade some natural light from the plants and therefore may be used almost as a substitute for natural light.

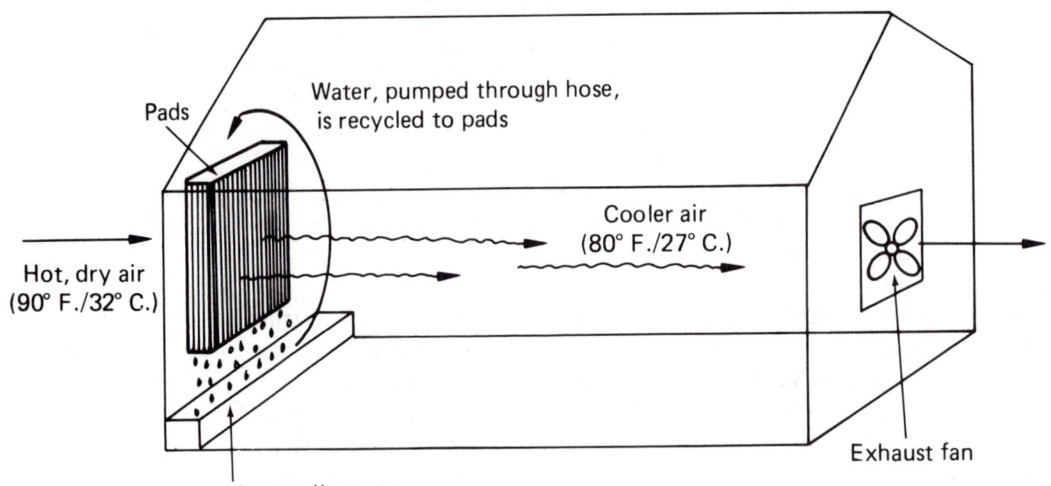


*FIGURE 9–9*
Drawing depicting evaporative cooling system for a greenhouse and how it works.

## WATERING METHODS

Watering is one of the most important aspects of growing plants in a greenhouse. To save labor and to conserve water, several innovations can be employed:

1. Bases of pots can be buried in pebbles or cinders in the benches.
2. Cuttings can be misted automatically (15 seconds every half-hour).
3. Potted plants, seedlings, and cuttings can be bottom-watered in a tray or pan.
4. Judicious use of shading and ventilating devices can be installed to reduce loss of water from the plants by transpiration.
5. Plastic wrap or pieces of plate glass can be placed over pots of germinating seeds.
6. The floor and the sides of the greenhouse can be watered to maintain the humidity.
7. Plants with high water requirements can be grown in special frames with glass sides and tops. These are essentially large terraria (Wardian cases) with tops that open in case of high temperatures.
8. The greenhouse bench area above the plants may be covered with shade cloth (coarse mesh plastic shading cloth) to keep them cooler. Plants that prefer indirect light are ferns, bromeliads, and many orchids.
9. Trickle irrigation can be employed, which uses a perforated plastic hose laid on the bench or soil in a floor bed. Or individual small tubes can be used to carry water from a main water line right into individual containers such as hanging baskets.

This last method puts the water directly at the roots of the plants, thus conserving water. Plants should be well watered at their bases. Sprinkling water over the tops is the least effective type of watering because soil is splashed out of the pots and this helps to spread many greenhouse pests.

## GREENHOUSE USES

Greenhouses usually have a combination of shelves and benches for growing plants. In fact, most of the vertical and horizontal space in a greenhouse can be effectively utilized by growing ferns and other shade-requiring plants under the benches, on the sides of the benches, on shelves mounted on the side walls above the benches, and in hanging baskets and pots. Many plants grouped together help maintain high humidity.

The greenhouse needs space to accommodate tools such as rake, shovel, trowel, scoops for handling soil and other media, tampers, broom and brush, labels, pencils, pots, sprayer, hose, watering can, fine spray misters for hose, newspaper, and plastic bags. Dry planting media can be stored out of the way under benches in metal or plastic garbage cans or in wooden bins. Common planting media include perlite, vermiculite, peat moss, sphagnum moss, sand, loam, clay, gravel, fir bark, osmuda fiber, and nutrient solutions for *hydroponic growth* (plants grown without soil in nutrient-containing water) (Figures 9–10, 9–11, and 9–12).

*FIGURE 9–10*
Plastic-covered greenhouse with benches of hydroponically-grown butter-crunch lettuce plants being grown in perlite-filled plastic gutter troughs. Large black pipe on left carries nutrient solution. (Photograph courtesy of John Clark.)

(A)

(B)

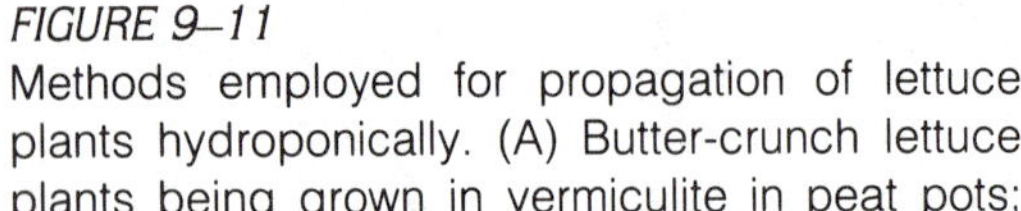

*FIGURE 9–11*
Methods employed for propagation of lettuce plants hydroponically. (A) Butter-crunch lettuce plants being grown in vermiculite in peat pots; these are watered with nutrient solution. (B) Peat pots with lettuce plants are now set in plastic troughs containing perlite. (Photographs courtesy of John Clark.)

(A)

(B)

*FIGURE 9–12*
Hydroponically-grown butter-crunch lettuce plants. (A) Plastic pipe with spouts directed to plastic gutter troughs filled with perlite and peat pots of lettuce plants. Nutrient solution travels under pressure through the large plastic pipe and is delivered to each gutter trough through the plastic spouts. (B) Close-up view of one of the lettuce plants grown in this system. It shows an extensive root system covered partially with perlite particles. These roots have developed into the perlite from the base of the peat pot in which the lettuce plant had its initial start (see Figure 9–11). (Photographs courtesy of John Clark.)

Once the greenhouse is built, insulated properly, and equipped with tools, media, and bench or bed space, it can be used for many purposes: growing food plants, including herbs; growing exotics such as aroids, cacti and succulents, ferns, orchids, and gesneriads; propagating cuttings; germinating seeds and growing seedlings for the garden outdoors in spring and summer; starting bulbs for forcing after cold treatment; forcing twigs; growing mushrooms; grafting potted trees; and growing bonsai.

## Dwarfing plants for bonsai

Bonsai is an art form in which a woody plant such as a tree, shrub, or vine is grown in a small container to represent a miniature landscape, dominated by an interesting tree form (Figure 9–13). It involves good horticulture and good design. Further, it suggests more than just a tree in a pot. It reflects nature rather than distorting it.[1]

The simplest way to start a bonsai is to obtain a container-grown plant from a nursery. Look for a specimen with a potentially good form, twisted

*FIGURE 9–13*
(A) A Bonsai hemlock tree (*Tsuga* sp.). This one clearly illustrates the one-sided balance typical of many bonsai plants. (B) Bonsai of a pine tree (*Pinus* sp.). Note the bonsai triangle formed by the branches. The three corners of the triangle represent heaven, man, and earth.

[1]These ideas are those of Jack Wikle, bonsai enthusiast and Director of Education at Hidden Lakes Gardens, Tipton, Michigan.

or dwarflike. Plants can also be grown from seed, from cuttings, or by layering. Another source is a natural habitat where plants have struggled to exist and may be dwarfed and distorted by environmental rigors.

Basically the objective of bonsai is to create in miniature the form of trees which have grown in particularly artistic shapes. The species should be one in which the flowers, fruits, and foliage remain in scale as the tree is dwarfed. The trunk is the central element of the bonsai; it should taper gradually toward the top. The roots are also important; they add to the character if partially exposed or trained over a rock.

With specimen in hand, the next step is to plan the form in which it is to develop. This can be done by using wire and judicious pruning. Most bonsai are grown in trays or shallow dishes; thus some root pruning and shaping of the root system is necessary, especially if the roots are to clasp a stone. Questions that should be answered in the initial planning are as follows:

1. Is the bonsai to be a single tree or several trees?
2. Is the trunk to be single or multiple stem?
3. What habit should the specimen have: upright, slanted, or cascade?
4. Is the general form to be straight or the trunk and branches to be twisted?
5. Is the base to rise from the soil, grow on a rock, or expose bare roots?

The three basic operations in bonsai design are pruning, wiring, and repotting (Figure 9–14). A tree usually requires one heavy branch pruning to establish its basic form; subsequently, the shaping is done by pinching back or nipping. Do not shear a bonsai. Repotting and root pruning should be done when the root system has become tightly packed and there is little space in the root mass for soil.

The bonsai must be watered regularly but should not be constantly saturated. Feeding is also essential; use any houseplant fertilizer but in a diluted form. Winter protection of most hardy species should be provided in a lath house or a cold frame. For the normal growth cycle they cannot remain indoors through the winter but must be gradually acclimatized to the changing seasons.

## FORM AND POSITION FOR BONSAI

### *Basic styles of bonsai*

1. Formal upright: the trunk of the tree stands upright, reminding us of a big old tree in a vast open area.
2. Slanting: the trunk of the tree is slanting to one side. This shape is found on a slope or in a windy place.
3. Coiled: the trunk is crooked. This shape reminds us of an aged tree "fighting for survival."

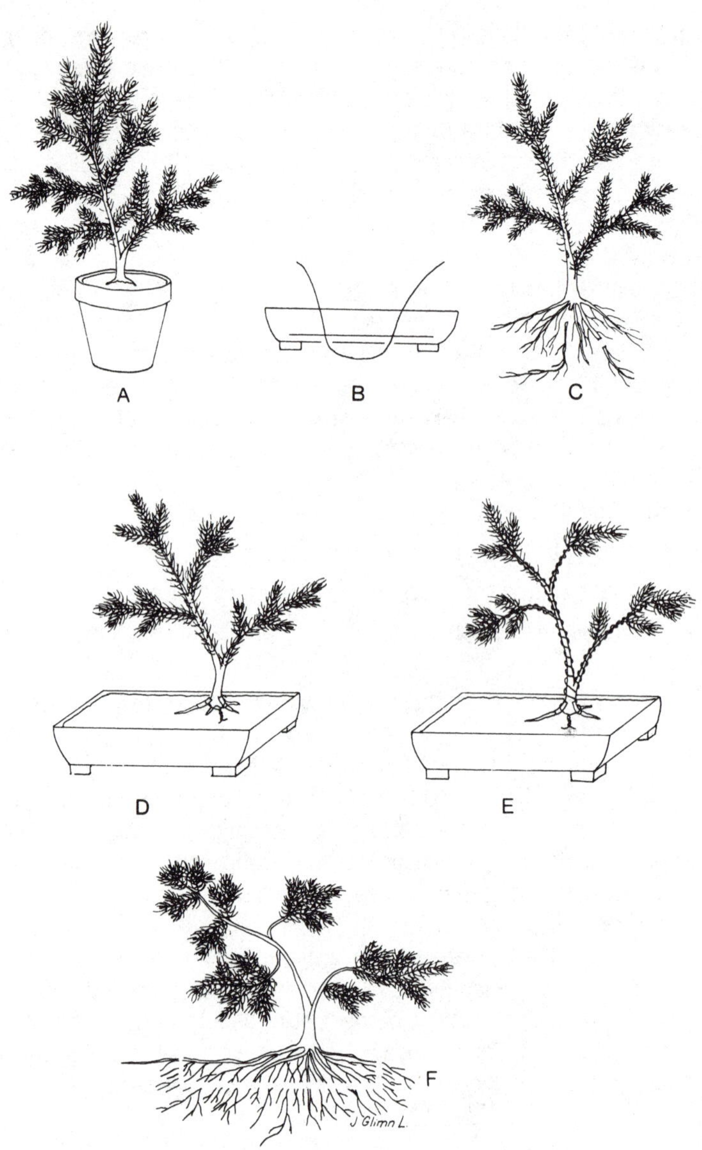


*FIGURE 9–14*
Primary steps used in preparing a bonsai of white spruce *(Picea glauca):* (A) select potted plant; (B) select bonsai pot with two holes in the bottom, place plastic screen over holes, and run wire through holes and screen; (C) prune leader and other undesired laterals, remove from pot and prune large, long roots; (D) place plant in bonsai pot, off-center, wire trunk in place using rubber pads to protect bark, then add soil and water; (E) week by week rewire branches to bend the desired angle; (F) three years later spread and trim roots to one-half inch less than pot size, rewire trunk into pot, and add fresh soil.

4. Twin trunks: two trunks from a single root system, not widely separated.
5. Cascade: the trunk curves down over the rim of the container as if it is hanging over a steep cliff in a remote mountain area.
6. Group planting: many trees, uneven number, planted in a single container in order to create a scene in nature such as a grove of trees in a forest or a seashore grouping.

### *Criteria for plant position*

1. Branches should reach out to the sides, not directly forward.
2. Bottom one-third of trunk should be bare of branches.
3. Place tree off-center in a pot with a long dimension and position so that 40 percent of the area is on one side of the tree and 60 percent on the other.
4. Base of the trunk must be near the center line of the container.
5. If the tree slants to the right, place the base of the trunk to the left of center; if the tree slants left, place the base of the trunk to the right of center.
6. A tree should have balance, depth, and appear stable in its container.

## PLANTS FOR BONSAI

While some species are more popular than others for training as bonsai, almost any woody plant can be made into an attractive form. Some representative genera suited for bonsai are listed in Table 9–1.

*TABLE 9–1*
*Plants that have been found to be especially suitable for bonsai*

| *Outdoor Hardy Bonsai Plants* | |
|---|---|
| *Acer* (maple) | *Malus* (flowering crab) |
| *Cedrus atlantica* var. *glauca* (Atlas cedar) | *Picea* (spruce) |
| *Chaenomeles* (flowering quince) | *Pinus* (pine) |
| *Cotoneaster* (cotoneaster) | *Pyracantha* (firethorn) |
| *Crataegus* (hawthorne) | *Rhododendron* (azalea) |
| *Fagus* (beech) | *Taxus* (yew) |
| *Juniperus* (juniper) | *Tsuga* (hemlock) |
| *Larix* (larch or tamarack) | *Wisteria* (wisteria) |
| *Indoor Nonhardy Bonsai Plants* | |
| *Ficus* (fig)—especially dwarf fig *(F. pumila)* | *Citrus* (orange) |
| *Hedera* (English ivy) | *Punica granatum* var. *nana* (dwarf pomegranite) |
| *Cuphea* (cuphea) | |

## PLANTING A BONSAI PLANT

Plan to start a bonsai plant in the spring after danger of frost is past. Select a woody plant in the 6 to 15 inch height range and a container. Wooden containers or a rock as a container are sometimes used. Pots with two holes can be used to wire plants into desired form.

### *Planting procedure*

1. Cover bottom of pot with layer of pea-size gravel.
2. Cover gravel with a layer of sand. Bonsai need perfect drainage.
3. Cover sand with a light layer of potting mix. Generally a good soil mixture is one-third potting soil, one-third sand, and one-third peat.
4. Remove soil from plant's roots and prune to fit pot.
5. Carefully, but thoroughly, adjust the roots in the pot, adding more potting mix a little at a time.
6. When planting is finished, soak in tray of water until soil surface is definitely moist. Then place plant in a protected area for at least one week.
7. Relocate plant where it will receive sun at least half of the day, and water when soil surface dries.

## BONSAI MAINTENANCE

Winter-hardy bonsai plants must be subjected to cool winter temperatures when grown in temperate regions. However, they need to be protected. This can be done by covering the plants with a styrofoam rose cone. Many people simply bury the bonsai pots in the soil and cover the plants with leaves or the styrofoam cone. They must be protected with a wire screen to prevent damage by rodents. In the spring the plants are dug up to put on display. Another alternative is to place the bonsai in a cold frame that is covered with lath or shade cloth to prevent the plants from overheating on sunny days. This can also be prevented by placing the cold frame on the north side of a house. The plants in a cold frame should be mulched for protection from cold and covered with a relatively fine-mesh screen to prevent damage by mice and rabbits. Hardy bonsai should not be stored over winter in a garage, as they will be too cold and dry out.

Bonsai, to be successful, must be fertilized relatively little. The rule of thumb is fertilize the trees once a week during periods of growth with a weak fertilizer solution. When the plants are actively growing, they should be watered as often as once a day. The soil mix should provide good aeration, containing ample sand and gravel.

Why do bonsai remain small? (Figure 9–15) It is seldom because they

*FIGURE 9–15*
Group display of small bonsai, three to nine inches (8 to 23 centimeters) in height, all grown under cool white fluorescent lamps. Names of plants, in order, from top, are as follows:
box honeysuckle *(Lonicera nitida)*
firethorn *(Pyracantha coccinea)*
creeping fig *(Ficus pumila)*
false heather *(Cuphea hyssopifolia)*
snow rose (*Serissa foetida* cv. 'Double')
dwarf Greek myrtle (*Myrtus communis* cv. 'Compacta')
and English ivy *(Hedera helix)*
(Photograph provided by Jack Wikle, Education Program Director, Hidden Lake Gardens, Michigan State University, Tipton, Michigan.)

are provided with a low level of essential nutrients. It is because the roots are given little space to grow and shoots are pruned regularly. Root pruning is done when the roots get too crowded in the pot, but avoid root pruning too often because it is disruptive. With deciduous plants, prune roots every second year and with conifers every three to five years. The shoots need to be pruned of new growth as it appears every year. This may need to be done several times during the growing season.

Growing bonsai indoors can be greatly improved by the use of a lighting system that employs fluorescent lamps. See Table 9–2 and Figures 9–15, 9–16A and B for plants successfully grown under lights.

*TABLE 9–2*
*Bonsai plants successfully grown under lights*

| | |
|---|---|
| Littleleaf boxwood *(Buxus microphylla)* | Juniper *(Juniperus)* |
| Glossy abelia *(Abelia grandiflora)* | Crepe myrtle *(Lagerstroemia)* |
| Natal plum *(Carissa grandiflora)* | Dwarf honeysuckle *(Lonicera nitida)* |
| Flowering quince *(Chaenomeles speciosa)* | Dwarf Norway spruce (*Picea abies* and its many cultivars) |
| Dwarf false cypress (*Chamaecyparis* and its many cultivars) | African boxwood *(Myrsine africana)* |
| Small-flowered chrysanthemum (*Chrysanthemum* and its many cultivars) | Dwarf Greek myrtle (*Myrtus communis* cv. 'Compacta') |
| Cotoneaster *(Cotoneaster)* | Indoor oak *(Nicodemia diversifolia)* |
| False heather *(Cuphea hyssopifolia)* | Olive *(Olea europea)* |
| Fukien tree *(Ehretia microphylla)* | Firethorn *(Pyracantha)* |
| Benjamin fig *(Ficus benjamina)* | Satsuki azalea *(Rhododendron lateritium)* |
| Mistletoe fig *(Ficus deltoides)* | Rosemary *(Rosmarianus officinalis)* |
| Creeping fig *(Ficus pumila)* | Chinese sweetplum *(Sageretia theezans)* |
| Dwarf creeping fig (*Ficus pumila* cv. 'Minima') | Snow rose (*Serissa foetida* cv. 'Double') |
| Oleander-leaved fig *(Ficus neriifolia)* | Small-leaved Oriental elms (*Ulmus* and its many cultivars) |
| English ivy *(Hedera helix)* | |

(A)

(B)

*FIGURE 9–16*
(A) Miniature bonsai of littleleaf cotoneaster *(Cotoneaster microphylla)*, two inches (five centimeters) high from rim of pot, about three years old from a cutting.
(B) Miniature bonsai of Chinese sweetplum *(Sageretia theezans)*, six inches (15 centimeters) high from rim of pot, about eight years old from a cutting.
Both plants were grown under cool white fluorescent lamps. Photographs provided by Jack Wikle.

## Growth chambers for plants

Growth chambers are extremely useful for growing plants in the laboratory under specifically controlled environmental conditions (Figure 9–17). They are widely used in classrooms, industry, and laboratories. They allow control of day-length cycles, light intensity and quality, air temperature, and humidity. With interval timers (usually with 15-minute intervals), a 24-hour period can be programmed for the lengths of light and dark periods and the temperatures desired for each. With movable tables or shelves, plants can be placed at different distances from overhead lamps mounted in the ceiling of the chamber. Cool white and warm white fluorescent lamps and low wattage incandescent (tungsten) lamps are used. Relative humidity can be set over a range of usually 20 to 90 percent. The controls are mounted on the outside of the chamber.

What have been the practical spinoffs from the use of plant growth chambers? In research, they have been largely responsible for telling us about the role of photoperiod and thermoperiod in controlling flowering in long-day and short-day plants (see Chapter 3). In teaching, they allow students to complete simple experiments on the role of day length or temperature in controlling plant growth. Teachers "house" their plant cultures such as freshwater algae, mosses and liverworts, and vascular plants in growth chambers. And in industry, they are widely used to grow plants commercially, especially in the culture of mericloned plants (see Chapter 6).

*FIGURE 9–17*
Controlled environment plant growth chamber.

## Terraria

Terraria date back as far as 1829 when Dr. Nathaniel Ward, a surgeon from London, placed a sphinx moth chrysalis for close observation in a closed glass container together with soil and plants. It turned out that he became more interested in the grasses and ferns which started to grow and develop inside this sealed glass container! Today, people use all kinds of transparent containers for terraria: candy jars, brandy snifters, wine jugs, carboys, aquarium containers, and even canning jars. The kinds of plants that can be grown and landscapes you can devise for plants are almost infinite.

### PREPARING THE TERRARIUM

Special tools can be purchased for digging, spreading roots, placing plants, and tamping the soil in terraria, or they can be fashioned from available materials. A long-handled spatula can be whittled from a piece of lath. A set of tongs can be made by splitting a piece of bamboo cane partway up the length, inserting a small wedge at the top of the split, then wrapping string above the wedge. A tamping stick can be made from a length of dowel with a piece of sponge tied to the end. A kitchen bulb baster is useful for watering.

A drainage layer of gravel or moss on the bottom of the terrarium is necessary to prevent waterlogging of the soil. Use pea-size gravel and some charcoal pieces to a depth of about one inch. Or use moss to cover the container bottom and to extend partway up the sides. A layer of moss can be placed green-side down on the container bottom and covered with soil.

The soil mixture depends on the kinds of plants you wish to grow. A mixture suited for most plants can be made of: 2 parts fibrous loam, 2 parts coarse sand, and 1 part flaky leaf mold. For bog plants and others which prefer an acid soil, substitute 2 parts peat for the leaf mold. A two- to four-inch layer of soil is adequate for most terrarium plants.

### TERRARIA PLANTS

Select plants of similar habitats based on requirements of soil type, soil pH, light, temperature, and humidity (Table 9–3). Some plants are unsuitable for terraria. Orchids require air movement that the closed environment of a terrarium does not provide. Succulents and cacti need air movement and sunlight and will not tolerate the high heat and stagnant air of a terrarium placed in the sun. But succulents can be grown together in a shallow bowl (uncovered) as a dish garden.

For woodland terraria experience shows that *Lycopodium, Coptis,* and *Chimaphila* will only last a little while; on the other hand, *Mitchella,* lichens, *Goodyera, Selaginella,* and ferns are long lasting and are thus excellent candidates for such terraria.

For the recommended carnivorous plants, it is advisable to buy only propagated specimens from nurseries, as these are most adapted to indoor

TABLE 9–3
*Plants suitable for different types of terraria*

| Tropical Foliage Plants for Terraria | |
|---|---|
| *Acorus* (sweet flag) | Grasslike plant, variegated, dwarf variety available |
| *Adiantum* (maidenhair fern) | Delicate wedge-shaped leaflets on dark stems; difficult as a houseplant because of humidity requirement but ideal in terraria |
| *Aglaonema* (Chinese evergreen) | Variegated leaves, capable of growing rather large, up to three feet |
| *Alternanthera* (alligator weed) | Small pink and red leaves; will remain small. Easy to grow |
| *Aucuba* (gold-dust tree) | Leaves speckled with yellow. Use young plants only |
| *Buxus* (boxwood) | Slow growing shrub with shiny dark green leaves |
| *Calathea* (peacock plant) | Red stalks, undersides of leaves red, grows to eight inches |
| *Callisia* (striped inch plant) | Can grow too fast, but colorful with red underside, green- and white-striped leaves |
| *Chamaedorea* (parlor palm) | Miniature, slow growing |
| *Codiaeum* (croton) | Most grow fairly large; have coarse texture. Need good light |
| *Cordyline, Dracaena* | Use young plants only. Palmlike foliage atop stiff stems |
| *Cymbalaria* (kenilworth ivy) | Dainty-leaved creeper suitable for ground cover |
| *Dizygotheca* (false aralia) | May outgrow terrarium. Likes moist soil and needs good light |
| *Ferns other than those listed* | Many kinds, many uses. (See also under woodland plants.) |
| *Ficus repens* (creeping fig) | Useful for ground cover; small leaves |
| *Fittonia* (mosaic plant) | Prominently veined leaves; ground cover |
| *Hedera helix* (English ivy) | Infinite varieties |
| *Helxine soleirolii* (baby tears) | Tiny dense leaves; will work as ground cover but spreads rapidly |
| *Hypoestes* (polka-dot plant) | Green leaves spotted with pink; grows fast—keep trimmed |
| *Maranta* (prayer plant) | Blotched oval leaves, folded in darkness |
| *Osmanthus* (false holly) | Slow-growing, hollylike leaves that are green or variegated |
| Palms other than those listed | Many miniature or young slow-growing kinds are possibilities |

*(Continued)*

TABLE 9–3 (Continued)

| | |
|---|---|
| *Peperomia* | Many kinds available; likes well-drained soil |
| *Philodendron* | Most are too large growing; keep trimmed, and use small-leaved types |
| *Pilea* (aluminum plant, artillery plant) | Some upright, others make good ground covers |
| *Pothos* (devil's ivy) | Easy, but may grow too fast |
| *Saxifraga* (strawberry begonia) | May spread too rampantly |
| *Schefflera* | Use young plants only; for larger bottles |
| *Selaginella* (spike-moss) | Many kinds; easy; fast growing |
| *Syngonium* (arrowhead vine) | Use young plants only; for larger bottles |
| *Tradescantia* (wandering Jew) | Many kinds; easy |
| *Zebrina pendula* (wandering Jew) | Many kinds; easy |
| *Flowering Plants for Terraria* | |
| *Begonia* | Wax begonia and dwarf varieties are best |
| *Crossandra* | Pastel, salmon orange flowers; keep trimmed |
| *Cuphea* (cigar plant) | Easy, shrublike |
| *Episcia* | May become too large and trailing; orange, pink, or white flowers |
| *Exacum* | Good for a short term |
| *Jasminum* (jasmine) | Slow growing with fragrant yellow and white flowers |
| *Oxalis* | Cloverlike leaves, with pink, red, yellow, or white flowers; may need occasional thinning |
| *Punica granatum* var. *nana* (dwarf pomegranite) | Blooms are orange-red and bears brownish-yellow to red fruit |
| *Saintpaulia ionantha* (African violet) | Many varieties; dwarf kinds are especially good in terraria |
| *Sinningia* (miniature gesneriad) | Only miniature sinningias are suitable; the larger species are too big |
| *Carnivorous Plants for Terraria* | |
| *Dionaea muscipula* (Venus-flytrap) | Hinged leaves open and close |
| *Drosera* (sundew) | Secretes sticky liquid that attracts and catches insects |
| *D. rotundifolia* (round-leaved sundew) | |
| *D. binata* (forked-leaf sundew) | |
| *D. capensis* (South African sundew) | |
| *D. intermedia* (intermediate sundew) | |
| *Pinguicula* (butterwort) | Leaves curl over and trap insects |
| *P. lutea* (yellow butterwort) | |

| | |
|---|---|
| *P. caerulea* (blue butterwort) | |
| *Sarracenia* (pitcher plant) | Leaves are pitcher shaped, inside are downward-pointing hairs and liquid which digests trapped insects |
| *S. purpurea* (purple pitcher plant) | |
| *S. psittacina* (parrot pitcher plant) | |
| *S.* hybrids: | |
| *S. minor* × *S. leucophylla* | |
| *S. purpurea* × *S. leucophylla* | |

### *Woodland Plants for Terraria*

| | |
|---|---|
| *Chimaphila* (pipsissewa, spotted wintergreen) | Plain or striped varieties, low growing |
| *Coptis trifolia* (gold-thread) | Dainty cloverlike plant |
| Ferns: | Many kinds, all interesting |
| *Asplenium* (spleenwort) | |
| *Adiantum* (maidenhair) | |
| *Camptosorus* (walking fern) and others | |
| *Goodyera pubescens* (rattlesnake plantain) | Woodland orchid with white-veined leaves; occasionally blooms |
| *Lycopodium* (club moss) | Good ground cover; many kinds |
| *Mitchella repens* (partridgeberry) | Small round leaves with red berries; creeper |
| *Selaginella uncinata* (spikemoss) | Almost iridescent "fern" with tiny blue green leaves |

### *Terrarium Plants for Different Environments*

**1.** These plants will do well in a room with a temperature in the 21.5° C (70° F) range.[1]

*Aglaonema commutatum* l
*A. costatum* l
*Alternanthera amoena* varieties l
*Anthurium scherzerianum* l
*Asparagus plumosus* var. *nanus* m
*Begonia foliosa* m
*B. imperialis* m
*B. imperialis* var. *smaragdina* m
*B. Rex-cultorum* m
*Billbergia nutans* m
*Calathea illustris* m
*C. roseo picta* m
*C. zebrina* (grows quickly) m
*Cissus (Vitis) rhombifolia* (florists' grape ivy) l
*Oxalis spp.* m
*Peperomia obtusifolia* var. *variegata* m
*P. rotundifolia* m
*P. obtusifoila* m
*P. sandersi* m
*Philodendron cordatum* m
*Pilea microphylla* m
*P. involucrala* m
*Saxifraga sarmentosa* m
*Scindapsus pictus* var. *argyraeus* l
*Selaginella emmellana* l
*S. draussinna* var. *browni* l
*Sinningia* m
*Syagrus (Cocos) weddeliana* m

*(Continued)*

[1]Light preferences are indicated by the following symbols in the table: l = low light; m = medium light; h = high light. These plants are available from florists or commercial greenhouses.

## *TABLE 9–3 (Continued)*

*Codiaeum (Croton)* varieties (need sunshine, quick growers) h
*Cryptanthus bivittatus* m
*C. zonatus* m
*Dieffenbachia seguine* (small plants only) m
*Dracaena godseffiana* m
*D. sanderiana* l
*Ficus pumila (repens)* l
*Fittonia argyroneura* l
*Helxine spp.* l
*Maranta arundinacea* var. *variegata* l
*M. leuconeura* var. *kerchoveana* l
Mosses and lichens will also do well.
*Tradescantia spp.* l
*Zebrina pendula* l
Ferns:
*Adiantum capillus-veneris* l
*A. cuneatum* l
*Asplenium nidus* m
*Davallia bullata* m
*D. pentaphylla* m
*Nephrolepis* (small varieties) m
*Polystichum tsus-sinense* m
*Pteris cretica varieties* m

**2.** Plants for the cool room or sunporch. Winter temperature 4° to 10° C (40° to 50° F).

*Acorus gramineus* vars. *pusillus* and *variegatus*
*Buxus sempervirens* var. *suffruticosa*
*Camellia japonica*
*Coprosma baueri variegata*
*Daphne odora*
*Euonymus fortunei* varieties
*Ficus pumila*
*Hedera helix* varieties *argenteo, variegata, conglomerata,* and 'Merion beauty'
*Pittosporum tobira variegata*
*Primula obconica*
*P. sinensis*
*Pteris cretica*
*Rhododendron* and azaleas (kurume varieties in small sizes)
*Saxifraga sarmentosa* (strawberry begonia)

conditions. Some of the native carnivorous plants are endangered or threatened, and it is illegal to dig them from the wilds and sell them. The California pitcher plant *(Darlingtonia)* should not be purchased for any purpose. Its roots cannot tolerate heat above 18° C (65° F). Other carnivorous plants are ideal for a well-lighted terrarium, such as *Drosera, Pinguicula* and *Dionaea.* The tall species of *Sarracenia* (pitcher plants) do very poorly, but *S. purpurea* (purple pitcher plant) and *S. psittacina* (parrot pitcher plant) are small and should do well if given bright light in an open-top terrarium.

A carnivorous plant terrarium is best made in an open aquarium with charcoal bits and gravel in the bottom (to absorb foul odors and excess impurities), covered with a generous layer of live or reconstituted whole-fiber sphagnum moss. Keep the plants quite moist and give them as much sun as possible. Do not cover the aquarium tightly. Keep the temperature below 32° C (90° F) and the plants should thrive. Never apply fertilizer to carnivorous plants; leave the terrarium open, and the plants will lure and catch their own insects (pieces of meat are usually too concentrated and will kill the leaves). Carnivorous plants may also be grown in individual plastic pots outdoors or under lights.[1]

## PLANTING AND MAINTENANCE

Terrarium soil should be moist before adding plants. Make a well in the center of the soil mixture and add water in small amounts until the soil is moist and crumbly. Decide on plant arrangement in the terrarium before transplanting from pots. Transfer plants with soil around the roots and space them with room to grow. For narrow-topped containers, grasping-tools are necessary for moving plants. Once the plants have been placed into depressions large enough to accommodate their roots, the soil should be firm around the roots. Whenever possible, plant roots with their original pot soil around them.

To maintain humidity in the terrarium, cover the opening with clear plastic or plate glass. Bottles with openings two inches or less in diameter usually need no cover, as evaporation takes place very slowly. Knowing when to water a terrarium is very important. If the container opening is large enough, touch the soil surface to determine its moisture content. If dry, mist the planting surfaces thoroughly. Ideally, terraria should not be allowed to dry out completely but be maintained at an evenly humid environment. Some moisture condensation on the walls is a good sign. In the event of excess humidity, which promotes the growth of molds and bacteria, terraria should be ventilated by removing the cover for a few hours.

For watering, tap water may be used if allowed to stand overnight in an open container; however, unpolluted rainwater or melted snow at room temperature is best. A bulb baster is a fine water device for terraria, and a funnel may serve the purpose for narrow-necked bottles. Watering is necessary when plants begin to look limp, soil looks light colored, and there are no drops of moisture on the interior of the bottle.

To maintain a terrarium, dead or decaying vegetation should be removed as soon as possible, and luxurious growth should be pruned judiciously to prevent overcrowding. Remember that overwatering is a dangerous practice when dealing with terrarium plants just as it is with potted plants. In spite of isolation, terraria are sometimes invaded by insect pests, which may be treated with aerosol insecticides. In case of fungus mold, Benlate or Captan may be used as directed, or powdered sulfur may be used as a control measure.

Fertilizer diluted to one-quarter normal strength may be applied when necessary to keep plants properly nourished. Avoid overfertilization as the plants quickly outgrow their confined space.

---

[1]Consult the quarterly *Carnivorous Plant Newsletter* (write c/o Fullerton Arboretum, Fullerton, CA 92634) for much interesting cultural information on carnivorous plants and source lists for plants (*Newsletter* 10:1, 1981).

The type of plants grown in a terrarium determines the proper amount of light. However, in general, terraria should not be placed in full sunlight because of excess heat buildup inside the closed container. Artificial light (either fluorescent, mercury vapor, or incandescent) may be used as a substitute for natural light or as a supplement to inadequate natural light.

## Aquaria

Submerged plants are used in aquaria as oxygenators. Floating aquatic plants should be used sparingly in aquaria because an open surface is most desirable. However, the majority of floating aquatic plants are suitable only for summer culture, and the water of a house aquarium is usually deficient in mineral nutrients necessary for growth of surface or floating plants.

Do not chill the plants by putting them in unusually cold water (less than 13° C or 55° F). Use a soil layer of loam covered with sand and a thin layer of stone chips or very small pebbles. Fertilization is not usually necessary. Do not place the aquarium in deeply shaded areas or overcrowd with plants. In their native habitats, aquatic plants grow in relatively shallow water where light is more readily available than at greater depths. Propagate aquatic plants by division, cuttings, or seeds. Plants for aquaria are listed in Table 9–4.

*TABLE 9–4*
*Plants suitable for aquaria*

|  |  |
|---|---|
| *Azolla* (mosquito fern) | Small floating aquatic fern which grows rapidly in the summer |
| *Cabomba* (Washington grass, fish grass) | Good oxygenator. Temperature range 13° to 27° C (55° to 80° F) |
| *Anacharis* (elodea, waterweed) | Rapid grower. Plants need clean, lime-free water. Thrives where some sunlight is available |
| *Lemna* (duckweed) | Floating aquatic; provider of shade |
| *Ludwigia* (ludwigia) | Plants should be grown earlier in the summer and transferred to aquaria as required |
| *Myriophyllum* (parrot feather) | Requires fairly strong sunlight and lower temperatures |
| *Utricularia* (bladderwort) | Carnivorous, submerged aquatic plant; shallow pools are natural habitat |
| *Vallisneria* (eel-grass, tape-grass) | Tends to die down in winter |

## REFERENCES

Behme, R. L. 1969. *Bonsai, saekei,* and *bonkei.* New York: W. Morrow and Co.

Brooklyn Botanic Garden. 1970. *Greenhouse handbook for the amateur.* Plants and Gardens Handbook 42.

Carnivorous Plant Newsletter. 1981. Carnivorous plant sources and books. *Carnivorous Plant Newsletter 10(1):23, 25.*

*Cathey, H. M. 1973. Growing bonsai.* USDA Home and Garden Bulletin No. 206. Washington, D.C.: U.S. Government Printing Office.

Chidamian, C. 1955. *Bonsai miniature trees.* Princeton, N.J.: Van Nostrand Co.

Elbert, V., and Elbert, G. A. 1972. *Fun with terrarium gardening.* New York: Crown Publishing Co.

Evans, C. M., and La Pliner, R. 1973. *The terrarium book.* New York: Random House.

Fitch, C. M. *The complete book of terrariums.* New York: Hawthorne Books.

Hull, G. F. 1964. *Bonsai for Americans.* Garden City, N.Y.: Doubleday and Co.

Kane, M. 1981. Solar greenhouses built to last. *Organic Gardening* 28(8):52–61.

Kramer, J. 1973. *The complete book of terrarium gardening.* New York: Charles Scribner's Sons.

———. 1974. *Underwater gardens: planning your aquarium.* New York: Charles Scribner's Sons.

———. 1975. *Your homemade greenhouse and how to build it.* New York: Cornerstone Library.

McCullagh, J. C. 1978. *The solar greenhouse book.* Emmaus, Pa.: Rodale Press.

McFarlane, J. C. 1981. Measurement and reporting guidelines for plant growth chamber environments. *Plant Science Bulletin* 27(2):9–11.

McPhillips, M., ed. 1979. *The solar age resource book.* New York: Everest House.

Miller, J. 1980. Growing carnivorous plants under lights. *Carnivorous Plant Newsletter* 9(4):94, 97.

Slack, A. 1979. *Carnivorous plants.* Cambridge, Mass.: M.I.T. Press.

Scheller, W. 1977. *Successful home greenhouses: building, stocking, using.* Farmington, Mich.: Structures Publishing Co.

Schnell, D. E. 1976. Carnivorous plants of the U.S. and Canada. Winston-Salem, N.C.: J. F. Blair.

Stowell, J. P. *Bonsai: indoors and out.* Garden City, N.Y.: American Garden Guild.

Sunset Books Editorial Staff. 1965. *Bonsai.* Menlo Park, Calif.: Lane Books.

Sunset Books and Sunset Magazine. 1973. *Terrariums and miniature gardens.* Menlo Park, Calif.: Lane Books.

Swenson, A. 1977. *Cultivating carnivorous plants.* Garden City, N.Y.: Doubleday and Co.

Wikle, J. 1980. Growing bonsai under indoor lights. *Light Garden Magazine,* June–July, 131–39.

———. 1982. How to grow a bonsai like this . . . under fluorescent light. *Bonsai Journal,* Fall, pp. 60–66.

Yashiroda, K. 1960. *Bonsai: Japanese miniature trees.* Newton, Mass.: C. T. Bradford Co.

---

*Note:* The references included here are self-explanatory; they cover the basics of setting up and maintaining cold frames, greenhouses, bonsai, terraria, and aquaria.

# Natural plant communities 10

We continually go back to natural and wilderness areas to get ideas for landscaping our homes, to photograph rare and endangered species of plants, to obtain recreation and exercise, to search for edible wild foods, and simply to enjoy seeing some of the earth's magnificent beauty. We cannot help but be impressed by such plants as the ancient bristlecone pines and giant redwoods in California, towering white pines in Michigan's Keewenau Peninsula, charming calypso orchids in the white cedar swamps of northern Michigan and Canada, the famous natural rhododendron gardens high in North Carolina's Appalachians, lush subtropical forests in Florida's fragile Everglades, and the immense Douglas fir trees in the Olympic National Park's temperate rain forest of Washington state.

These areas, and others like them worldwide, are important because they contain an irreplaceable gene pool of wild species and hence can provide for us new, undiscovered plants as a source of energy, food, building materials, fiber, craft materials, and aesthetic enjoyment. By knowing something about the natural environment of a plant, we are better able to grow it in cultivation. For example, some plants are sensitive to soil pH (acid-loving heaths), some to variations in light and temperature throughout the year (temperate versus tropical plants), and others are regulated by certain environmental stimuli that induce flowering (desert plants which bloom after heavy rains).

Large, relatively undisturbed wilderness or natural areas provide a base line of environmental quality that allows us to monitor subtle changes in overall environmental conditions. They provide the opportunity for scientific study of the complex interactions between organisms, leading to a better understanding of species' adaptations to the environment. The endangered plants and animals that may be preserved are symbolic of all life that has evolved together on the planet Earth, and their survival should be taken seriously. Thus, it is important to understand why natural and wilderness areas are so fragile as ecological entities, what types of plants make up the various communities, and which plants are now considered to be rare and endangered species.

A *wilderness area* is defined as a large primitive landscape of rugged or undeveloped land and water that preserves wildlife, plants, and geological formations in an undisturbed setting and that allows people to experience the challenges of a natural environment and opportunities for solitude or escape from our crowded urban environment. A *natural area* is a relatively unspoiled piece of land that encompasses an entire ecosystem or habitat type containing a community of plants and animals which represent a typical, or often unique and unusual, association of recognized scientific, cultural and aesthetic values. A natural area, diverse in landscape, can be maintained against outside development and is capable of being managed properly. It may contain unusual geological formations or harbor rare and endangered species which cannot survive except in this environment. Natural areas may be as small as a privately owned 2.02 hectare (5 acre) wood lot in an urban setting or may be thousands of acres of state-owned land in an area where recreation, nature study, management of rare species, and scientific research can all coexist. The key to recognizing a natural area is that it contains land, water, rocks, plants, and animals worth preserving; in other words, our "natural heritage."

We have recently come to a greater awareness of efforts to maintain environmental quality and preserve significant pieces of natural landscapes for future generations. Environmental quality is a direct indicator of the quality of our own lives and the strength of our society. For the future, we must be concerned with the preservation of natural diversity today in as many ways and places as we can. Since we can never save every desirable piece of land and its living creatures, nor stop the ever-increasing destruction that civilization demands, we must be very careful to select the best representative natural areas to be set aside for scientific research, educational activities, and simply as preserves of our unique and varied natural heritage. During the past twenty years, many states have established programs to identify, inventory, select, dedicate, and maintain significant natural areas (sometimes as parks) for a variety of purposes. In addition, the Nature Conservancy, a private conservation organization, has worked extremely hard to purchase lands that are outstanding examples of our natural heritage and see to it that they are properly maintained. We can never retreat from our efforts to support the cause of conservation because it may soon be too late for some specialized habitat or endangered species. We all lose a little of ourselves whenever we lose a species with which we have shared the environment of the earth.

Many species of plants and animals are now recognized officially as being rare, threatened, or endangered. The same applies to wilderness and natural areas. Preserving such plants and animals and the areas in which they thrive is essential. They must not only be preserved; they must also be maintained as such, so that they are not destroyed in the long run. Processes such as natural succession must be allowed to continue. Human pressures must be minimized in natural and wilderness areas. If we save these areas with their diverse plant communities and the rare and endangered plants and animals that exist in them, we shall have all these creatures around for future generations to enjoy.

The following outlines, prepared by the state of North Carolina and the state of Illinois are examples of efforts to produce guidelines for the establishment and maintenance of natural areas.

## Principles Governing the Establishment, Extension, and Development of State Park Natural Areas of the State of North Carolina: 1963[1]

*Purpose*

The purpose of the North Carolina State Parks Natural Areas System shall be:

TO SERVE THE PEOPLE OF NORTH CAROLINA AND THEIR VISITORS by:

1. Preserving and protecting natural areas of scientific, aesthetic, or geologic value not only for the knowledge and inspiration of the present generation; but also, for generations to come.
2. Portraying and interpreting plant and animal life, geology, and all other natural features and processes in the various Natural Areas.

*Quality*

Sites selected as Natural Areas shall be those which (1) best portray the natural processes that have formed the earth and its plant and animal life; (2) portray some specific natural process so dramatically and arrestingly as to be unique and of sufficient importance to be worthy of statewide interest; (3) contain some outstanding examples of native plant and animal communities or other outstanding sufficient natural objects, conditions, and phenomena.

*Maintenance and Protection*

In order to protect the quality of Natural Areas, the basic principle of maintenance will be to preserve the area in its natural condition.

Paths or trails will be opened only in a manner compatible with the basic principle of maintenance.

---

[1]Adopted by the Board of Conservation and Development, 23 April 1963.

No roads will be constructed, opened, or maintained except those required for protection and minimum maintenance.

No buildings of any sort will be erected except those required for interpretation, protection, and minimum maintenance.

Agricultural operations of any kind will not be permitted.

Natural Areas, being units of the Division of State Parks, will be subject to State Park Regulations as authorized and directed by Chapter 113, Section 35, of the General Statutes of North Carolina.

*Size*

Sites selected as Natural Areas shall be large enough to (1) completely include the natural features the area is established to preserve and protect; (2) provide sufficient buffer area to protect the natural features from outside influences or encroachments; (3) permit the development of interpretive devices if these can be provided without damage or impairment of the primary purpose of preserving the natural features.

*Location*

Intrinsic values will determine the location of Natural Areas and they must be located where existing values justify their acquisition regardless of geographic location in relation to population.

*Access*

The boundaries of all Natural Areas should be accessible over public roads or byways.

*Cost*

The cost of land should be reasonable, considering values in the section of the State in which the Natural Area site is located. No property shall be purchased or accepted as a gift unless it meets all the requirements herein set forth for establishing Natural Areas and involves no commitments, privileges, or conditions except a condition requiring that the property be used only as a Natural Area.

*Development*

Natural Areas shall receive only such development as is necessary to preserve and protect their natural values, protect public health, and provide adequate interpretive programs. All developments shall be planned and executed so as to in no way impair, damage, or detract from the natural values for which the areas were established to preserve and protect. Development of recreational and public use facilities which provide for organized sports or contests, swimming, camping, picnicking, and the like shall not be provided in Natural Areas.

## Objectives of the Illinois Natural Preserve System

*Acquisition*

1) To preserve adequate samples of all natural land types occurring in the state.
2) To preserve natural areas in all portions of the state.
3) To preserve unique and outstanding natural areas.
4) To preserve wilderness remnants.

5) To preserve habitats for rare or endangered species of plants or animals.

*Maintenance*

6) To provide perpetual protection for natural preserves against external intrusions.
7) To provide management of nature preserves which will assure their perpetual maintenance as nearly as may be in their natural condition.

*Records*

8) To provide for the accumulation of knowledge concerning features and conditions within nature preserves.

*Use*

9) To allow and facilitate the conduct of research studies in nature preserves in such manner and to such degree as will not modify natural conditions.
10) To allow and facilitate the visiting of nature preserves for purposes of observation and study for education and pleasure in such manner and to such degree as will not modify natural conditions.
11) To provide for the interpretation of nature preserves to visitors to enhance their understanding and enjoyment.

# Basic ecological concepts

## ORGANISMS LIVING TOGETHER

Plants typically live in *communities*, varying from magnificent clones of aspen and sumac to large swamps of mangrove, relatively homogeneous communities. Others are more mixed communities: beech-maple forests, oak-hickory forests, beach grass-juniper dune formations, cattail-sedge marshes, and others. These communities are denoted by the dominant plant species in the association. There are, of course, many other important species present, but in smaller numbers or biomass. A *community*, then, may be defined as a group of species living together in a relatively distinctive association under a similar set of influential environmental conditions such as soil type, soil pH, soil moisture level, temperature, and so forth. Communities may be relatively uniform, or they may be ever changing.

What happens in plant communities? They are clearly not static entities. One basic, dynamic process is called *succession*. In disturbed agricultural or urban areas, or in land newly exposed such as roadsides, volcanic areas, and hillsides after landslides, *pioneer species* such as mosses, lichens, and weeds become established on the bare soil. These are followed by other species of perennial herbs, shrubs, then trees. In a given area the continual change in species composition through time is called *succession*. The final stage is called a *climax* community of plants. Examples of climax forests are oak-hickory or beech-maple associations in northern temperate forests. Theoretically, in a climax community, a condition of relative stability is achieved for a given set of environmental factors, especially moisture and temperature and soil type, and the species composition appears to remain the same over a very long period of time.

Within a plant community, we can observe many interesting types of relationships which help determine the exact composition of the community. One factor is *competition,* where plants in association with each other compete for light, water, and mineral nutrients. In old farm fields black mustard may outcompete barley or wheat, and in your garden, lambs-quarters may take over lettuce, peas, and radishes. Most wild plants have evolved adaptations to help them avoid or cope with competition. However, in the final analysis, some species may be more aggressive and competitively successful than others at reproducing and finding niches. Those successful species will become more common, while others may become rare.

Another relationship is called *mutualism,* where two species of plants coexist, each one benefiting from the other. Examples include *mycorrhizae* (fungi growing in association with the roots of vascular plants), *lichens* (algal and fungal species living together as one plant body), and *symbionts* (such as blue-green algae or certain species of bacteria living in association with plant roots or leaves and fixing nitrogen). Both mutual partners benefit in terms of water, nutrition, light, or structural support. In mycorrhiza, for example, the fungus obtains water and some nutrients from the roots. The roots, in turn, derive a far greater nutrient supply as the fungus breaks down organic matter in the soil and transports it as usable compounds into the plant cells.

*Commensualism* is a relationship where one member of the association definitely benefits, while the other receives no benefit but is relatively unharmed. *Epiphytes,* plants that grow upon other plants, are good examples of commensualism. Some epiphytes are Spanish moss on live oak trees in the southeastern United States; lianas or vines growing on forest trees; and ferns, orchids, and bromeliads growing on tropical forest trees. In the tropics whole communities of ecologically similar plants exist as epiphytes, literally covering the branches of tall rain forest trees. These epiphytes are "air plants," deriving their nourishment from the air, from chemical-laden rainwater that drips from the leaves of the "host," and from decaying organic matter that accumulates around their root masses. We should emphasize that epiphytes are not parasites. At worst, the trees in a rain forest become top-heavy from their "load" of epiphytes and are thus more susceptible to being blown down by the wind.

A definitely one-sided relationship in plant communities is called *parasitism,* where one member such as fungus, virus, bacterium, or higher plant derives its nutrition at the expense of the other member, the host. Parasite examples are disease-causing microorganisms, gall insects, and vascular plants such as dodder, mistletoe, witch-weed *(Striga),* and broomrape *(Orobanche).* The latter two types can be serious economic pests of agricultural crops in the southeastern United States and Europe, respectively. When an organism grows in nonliving plant or animal material, it is called a *saprophyte.* Mushrooms (fungi) growing on dead limbs or tree trunks, or in the rich organic matter of the forest floor are typical examples.

In plant communities, we find *food chains* operating. The green plants in all types of food chains are called the *primary producers,* as they are the only organisms capable of utilizing the sun's energy to produce carbon compounds (sugars) from inorganic raw materials ($CO_2$ and $H_2O$). Green plants are, in turn, eaten by animals (herbivores) which are eaten by other animals (carnivores). In such systems composed of the different trophic levels of primary producers, primary consumers, and secondary consumers there is approximately a 90 percent loss in energy in each move "up" to the next trophic, or consumer, level. Also, pesticides such as DDT accumulate in increasing amounts at successively higher trophic levels. Thus, carnivorous birds, especially falcons and eagles, can accumulate extremely high levels of pesticides in their fatty tissue. Pesticide accumulation is responsible for the death of many birds at the "top" of the food chain.

Primary producers also accumulate toxic compounds, especially heavy metals such as mercury and lead. This capability of "luxury consumption" by plants of heavy toxic metals is being used today to depollute our rivers, lakes, and waterways. The classic example is water hyacinth *(Eichhornia crassipes)* which grows as an introduced aquatic weed in the southern United States.

## Types of plant communities

### FOREST COMMUNITIES IN TROPICAL AND TEMPERATE REGIONS

Tropical forests (Figures 10–1 and 10–2) can be extremely diverse in numbers of species, while forests of temperate regions (Figure 10–3) contain relatively few dominant tree species. In temperate regions of the world, much of the original forest has disappeared due to agriculture, lumbering, grazing, firewood cutting, and urbanization. Some of the forest land that was cut over for fuel or lumber has been replaced by secondary or tertiary growth of new trees through succession, regrowth of stump sprouts, or through reforestation programs. Thus, today there are only small relic stands of original climax forests in national parks, state parks, private natural and wilderness areas, and some national and state forests.

The same type of destruction of primeval forests is going on today in the tropics worldwide at an alarming rate. One of the largest tropical forest ecosystems is in the Amazon basin in Brazil. It is here that agricultural, industrial, and urban "development" is threatening the very existence of these tropical forests. It is estimated that at the present rate, the tropical forests in the Amazon could be wiped out within thirty years! The effects on rainfall, plant-animal interactions, and human existence would be catastrophic. The tropical soils are very fragile, and forests do not regenerate as easily as in the temperate regions. Ironically, tropical soils are very poor under intensive long-term agricultural production.

Fortunately, some wilderness or natural areas have been set aside for preservation, preventing their destruction and allowing us and future generations to enjoy them. More worldwide forests in wilderness and natural areas must be set aside and preserved. Many forested areas have been set aside but are not maintained to be preserved in their natural condition. The pressures of maintenance are immense. Some forested areas are overused

*FIGURE 10–1*
Secondary growth in a diverse tropical forest. (Monteverde Biological Reserve, Costa Rica)

*FIGURE 10–2*
Humid tropical cloud forest. Notice the lush growth of mosses and other epiphytes on the dwarf trees in this high elevation rain forest. Monteverde Biological Reserve, Costa Rica. *(Photographed by K. S. Walter)*

*FIGURE 10–3*
Beech-maple woods. Note the lush ground cover of spring herbs characteristic of such rich forests. (Irwin's Woods, Washtenaw County, Michigan)

by campers and backpackers. Some forests have multiple uses, such as snowmobiling, selective cutting, camping, and picnicking. Some evidence the destruction caused by man-made forest fires. We can add to this list the forests in North India, overcut for firewood, and land in North Africa, overgrazed. In the United States, all of the above uses, abuses, and destruction have occurred to an excessive extent.

The solutions for this overuse and overexploitation include the following: increasing the number of desirable areas available for use to spread out the impact; reforestation and proper management; limiting recreational uses of forests by controlling the numbers of visitors; limiting the kinds of multiple-use activities that are allowed; prosecuting violators who inadvertently or willfully cause destruction of forested areas; and increasing our use of alternative energy sources such as wind, geothermal, photovoltaic, methane, and solar. We must also try to prevent the introduction of disease-causing pathogens that could destroy certain tree species. Already we have witnessed the destruction caused by American chestnut blight and Dutch elm disease. These two devastating diseases were inadvertently introduced into America on imported plants just after the turn of the century. Their presence was not recognized until it was too late to stop their spread.

In the northeastern United States our forests have been divided into three basic categories based on the most typical dominant tree species:

1. The beech-maple forest type (American beech and sugar maple). Occurs on the most moist (mesic) sites and has the richest collection of spring wild flowers (Figures 10–3 and 10–4).
2. The oak-hickory forest type (red oaks, chestnut oak, and various hickories). Occurs on drier sites and may contain shrubs and some understory herbs.
3. The evergreen spruce-fir forests of high elevations in the Appalachians and to the north in Canada (boreal forests) (Figure 10–5). Forests are dominated by evergreen species.

In all of these forests, there may occur many other species besides the characteristic dominants. For example, white ash, basswood, black cherry, tulip poplar, and white oak are commonly found in the beech-maple association. Dogwood, sassafras, black gum, black oak, and red maple may be seen in the oak-hickory forest. Birches, cherries, and maples are to be found as deciduous species in the spruce-fir communities.

## ALPINE COMMUNITIES: KRUMMHOLZ AND TUNDRA

Alpine communities contain plants that have very short growing seasons and have become adapted to very harsh and changing environments (Figures 10–6 and 10–7). The most important environmental factor for identifying alpine

*FIGURE 10–4*
Beech-maple woods herbaceous ground cover plants, including rue anemone *(Anemonella thalictroides)* right and trout lily *(Erythronium americanum)*. (Horner Woods, the University of Michigan Matthaei Botanical Gardens)

*FIGURE 10–5*
Eastern Appalachian deciduous forest. Vegetation here exhibits a transition from rich moist coves at lower elevations, to drier slopes, to spruce-fir forests at higher elevations. (Great Smoky Mountain National Park, S.E. of Clingman's Dome)

*FIGURE 10–6*
Subalpine meadow; trees shown are Engelman spruce *(Picea engelmannii).* Evergreen conifers share this rugged terrain with a lush herbaceous flora. (Mt. Audubon, Rocky Mountains, Colorado)

*FIGURE 10–7*
Krummholz alpine community. Harsh environment and drying winds prevent these trees from ever growing more than a few feet tall at elevations above 3,000 meters (10,000 feet). (View to east plains from Mt. Audubon, Rocky Mountains, Colorado)

regions is the low temperatures year-round. Tundra is characterized as an undulating, treeless plain, above timberline, that is typical of arctic, subarctic, and mountainous areas worldwide. Krummholz occurs at and just below timberline in mountain alpine areas and is characterized by having stunted, windswept woody vegetation (Figure 10–7). Both types of areas are very fragile and can easily be destroyed by fire, disease, overgrazing, and avalanches.

## PRAIRIE COMMUNITIES

Prairie communities once dominated the midwestern United States but are now largely lost to agricultural, urban, and industrial development. Prairie regions were grasslands, defined by an average annual rainfall of between ten and thirty inches. However, a few relic prairies remain along railroad rights of way and in a number of recognized natural areas. One large prairie still exists at Ft. Custer near Battle Creek, Michigan, a former military base where minimal disturbance has occurred in the fields. And, one can find plants that are characteristic of prairies in old field communities throughout the midwest, anywhere that soil moisture is the limiting factor.

What are these plants that we call prairie indicator species? The dominant plants of a prairie are grasses with certain indicator or "marker" species (Figure 10–8). The short-grass prairies of the western great plains are characterized by the presence of buffalo grass *(Buchloe dactyloides)*, blue grama *(Bouteloua gracilis)*, and hairy grass *(Bouteloua hirsuta)*. The tall-grass prairies of the midwest contain big bluestem *(Andropogon gerardii)*, little bluestem *(Andropogon scoparius)*, Indian grass *(Sorgastrum nutans)*, smooth dropseed *(Sporobolus heterolepis)*, tall panic grass *(Panicum virgatum)*, slough grass *(Spartina michauxiana)*, and porcupine grass *(Stipa spartea)* (Figure 10–9).

Many nongrass types of flowering plants are also considered to be prairie indicator species, such as the locoweed *(Astragalus)*, rattlesnake master *(Eryngium)*, colicroot *(Veronicastrum)*, puccoons *(Lithospermum)*, prairie dock *(Silphium)*, lupine *(Lupinus)*, false indigo *(Baptisia)*, and brown-eyed susan *(Rudbeckia)*. Many of these nongrasses are in the pea family *(Fabaceae)* and the aster family *(Asteraceae)*. We should mention that in order to maintain a prairie, the land must experience periodic burning. The short- and tall-grass prairies of the great plains and midwest were periodically burned as a result of lightening-caused fires. Today, prairies are maintained as "managed tracts," by burning-over every year, under controlled conditions.

## COASTAL COMMUNITIES

Marine habitats, like fresh water habitats, are very fragile. They have been destroyed by oil dumping, chemical wastes, garbage, dredging, development of marinas and housing, herbicides, and erosion. Plants growing in marine

*FIGURE 10–8*
Prairie grass "marker" species, blue grama grass *(Bouteloua gracilis).* (Guernsey Campground, Wyoming)

*FIGURE 10–9*
Larry Mellichamp holding prairie grass "marker" species, big bluestem *(Andropogon gerardii)* together with the giant dicot, rosinweed *(Silphium),* another prairie plant. (Photographed at the University of Michigan Matthaei Botanical Gardens by K. S. Walter)

environments are important because they are among the producers at the base of the food chain, create a stable environment for planktonic organisms, and provide a breeding ground for marine life. For us, nonvascular marine plants such as red algae and kelps are important sources of food, potash fertilizers, iodine, agar, emulsion stabilizers, and insulation (diatomaceous earth). Marine communities also are used for aquaculture or mariculture.

The pressure of people on marine habitats is increasing. Most of the world populations live near or on sea coasts. Many marine habitats have been utterly destroyed where people live or have their recreation spots. It is only by setting aside and permitting no further development of still existing areas that they will be saved. In the United States, California, Oregon, Washington,

Maine, and the Carolinas have done a great deal to save their remaining marine coastlines (Figure 10–10). One area in the world where an entire marine ecosystem of mangrove was wiped out by herbicides (agent orange and cacodylic acid) is in South Vietnam. It is estimated that it will be at least 200 years before this ecosystem will become reestablished, and many changes can occur during that time. A typical red mangrove coastal swamp exists along the coasts of Florida (Figure 10–11).

*FIGURE 10–10*
Salt marsh grass vegetation along coastline of North Carolina. (Figure Eight Island)

*FIGURE 10–11*
Red mangrove swamp *(Rhizophora mangle).* Note the abundant production of aerial roots holding the main body of the plants above the water line. (Everglades National Park, Florida)

## LAKE COMMUNITIES

Lake communities are marked by having aquatic plants growing around their fringes (emergent aquatics), on their surfaces (floating aquatics), and below the water (submerged aquatics). Many of these plants, such as the emergent aquatic plants, cattail *(Typha)* and crack willow *(Salix fragilis)*, and the submerged aquatic plants, pondweed (*Potamogeton* spp.) and *Vallisneria*, are efficient at removing heavy metal pollutants from the waters of lakes. This was shown at Saginaw Bay, Michigan, one of the most polluted areas of the Great Lakes. Aquatic plant communities are also important because they are significant habitats for birds and other wildlife. They also serve as breeding areas for fish and microscopic plankton at the base of the aquatic food chain and a source of edible wild plants (cattail, arrowhead, watercress) in nonpolluted waters. It is interesting that lakes with various types of aquatic plants are now being used to remove pollutants from sewage (as is being done for East Lansing, Michigan). The species of diatoms that grow in particular lakes can tell us the extent of pollution by nitrogen and phosphorus compounds. Nitrogen and sulfur oxides form, respectively, nitric and sulfuric acids and, of course, most of us are now acquainted with the drastic effects of acid rain from snow, dust, and air pollution, on plants and aquatic animal life in lakes of Scandinavia, the Adirondacks of New York, and Ontario. Now, you can appreciate the fact that our lakes are indeed extremely fragile ecosystems.

## BOG AND FEN COMMUNITIES

A bog is an inadequately drained area rich in organic plant remains, usually acidic (pH range of 4 to 5), frequently surrounding a body of open water, and having a characteristic flora. Examples of plants are pitcher plants, sundews, sphagnum moss, black spruce, tamarack, cranberries, and other members of the heath family *(Ericaceae)*. A bog starts out as a small glacial lake. A mature bog mat, composed mainly of sphagnum moss (Fig. 10–12).

*FIGURE 10–12*
Bog plant, sphagnum moss (*Sphagnum* sp.). This plant is responsible for creating the organic, acid environment characteristic of the bog. It provides the substrate in which many other species of plants grow. (Mud Lake Bog, Washtenaw County, Michigan)

grasses, and sedges, usually forms around the margin of the lake and gradually fills in over the open water (Figures 10–13 and 10–14). As decomposed plant material accumulates (due to high acidity and anaerobic conditions) and sinks to the bottom, a more or less solid bed of organic matter forms. Natural succession then brings about invasion of the mat by shrubs and trees to establish a new forest community on relatively solid ground. This may take thousands of years.

In contrast to a bog, a fen consists of low, moist land, partly or wholly covered by water, is usually alkaline in pH, and has a characteristic flora. Examples of plants are sedges, potentilla, gentians, valerians, and algal stoneworts *(Chara, Nitella).*

*FIGURE 10–13*
Livingston Bog, the University of Michigan Biological Station, Pellston, Michigan. Notice the transition from open water, onto the narrow bog mat with shrubs, to the surrounding forest dominated by conifers (spruce, larch, and *Thuja*).

*FIGURE 10–14*
Leatherleaf Bog *(Chamaedaphne calyculata),* near Traverse City, Michigan. An evergreen shrub bog completely overgrowing an old filled-in pond.

Both bog and fen types of communities are very fragile. Excessive tramping around them destroys the substrate and many of the plants. Both communities contain many rare and endangered species of plants such as gentians, terrestrial orchids, and some of the insectivorous plants (pitcher plants, butterwort, sundews) (Figures 10–15 and 10–16). Collecting these plants in bogs and fens is discouraged. Already, many of these rare plants have disappeared due to collecting for commercial purposes. Seeds, however, can be collected and plants grown to become available to collectors. The terrestrial native orchids are almost impossible to grow from seeds and cannot live without their fungal mycorrhizal associations.

*FIGURE 10–15*
Bog plant, pitcher plant *(Sarracenia purpurea)* growing with leather-leaf and cranberry. These plants thrive in the acidic, moist, nitrogen-poor, open environment of the sphagnum bog. (Mud Lake Bog, Washtenaw County, Michigan)

*Figure 10–16*
Bog plant, pitcher plant *(Sarracenia minor),* showing both flowers and leaves. The southern bogs are more like moist, grassy meadows than the sphagnum bogs of the north. Both are moist, acidic, and open (few trees) and so provide a suitable habitat for similar plants such as carnivorous plants, orchids, and members of the heath family *(Ericaceae).* (Southern North Carolina)

## SWAMP AND MARSH COMMUNITIES

Swamps and marshes are wet places, where the water table is at or above the ground surface for much of the year. Swamps are characterized by forest communities dominated by tree species such as black ash, red maple, and yellow birch. Marshes are communities dominated by grasses and sedges or low shrubs (no trees), such as cattails, sedges, giant reed, and willows. They are frequently associated with lakes, rivers, streams, and springs. These habitats are very important for the survival of water fowl, amphibians, turtles, reptiles such as alligators and water snakes, many insects, and microscopic creatures. The plants in swamps are also very characteristic, as witnessed by the baldcypress swamps of the southeastern United States, cattail-sedge swamps of the Great Lakes region, tule (a sedge) swamps in California, and sawgrass marshes in the Florida Everglades. Their fragility is not due to the delicate nature of the plants (which are quite tough), but to the fact that their natural environment is so drastically altered when the wetlands are drained for commercial purposes. Swamps and marshes can be wiped out by a long-term lowering of the water table as a result of drought, drainage, and overuse of water (Figures 10–17 and 10–18).

*FIGURE 10–17*
Swamp community, Cedar Lake, Michigan. A swamp is a wet woods dominated by characteristic trees and shrubs. (Photographed by J. G. Bruce)

*FIGURE 10–18*
Freshwater marsh. Graminoids such as cattails, sedges, rushes, and grasses dominate this vast area of shallow water that is too wet for trees. (Whitefish Point, Lake Superior, Michigan)

## DESERT COMMUNITIES

Desert communities are very interesting ecosystems. They contain plants which are drought-resistant or drought-escaping (Figure 10–19). *Drought-resistant* plants are able to withstand water stress because of a combination of certain features which may include deep root systems, small leaf surface, sunken stomata, and succulent water-storing roots, leaves, and/or stems. Desert plants which drop their leaves during dry periods and produce new crops of leaves during wet seasons are called *drought-escaping* plants. Those desert annuals whose seeds germinate only after heavy spring rains and complete their life cycle by flowering and producing seeds in a matter of a few weeks during the rainy season are also called drought-escaping. During the dry season, these plants survive by dormant seeds present in the sandy soils of the desert.

Desert communities are very fragile. They contain dozens of rare and endangered species of cacti and other types of plants. These are in danger of being lost due to overcollecting by specialty collectors or commercial dealers. Probably more than any other group of plants, cacti and succulents are overexploited. Their odd shapes, extreme age, and ability to withstand harsh shipping and handling conditions have contributed to this predicament. The deserts of the southwestern United States, Mexico, and South Africa are particularly rich in endangered succulents. International laws now protect dozens of species from being transported out of their native lands. Whole desert

*FIGURE 10–19*
Desert community. This community is dominated by cacti and other xerophytes. Lack of water, high light intensity, and extreme heat are the limiting factors in this environment. Plants have developed water conservation measures over their long period of evolution. (Sonoran Desert Museum near Tucson, Arizona)

communities are being destroyed by subdivisions, strip mining, highways, and off-road vehicles. It is fortunate that a few of our desert communities have been set aside for permanent preservation, such as in the Saguaro National Monument and Organ Pipe Cactus National Monument, Arizona, and Big Bend National Park, Texas.

## Practical plant ecology

In a practical sense, learning about natural plant communities shows that plants have preferences when it comes to environmental conditions. For example, many ferns grow best in the shade, dogwoods in part shade, and geraniums in full sun. In addition, bald cypress and black ash like "wet feet"; whereas other plants, such as strawberries, melons, peaches, and grapes, prefer well-drained, porous soils. If you wonder about soil or light preferences for cultivated plants, it is useful to see where the wild progenitors of our garden plants grow best in their native habitats.

It is important to know that the seedling stage is the most critical time in the plant's life cycle because while becoming established, it is most vulnerable to environmental conditions, predation, and competition. The seed must land in a suitable habitat and be able to grow in order to become a mature plant. Thus, plants occur in nature only where their seeds are dispersed and seedlings can grow! This explains many plant distribution patterns in relationship to moisture or temperature tolerances. For example, baldcypress *(Taxodium)*, sweet gum *(Liquidambar)*, rhododendrons, silver bell *(Halesia)*, shortia *(Shortia galacifolia)*, and many other well-known northern garden plants grow wild as native plants in the southeastern United States but not in the northeastern region. However, they do quite well in northern climates, far away from their native homes, when transplanted as saplings or mature specimens or are carefully grown from seed in nurseries. They can grow in cold climates, but not as seedlings! Of course, other factors, such as moisture and soil type, must be considered when explaining why we can grow a plant where it would not grow on its own. The controlling factors are usually a combination of conditions.

There are also examples of plant species that, when successfully established as garden plants far from their native haunts, reproduce and become locally abundant as if they were naturally growing there. Their spread may be due to establishment of seedlings or vegetative propagation. Such plants include goats-beard *(Aruncus)*, some ferns, many alpine plants grown in rock gardens, and many wildflowers like columbine *(Aquilegia)*, coneflower *(Echinacea)*, sundrops *(Oenothera)*, and phlox. Occasionally cultivated plants are said to become "pests" when they vigorously reproduce in the garden and overgrow and outcompete the less aggressive types. Some well-known examples of garden plants which are aggressive to the point of taking over all available space at the expense of more delicate species include: bishop's weed *(Aegopodium podagraria)*, lily-of-the-valley *(Convallaria majalis)*, horsetail and scouring rush *(Equisetum arvense* and *E. hyemale)*, English ivy *(Hedera helix)*, day lily *(Hemerocallis fulva)*, Japanese honeysuckle *(Lonicera japonica*, a devastating pest in native vegetation whenever it escapes cultivation), mints (*Mentha* species), Mayapple *(Podophyllum peltatum)*, bracken fern *(Pteridium aquilinum)*, lungwort *(Pulmonaria officinalis)*, and alder-buckthorn (*Rhamnus frangula*, which was introduced for use as a hedge plant but has escaped and is rapidly becoming a serious pest in natural areas in the northeastern U.S.), and moneywort *(Lysimachia nummularia)*. Also, we cannot forget kudzu *(Pueraria lobata)*, that aggressive vine introduced into the southern United States in 1876 as an ornamental plant. During the Great Depression it was widely planted throughout the South as an erosion control. It holds the soil, adds nitrogen (it is a legume), and provides fodder for cattle. However, it grows so fast that it takes over the field, forest, and home. It is difficult to manage and has fallen into disfavor since the late 1940s. Today, while considered nothing more than a blight on the countryside, little effort is being made to subdue or destroy it in most areas.

Many of our houseplants that have not been highly bred come directly from propagules obtained from plants collected from the wild in the tropics. If it is not possible to make firsthand observations or to find information in books, then it is best to consult a good field botanist or a gardener who knows plant habitat preferences. With houseplants, you quickly learn that coleus needs some light, but not too much; that grape ivy can grow in the shade of north windows; and that cacti and succulents obviously need sun. Those who cultivate aroids, bromeliads, and ferns know that some of them grow best in fairly shady locations with high atmospheric humidity. Connoisseurs of insectivorous plants such as Venus flytrap and pitcher plants know that they grow best in sphagnum moss and that they should be grown in full sun and watered with distilled water. The temperate pitcher plants and sundews need to experience a cold period during the winter months just as in their native habitats. (An annual six to eight weeks of dormant rest at 4° C or 40° F is sufficient to rejuvenate them and keep them growing for years.) Winter-hardy bonsai of conifer and deciduous plants must be left outdoors during the fall and chilled during the winter months (the dormant period) if they are to grow normally again in the spring.

From studying plant communities, we learn that in some ecological niches plants have mycorrhizal (fungus-root) associations in order to grow best. Examples are orchids, members of the heath family (blueberries and rhododendrons), and many forest trees. We have learned that unsterilized forest humus is a good source of these fungi, which form a mutualistic association with plant roots. The fungus helps make nutrients from the soil available to the plant roots, and thus provides a necessary function that allows the plant to thrive. We are just beginning to understand the significance of this widespread phenomenon.

## REFERENCES

Braun, E. Lucy. 1950. *Deciduous forests of eastern North America.* Philadelphia: Blakiston Co. (A classic on the ecology of eastern North America.)

Carpenter, P. L.; Walker, T. D.; and Lanphear, P. O. 1975. *Plants in the landscape.* San Francisco: W. H. Freeman and Co. (A landscape architecture textbook.)

Curtis, J. T. 1959. *The vegetation of Wisconsin.* Madison, Wisc.: University of Wisconsin Press. (Classification of habitats and vegetation.)

Gleason, H. A., and Cronquist, A. 1964. *The natural geography of plants.* New York: Columbia University Press. (An interesting and authoritative account of the biology of plants and their adaptations to the environment; excellent text and photographs of the major types of vegetation in North America. Highly recommended.)

Humke, J. W.; Tindall, B. S.; Jenkins, R. E.; Wieting, H. L., Jr.; and Lukowski, M. S. 1975. The preservation of natural diversity: a survey and recommendations. Arlington, Va.: The Nature Conservancy. (A conservation proposal.)

Kaufman, P. B., and LaCroix, J. D. 1979. *Plants, people, and environment.* New York: Macmillan Co. (Diverse topics on plants and their uses.)

Prance, G. T., and Elias, T. S., eds. 1977. *Extinction is forever.* Millbrook, N.Y.: New York Botanical Garden. (Collection of articles on the conservation of endangered plants.)

Wells, J. R.; Kaufman, P. B.; and Jones, J. D. 1980. Heavy metal contents in some macrophytes from Saginaw Bay (Lake Huron, USA). *Aquatic Botany* 9:185–93.

Winberry, John J., and Jones, David M. 1973. Rise and decline of the "miracle vine": kudzu in the southern landscape. *Southeastern Geographer* 13:61–68. (An interesting account of the history and impact of kudzu.)

# Edible wild plants 11

Even before civilizations had cultivated food plants, edible wild plants had been staple items in man's diet. The American Indians relied on locally available native plants as far back as their cultures have been known. The first European settlers in the Americas relied, to a large extent, on native plants for their survival in the wilderness.

Since then, people in Appalachia have made widespread use of native wild plants for sustenance. And now more people have a renewed interest in edible native plants, as well as the poisonous ones. This interest has gone hand in hand with our attempt to regain our roots in the natural environment by appreciating and saving wilderness and natural areas, in organic gardening outdoors, and in growing plants indoors. In this chapter we take a close look at edible wild plants—where you can find them, when to collect them, which parts are or are not edible, and some of their culinary uses.

Euell Gibbons probably rekindled our interest in edible wild plants with his famous book, *Stalking the Wild Asparagus* (1962). In fact this charming, folksy book marks the beginning of the "back-to-nature" movement of the early 1960s. However, scientific and popular writings dealing with edible wild plants in America go back even further—to M. L. Fernald (1943), one of the country's foremost botanists, and O. P. Medsger (1940). These workers synthesized the scattered knowledge about edible wild plants and made it available to the general public. The uses of wild plants as herbal remedies by European settlers date back as far as the seventeenth and eighteenth centuries in this country.

Why are edible wild foods in such fashion today? Let's list a few reasons:

1. They are excellent snack foods.
2. They serve as survival foods.
3. They are not contaminated with food additives or pesticide poisons (if collected in pesticide-free areas).
4. Many have especially high levels of vitamins and mineral nutrients.
5. They "spice up" your diet, providing some new alternatives to white bread, wilted lettuce, polluted potatoes, and putrid peas!
6. They allow you to try new recipes.
7. They are a source of teas and other beverages, including fermented drinks.
8. They cost less.
9. They are fun to collect.
10. They help you "get back to nature." All of our cultivated food plants began as edible wild plants.

## Habitats for edible wild plants

Edible wild foods can be found in a variety of habitats (Figures 11–1, 11–2, and 11–3). The primary ones[1] include:

1. Railroad banks and roadsides.
2. Old fields and meadows, unweeded gardens, newly abandoned farmland, old orchards.
3. Deciduous forests such as oak-hickory and beech-maple woods.
4. Marshes and swamps.
5. Streams, lakes, ponds, and bogs.

### PRECAUTIONS

A word of caution should be said about these sites. Do not collect edible wild plants to eat in any areas that have been sprayed with insecticides, fungicides, or herbicides. Do not collect in areas known to be polluted with heavy metals (as in wet habitats) or other industrial pollutants (PBB = polybrominated biphenyls, PCB = polychlorinated biphenyls, and PCP = pentachlorophenol). Dusty roadsides and oil-covered railroad embankments are also places to avoid. Gain permission to collect on private property. Last year's dried stalks indicating a rich bed of wild asparagus may not be available to a passer-by.

---

[1]See Weatherbee in Kaufman and LaCroix 1979.

*FIGURE 11–1*
Edible wild plant habitat along railroad tracks in a prairie community. (Pierce Road near Waterloo Recreation Area, Washtenaw County, Michigan)

*FIGURE 11–2*
Edible wild plant habitat, roadside bank. Here are seen "wineberries" *(Rubus phoenicolasius)* and various weeds. (Photographed near Lancaster, Pa.)

*FIGURE 11–3*
Edible wild plant habitat, old field. (Sanford Field, the University of Michigan Matthaei Botanical Gardens)

Identifying, collecting, and utilizing wild plants as supplements to your routine diet can be an interesting and rewarding activity. It would be misleading, however, to assume that you can obtain large quantities of free food from the wild with little work. Collecting and preparing the plants for consumption are enjoyable but can be hard work and take a considerable amount of time.

Also, do not expect to be able to just walk out and identify all edible plants easily, distinguishing them from poisonous plants. It can be very dangerous to eat something that you are not absolutely sure is safe. In fact, there have been cases of people getting sick from eating normally safe plant parts which were contaminated with herbicides, or other chemicals, or were an unusual strain of plant. Learning to identify wild plants and their edible parts requires time, patience, practice, and above all, experience. Cultivated food plants have poisonous parts which we avoid and normally do not think about such as tomato and potato leaves and potato tubers which have turned green. The same is true of edible wild plants.

The best ways to learn about wild plants and what and when their parts are edible are to consult a good book with clear illustrations,[2] take a course, have guidance from an authority, and simply begin to look at wild plants as food sources. *Be very careful that you identify all plants, plant parts, and mushrooms accurately.* Fatal mistakes occur every year when people think a certain plant is edible and it is not or when they prepare plants improperly.

These precautions are not meant to scare you away, only to warn you to proceed with care (but with no less enthusiasm), take your time, develop your skills, and learn from others. The rewards will be well worth the effort.

## Seasonal collecting

Edible wild plants can be collected year-round at most latitudes, but in temperate regions, the best collecting times are in spring and autumn (Figure 11–4). After the snow melts in the spring, start by harvesting the tender parts of dandelions, watercress, mints, and wild garlic. By late April and May, you can gather pokeweed and cattail shoots, young nettles, fern fiddleheads, milkweed shoots, dandelion flowers for wine, and wild strawberries. Maple sap is obtained earlier for boiling down into maple syrup and sugar. In some areas, this is morel mushroom collecting time, when droves of people are out scouring the woods for these delicacies (Figure 11–5).

With summer, you can go right to your garden to harvest greens, such as lamb's quarters, rough pigweed, and purslane. This is the time to harvest enough, both to eat fresh or cooked and to freeze for winter use. Young plants are the best to collect. Later in summer, it is time to harvest young milkweed pods, elderberry flowers, monarda or bee balm for tea, and young cattail spikes to roast.

---

[2]Peterson 1977; Weatherbee and Bruce 1979.

*FIGURE 11–4*
Spring edible wild plants class near Plymouth, Michigan in May.

*FIGURE 11–5*
Spring edible wild mushrooms, morels *(Morchella esculenta)* collected in North Carolina in May.

Fall brings a plethora of edible wild foods to the landscape. There are hog peanut tubers, Jerusalem artichoke tubers, oak acorns, black walnuts, hickory nuts, rose hips, wild apples, arrowhead tubers, giant puffballs, elderberries, blueberries, cranberries, May apples, and wintergreen berries. Our students in Practical Botany at the University of Michigan prepare over 100 different dishes and beverages from edible wild plants collected in the fall. Furthermore, they only collect in safe, unpolluted areas and practice conservation by not overcollecting in any given area. Ellen Elliot Weatherbee, noted authority on edible wild foods, teaches courses at the University of Michigan Botanical Gardens. During the midwinter course she and her students find fruits, nuts, and other plant parts that are edible and can be used as survival foods. During three-day camp-outs in northern Michigan during January, the students use these foods for sustenance.

## Nutritional values

Edible wild plants are just as nutritious as cultivated plants, and some are superior as far as certain minerals and vitamins are concerned (Table 11–1). Lamb's quarters and dandelions are very high in calcium. Dandelions, lamb's quarters, mints, nettles, pokeweed, and violet leaves are high in vitamin A. Vitamin C is relatively high in highbush cranberry fruits and violet blossoms, and very high in strawberry leaves and violet leaves. One further benefit is that many edible wild plants are low in calories, per serving only 15 calories for purslane, 20 for pokeweed, 20 for asparagus, and 33 for dandelions. The Jerusalem artichoke is high in the storage carbohydrate, fructan (made up of units of the sugar, D-fructose), in contrast to potatoes which contain starch (made up of units of the sugar, D-glucose). Both are easily digested by humans and serve as rich carbohydrate sources. Finally, seeds of sunflowers are excellent sources of fats, while seeds of legumes are among the richest sources of protein (30 to 70 percent by weight). In contrast, seeds of grasses are good carbohydrate (starch) sources and only fair sources of proteins (6 to 16 percent by weight).

*TABLE 11–1*
*Nutritional values of wild plants and some domestic foods*

| | ***Calories*** *100 Grams of Edible Portion*[1] |
|---|---|
| Hickory nuts | 673 |
| Walnuts | 654 |
| Chocolate cake | 339 |
| Burdock | 94 |
| Black raspberries | 73 |
| Pears | 61 |
| Apples | 58 |
| Dandelion greens | 45 (33) |
| Lamb's quarters | 43 (32) |
| Peppergrass | 41 |
| Amaranth | 36 |
| Sheep sorrell | 28 (19) |
| Asparagus | 26 (20) |
| Spinach | 26 (23) |
| Pokeweed shoots | (20) must be cooked |
| Purslane | 21 (15) |
| Elderberries | 21 |
| Chicory greens | 20 |
| Watercress | 19 |
| Lettuce | 18 |
| Jerusalem artichokes range from 7 cal. when freshly harvested to 75 cal. after being stored for "a while" | |

[1]All figures given are for 100 grams of edible portion. Figures in the column at right are for cooked portions.
*Source:* USDA. *Composition of foods.* Agriculture Handbook, No. 8. A few figures are from Gibbons' *Stalking the Healthful Herbs.*

| **Vitamin C** *Milligrams/100 Grams of Edible Portion* | |
|---|---|
| Strawberry leaves | 229 |
| Violet leaves | 210 |
| Wintercress buds | 163 |
| Wintercress leaves | 152 |
| Violet flowers | 150 |
| Mustard greens | 102 |
| Highbush cranberries | 100 |
| Catnip leaves | 83 |
| Pokeweed shoots (cooked) | 82 |
| Amaranth | 80 |
| Lamb's quarters | 80 |
| Watercress | 79 |
| Nettles | 76 |
| Strawberries | 60 |
| Spinach | 51 |
| Oranges | 50 |
| Dandelion greens | 35 |
| Purslane | 25 |
| Lettuce | 18 |

| **Vitamin A** *International Units/100 Grams of Edible Portion* | |
|---|---|
| Lamb liver (broiled) | 74,500 |
| Dandelions (cooked) | 11,700 |
| Carrots (raw) | 11,000 |
| Lamb's quarters | 9,700 |
| Pokeweed (cooked) | 8,700 |
| Violet leaves | 8,258 |
| Spinach (cooked) | 8,100 |
| Nettles | 6,566 |
| Amaranth (raw) | 6,100 |
| Wintercress leaves | 5,067 |
| Elderberries | 3,300 |
| Purslane (cooked) | 2,100 |
| Highbush cranberry | 2,100 |
| Lettuce | 1,900 |
| Green peppers | 420 |

| **Calcium** *Milligrams/100 Grams of Edible Portion* | |
|---|---|
| Lamb's quarters | 309 |
| Amaranth | 262 |
| Mustard greens | 220 |
| Dandelion greens | 187 |
| Watercress | 151 |
| Milk | 120 |
| Purslane | 103 |
| Spinach | 93 |
| Lettuce | 68 |
| Burdock | 60 |
| Pokeweed shoots (cooked) | 53 |

*(Continued)*

*TABLE 11–1 (Continued)*
*Nutrition values of wild plants and some cultivated counterparts*

| **Iron** *Milligrams/100 Grams of Edible Portion* | |
|---|---|
| Pork liver (fried) | 29.1 |
| Amaranth (raw) | 3.9 |
| Purslane (raw) | 3.5 |
| Raisins | 3.5 |
| Dandelion greens (raw) | 3.1 |
| Spinach (raw) | 3.1 |
| Dandelion greens (cooked) | 1.8 |
| Pokeweed shoots (cooked) | 1.7 |
| Elderberries | 1.7 |
| Watercress | 1.7 |
| Lettuce | 1.4 |
| Purslane (cooked) | 1.2 |
| Lamb's quarters (raw) | 1.2 |
| Lamb's quarters (cooked) | .7 |
| **Protein** *Grams/100 Grams of Edible Portion* | |
| Lamb liver (broiled) | 32.3 |
| Walnuts | 15.0 |
| Nettles | 6.9 |
| Lamb's quarters (raw) | 4.2 |
| Amaranth | 3.5 |
| Lamb's quarters (cooked) | 3.2 |
| Pokeweed shoots (cooked) | 2.5 |
| Jerusalem artichokes (raw) | 2.3 |
| Watercress (raw) | 2.2 |
| Purslane (raw) | 1.7 |
| Purslane (cooked) | 1.2 |
| Lettuce | 1.3 |
| Oranges | 1.0 |

## Uses of edible wild plants

### TEA PLANTS

Leaves for tea are best collected when they are just fully expanded, not when they are old and mature. Fruits for tea are collected when they are mature and ripe. The collected leaves and fruit can be dried on newspaper or in a special fruit drier and then stored for later use in glass or plastic bottles with tight lids. When mint plants are actively growing, the leaves can be collected fresh and placed in hot water to steep. Strawberry leaves make a delicious tea, but must be crinkly dry before using. Collect fuzzy clusters of red sumac berries after a hard autumn frost. Then wash and place them in cheesecloth or muslin and tie closed. This strains out the hairs and other parts while the berries are steeping in warm water to extract the flavorful constituents. The tart "sumac-ade" or "rhus juice" can be sweetened with honey, chilled, and served like lemonade.

Many of the teas made from edible wild plants have been found to be useful as "spring tonics," caffeine-free tea substitutes, and even "herb teas" (Figure 11–6). Some common edible wild plants used to make teas are listed in Table 11–2. Several cultivated plants can also be used to make teas other than the tea of commerce *(Camellia sinensis).* These include the dried leaves of the cultivated raspberry and the fresh leaves of violets. Bark from the roots of the sassafras tree *(Sassafras albidum),* so often used to make tea, is no longer recommended because of its documented carcinogenic effects.

*FIGURE 11–6*
Wild tea plants: mint, sassafras, juniper, sumac, elderberry.

## *TABLE 11–2*
## *Edible wild plants used to make teas*

| *Name of Plant* | *Part Used* |
|---|---|
| Wild strawberry *(Fragaria virginiana)* | Leaves (dried) |
| Wild rose *(Rosa)* | Fruits (hips) |
| Violets *(Viola)* | Leaves |
| Clover *(Trifolium)* | Flower heads |
| Bee balm *(Monarda)* | Leaves |
| Mints *(Mentha)* | Leaves |
| Staghorn sumac *(Rhus typhina)* | Berries |
| Labrador tea *(Ledum)* | Leaves |
| Mexican tea *(Chenopodium ambrosioides)* | Leaves |
| Basswood *(Tilia)* | Flowers |
| Wintergreen *(Gaultheria procumbens)* | Leaves |

## POTHERBS

Potherbs are plants, edible wild or cultivated, cooked as green vegetables. Many wild edibles used for this purpose are what we recognize as weeds in the garden! This includes pigweed, lamb's quarters, purslane, and nettles. In Amerindian cultures, these weeds are respected and remain in the garden for harvesting as edible wild foods.[3] Today, we, too, are beginning to recognize that these are valuable plants worth saving for harvest. They are valuable as rich sources of vitamins or essential mineral nutrients.[4] Nettles and lamb's quarters are very high in Vitamin A, and both lamb's quarters and pigweed contain large amounts of calcium. Even dandelion greens are very rich in Vitamin A.

Lamb's quarters can be substituted for spinach in any recipe. But a few of the potherbs must be treated in a special way before they can be eaten. Milkweed shoots are very bitter if eaten after a single boiling in water because they contain latex. To make palatable, place the shoots in boiling water for one minute, then pour off the water. This is then repeated two more times. After this the shoots should no longer be bitter to the taste. Another potherb is pokeweed whose mature shoots contain a poison which must be avoided. The roots and berries also contain copious amounts of this toxin. But young shoots are low in toxin. To be safe, boil the young shoots in several waters to remove any toxin that may be present. A word of caution then, only the young shoots of pokeweed can be eaten, those which have just emerged from the ground; all other parts of the plant, *the roots, fruits, and mature shoots should not be harvested.*

A number of the potherbs can be harvested when young in summer, then frozen for winter use. Especially good for this purpose are lamb's quarters, purslane, and nettles. Nettles have stinging hairs, so wear gloves when harvesting. Boiling in water removes the irritants (histamine, serotonin, and acetyl choline) that occur in the stinging hairs of nettle shoots. Table 11–3 lists a number of common edible wild plants that are useful as potherbs.

*TABLE 11–3*
*Edible wild plants used as potherbs*

| *Name of Plant* | *Part Used* |
|---|---|
| Lamb's quarters *(Chenopodium album)* | Young shoots |
| Rough pigweed *(Amaranthus retroflexus)* | Young shoots |
| Common nettle *(Urtica dioica)* | Young shoots |
| Purslane *(Portulaca oleracea)* | Young shoots |
| Pokeweed *(Phytolacca americana)* | Young shoots only |
| Asparagus *(Asparagus officinalis)* | Young shoots |
| Milkweed *(Asclepias syriaca)* | Young shoots |
| | Young pods |
| Russian thistle *(Salsola kali* var. *tenuifolia)* | Young seedlings |

[3]See Ford in Kaufman and LaCroix 1979.

[4]See Weatherbee in Kaufman and LaCroix 1979, Table 17–1.

## BERRIES

Edible wild plants are the source of many delicious fruits which can be eaten fresh, frozen, or used to make wines. They are perhaps best known as snack foods for the hiker and backpacker. When in season, many fruits such as black raspberries or wild blueberries are collected in abundance to eat fresh. Many of these fruits from edible wild plants have different and better flavors than their cultivated counterparts. Further, they can be very rich in vitamins. Wild strawberries, for example, are high in vitamin C. One of the nicest treats is the wild blackberry, both the large-fruited one (fruits two to three centimeters long) and the small-fruited one (fruits one to two centimeters long). With the addition of sugar to the small fruits of the wild blackberry, you can make the best pie ever tasted.

Remember that the fruits of some plants are poisonous. Berries that may be edible to birds may be poisonous to humans, so do not use this as a guide. This includes pokeweed berries, which can be used as a source of ink or natural plant dye, but should *never* be eaten! Table 11–4 lists a number of relatively common edible wild plants that yield excellent ripe fruits, which can be eaten fresh or made into jams and wines.

## SALAD PLANTS

Throughout the growing season, you can collect edible wild plants for salads. This includes greens, starchy tubers or tuberous roots, and flavorings. In winter, when these wild edibles are either impossible or difficult to obtain, you can switch to sprouts (seedlings) of cultivated alfalfa, mung bean, sunflower, and wheat by growing them indoors.

Collect edible wild plants for salads from sites where no pesticides have been used. Very commonly, herbicides are used in lawn areas or along roadsides; avoid these sites. The same is true for collecting water-growing wild edibles such as cattail and watercress; avoid polluted ditches, streams, rivers, and lakes. Even when you know that the site is safe for collecting, the plant

*TABLE 11–4*
*Edible wild plants that yield excellent fruits*

| *Name of Plant* | *Part Used* |
|---|---|
| Wintergreen *(Gaultheria procumbens)* | Fruits |
| Wild strawberry *(Fragaria virginiana)* | Fruits |
| May apple *(Podophyllum peltatum)* (*all parts are poisonous,* except ripe fruits) | Ripe fruits *only* |
| Blueberry *(Vaccinium)* | Fruits |
| Bog cranberry *(Vaccinium macrocarpon)* | Fruits |
| Thimbleberry *(Rubus parviflorus)* | Fruits |
| Black raspberry *(Rubus allegheniensis)* | Fruits |
| Elderberry *(Sambucus canadensis)* | Fruits |
| Highbush cranberry *(Viburnum trilobum)* | Fruits |
| Red mulberry *(Morus rubra)* | Fruits |

TABLE 11–5
*Edible wild plants that provide especially good salad materials*

| Name of Plant | Part Used |
|---|---|
| Violet *(Viola)* | Leaves, flowers |
| Jerusalem artichoke *(Helianthus tuberosus)* | Tubers |
| Groundnut *(Apois americana)* | Tubers |
| Watercress *(Nasturtium officinale)* | Young shoots |
| Dandelion *(Taraxacum officinale)* | Young shoot rosettes in early spring or late fall (with tight flower buds) |
| Cattail *(Typha)* | Young white shoots (outer green leaves removed) |
| Peppercress *(Lepidium)* | Young leafy shoots |
| Sheep sorrel *(Rumex acetosella)* | Young leafy shoots |

parts should be thoroughly washed several times in water. Table 11–5 lists some of the common wild edibles that are especially good in salads.

## PLANTS FOR OTHER CULINARY USES

Additional uses of edible wild plants are in soups, as desserts, as snack foods, and in stews. We mention only a few examples. Some snack foods are beechnuts *(Fagus)*, hazelnuts *(Corylus)*, hickory nuts *(Carya)*, black walnuts *(Juglans nigra)*, chestnuts *(Castanea)*, as well as candied underground stems of wild ginger *(Asarum canadense)*. Wild plants used in soups are wild garlic, wild onions, and wild leeks *(Allium)*, as well as wild asparagus *(Asparagus officinalis)* and nettles *(Urtica dioica)*. In stews use Jerusalem artichoke *(Helianthus tuberosus)*, groundnut *(Apios americana)*, and edible mushrooms such as giant puffball *(Calvatia gigantea)*, morels *(Morchella)*, honey mushroom *(Armillaria mellea)*, and oyster mushroom *(Pleurotus ostreatus)*. Desserts are made from berries listed in Table 11–4. Finally, we should mention that a number of edible wild plants are important in making breads: oak acorns *(Quercus)* cattail *(Typha)* pollen, wild rice grains *(Zizania aquatica)* and grains of many other wild grasses, and blueberries *(Vaccinium)* in muffins. Violet flowers make a colorful, delicate jam.

# REFERENCES

Angier, B. 1969. *Feasting free on wild edibles.* Harrisburg, Pa.: Stackpole Books.

Berglund, B., and Bolsby, C. E. 1971. *The edible wild.* New York: Charles Scribner's Sons.

Brackett, B., and Lash, M. A. 1975. *The wild gourmet.* Boston: Godine Press.

Elliot, D. B. 1976. *Roots: an underground botany and forager's guide.* Old Greenwich, Conn.: Chatham Press. (Excellent drawings)

Fernald, M. L., and Kinsey, A. C. 1958. *Edible wild plants of eastern North America.* New York: David McKay Co. (A classic)

Gibbons, E. 1962. *Stalking the wild asparagus.* New York: David McKay Co.

———. 1966. *Stalking the healthful herbs.* New York: David McKay Co.

Hall, A. 1973. *The wild food trailguide.* New York: Holt, Rinehart, and Winston.

Harrington, H. D. 1976. *Edible native plants of the Rocky Mountains.* Albuquerque: University of New Mexico Press.

Hatfield, A. W. 1971. *How to enjoy your weeds.* New York: Macmillan Co.

Harris, B. C. 1971. *Eat the weeds.* Barre, Mass.: Barre Publishing Co.

Kaufman, P. B. and LaCroix, J. D. 1979. *Plants, people and environment.* New York: Macmillan Publishing Co.

Kingsbury, J. M. 1964. *Poisonous plants of the U.S. and Canada.* Englewood Cliffs, N. J.: Prentice-Hall.

Kingsbury, J. M. 1965. *Deadly harvest.* New York: Holt, Rinehart, and Winston.

Kirk, D. R. 1970. *Edible wild plants of the western United States.* Healdsburg, Calif.: Naturegraph Pub.

Martin, A. C.; Zim, H. S.; and Nelson, A. L. 1961. *American wildlife and plants: a guide to wildlife food habitats.* Reprint. New York: Dover.

McKenny, M. 1971. *The savory wild mushrooms.* Seattle, Wash.: University of Washington Press.

Medsger, O. P. 1940. *Edible wild plants.* New York: Macmillan Co.

Peterson, L. 1977. *A field guide to edible wild plants.* Boston: Houghton Mifflin Co.

Smith, A. H. 1963. *The mushroom hunter's field guide.* Ann Arbor, Mich.: University of Michigan Press.

Tatum, B. J. 1976. *Billy Joe Tatum's wild foods cookbook and field guide.* New York: Workman Publishing Co.

Weatherbee, E. E. 1982. Myth and fact in determining edibility of wild plants. *The Michigan Botanist* 21(3):99–102.

Weatherbee, E. E., and Bruce, J. G. 1979. *Edible wild plants of the Great Lakes region.* P.O. Box 8253, Ann Arbor, Michigan, 48107. (Excellent guide)

---

*Note:* The references included here are self-explanatory; they cover the basics of foraging for edible wild foods and explain how to prepare them in the field or the kitchen.

# Spice, herb, and drug plants 12

Although both spices and herbs are used for flavoring, and some think of herbs as a group of spices, a distinction can be made between these plants. Spices are usually tropically grown, being either woody, perennial shrubs or trees. The commercial product is generally derived from roots, bark, flowers, fruits, or seeds, and the plants are difficult to grow in home gardens. Herbs, on the other hand, are more often temperate plants, soft stemmed or nonwoody (herbaceous), and may be annuals, biennials, or perennials; it is usually the leaves and flowers that contain an aroma or flavor, and they are easily grown in home gardens.

## The role of spices in world trade

Spices have played a major role in world history and commerce; and it would be difficult, if not impossible, to pinpoint the exact moment in time when man first realized how valuable were these gifts of nature. We traditionally think of them as additives to our food—flavor enhancers—that do much to increase our enjoyment of food, and for many centuries they have been used in that way. However, the value of spices has never been limited to the kitchen, and history records that they probably were first used in perfumes and to disguise and cover up the many unpleasant odors that people encountered as they went about their daily lives.

The ancient Egyptians used sweet and strong smelling spices and herbs in their elaborate burial rituals. The bodies of the pharaohs and other nobles were treated with such substances, probably to counteract the odor of putrefaction that would so quickly develop in that hot climate. Spices and herbs were also buried along with the corpse in the tomb so that the departed individual could enjoy them in his other "afterlife." Even in the Bible we find reference made to the use of these aromatic substances as in the annointing of the body of Christ with spices and herbs before He was laid in the tomb. In the account of His birth we read of the journey of the three Wise Men from the East who brought gifts of frankincense and myrrh.

The body odors of life as well as those associated with death have also been masked by the use of such spices as cinnamon, cloves, anise, and cassia. Ancient Chinese courtiers would chew on cloves when speaking to the emperor so as not to offend him with foul mouth odors; and the early Romans as well as Europeans of the Middle Ages would use heavy applications of pungent spices and herbs to negate body odors. Today we rely on soap and water, but we can certainly draw a parallel between modern and early times in our use of fragrant soaps and body and mouth deodorizers that derive their pleasant aroma from spices and herbs.

Certain aphrodisiac powers have also been attributed to spices and herbs by earlier civilizations, and we find that they have been used in religious ceremonies by various cultures through the years. Because of their great value, they have even been used as money—the invading Goths demanded 1,350 kilograms of black pepper from the Romans as part of a ransom in A.D. 408, and peppercorns served as payment for rent and taxes in medieval England.

Spices and herbs, when added to food, did more than improve taste and cut the bland flavor that was characteristic of much of what was consumed during the Middle Ages and before. They also served to make palatable food that was long past its prime. These aromatic additives covered the strong, foul odors of meats and other foods that spoiled quickly before man had refrigeration.

The impact of spices on the political history of mankind cannot be understated. Countries have risen and fallen; wars have been fought; great personal fortunes have been made; and the explorations of the fifteenth and sixteenth centuries occurred because of the demand for these highly prized "bounties of nature."

The Arabs supplied the ancient Egyptians, Romans, and Greeks with spices and, even as late as the fourteenth century, they were the chief providers of these desired products to the Western World. They carefully guarded their sources and knowledge of these plants and even went so far as to spread false rumors as to their origin. The early spice trade centered in China, Malaysia, and India, and although other countries tried to break the Arabian dominance of this lucrative trade, it was not until after Marco Polo's expedition to the Far East (1271–1297) that the door was opened to competition from Venice and Genoa.

In the fifteenth century the great sea powers, Portugal and Spain, entered into the period of exploration and quest for spices that sent out such men as Ferdinand Magellan, Vasco da Gama, Bartholomeu Diaz, Hernán Cortez, and Christopher Columbus. Columbus failed to find the spice lands but did discover America. For one hundred years Portugal held sway over the spice trade because of its dominance of the Indian Ocean, even extending control of this trade as far as China. All the spices from the East Indies were carried on Portuguese ships.

By the late sixteenth and early seventeenth centuries, Holland and England were challenging Portugal's command of the spice trade, and they sought to gain their share of the vast wealth emanating from this commerce. After Sir Francis Drake reached the spice lands of the East Indies in 1579 and destroyed the Spanish Armada in 1588, the English established the famous East India Company and soon controlled India and the Spice Islands. The Dutch formed the Dutch East India Company and, for a short time, monopolized the spice trade. Eventually, England and Holland divided the spoils between themselves, and their colonial empires were to last well into the twentieth century.

The spice trade also did much to enrich the young United States of America, and such New England seacoast towns as New Bedford and Boston gained much from trade with the Orient. Elihu Yale, founder of Yale University, made his fortune through the spice trade.

Although interest in spices and herbs has waned periodically, there is currently a renewed fascination with these plants. In 1978, U.S. imports of condiments, seasonings, and flavoring materials reached record levels with a total tonnage of 149,506 metric tons and a total value of $197 million. Contrary to popular opinion, the U.S. also exports substantial amounts of spices, although it is far from being a major spice-producing country. In 1979, approximately 11,000 metric tons, valued at nearly $20 million, were exported. Canada was the largest consumer, taking almost $3.4 million, which was approximately 18 percent of the total value. United States' exports of condiments and seasonings include black and white pepper, red peppers, cassia and cinnamon, mustard flour and prepared mustard, and mustard seed.

## Common spices

Although the list of spices is a lengthy one, this discussion will center on the most important ones, based on current volume of world trade (Table 12–1).

Allspice is derived from the unripe, dried fruits (berries) of *Pimenta dioica,* a small tree, a relative of eucalyptus, and very common in parts of South and Central America and the West Indies; the major producing country is Jamaica. It is an ingredient in pickles, meats, baked goods, and drinks including some liqueurs; the oil obtained from the leaves and fruits is used in perfumes. Allspice, capsicum peppers, and vanilla are the only important spices originally native in the New World (Western Hemisphere).

Capsicum (red pepper) is obtained from the fruits (which change in color from green to yellow to red) of a species of *Capsicum.* Different varieties supply us with pimento, used in cheeses and stuffed olives, and chili

*TABLE 12–1*
*Some common spices and their geographic sources*

| *Common Name* | *Scientific Name* | *Family* | *Part* | *Major Geographic Source* |
|---|---|---|---|---|
| Allspice | *Pimenta dioica* | Myrtle | Fruit | Jamaica |
| Capsicum (Red pepper) | *Capsicum* | Nightshade | Fruit | Mexico, India, China, Pakistan |
| Caraway | *Carum carvi* | Carrot | Seed | Europe |
| Cassia | *Cinnamomum cassia* | Laurel | Bark | Indonesia |
| Cinnamon | *C. zeylanicum* | Laurel | Bark | Sri Lanka |
| Cloves | *Syzygium aromaticum* | Myrtle | Flower Bud | Madagascar |
| Cumin | *Cuminum cyminum* | Carrot | Seed | India, Iran |
| Ginger | *Zingiber officinale* | Ginger | Rhizome | India, Fiji, China |
| Mace | *Myristica fragrans* | Nutmeg | Aril | Indonesia, Grenada |
| Mustard | *Brassica* | Mustard | Seed | Canada |
| Nutmeg | *Myristica fragrans* | Nutmeg | Seed | Indonesia, Grenada |
| Pepper (Black) | *Piper nigrum* | Pepper | Fruit | Indonesia, Malaysia, Brazil, India |
| Vanilla | *Vanilla planifolia* | Orchíd | Fruit | Madagascar |

powder, a common ingredient in pizza, tacos, some meats including sausage, and the famous or infamous (depending on the amount of spice used) concoction well-known in Texas and other parts of the country, chili.

Caraway *(Carum carvi)* produces spicy, aromatic seeds used in baked goods, such as certain kinds of rye bread and the liqueur, kummel. In 1978–79, the United States imported over 6,500 metric tons of caraway seed, primarily from the Netherlands.

Cassia (cinnamon) and cinnamon (true), derived from *Cinnamomum cassia* and *C. zeylanicum*, respectively, are important and much utilized spices so common in bakery products. Quills of cinnamon are formed from the rolled, inner bark which is allowed to dry. Most of the cinnamon imported into the U.S. is cassia (Figure 12–1).

Cloves are the dried, unopened flower buds of the clove tree *(Syzygium aromaticum)*. They are added to ham for flavor and to preserves such as canned pears and other fruits. Much of the clove supply is used in the production of Indonesian kretek cigarettes. It is also important as a clearing agent in the preparation of plant tissues.

Cumin *(Cuminum cyminum)* is a small annual, approximately 30 centimeters in height, with aromatic seeds. It is a common constituent, along with other spices, of curry powder which is used as a seasoning.

Ginger is the most important spice obtained from a rootstock (also referred to as an underground stem or rhizome) of *Zingiber officinale*, which is also related by family to another spice, cardamom. Ginger is widely cultivated, mainly in India, and has many uses such as an ingredient in candy, beverages, and curry powder.

Mustard *(Brassica)* is an annual which produces both black mustard seeds that are small, dark brown with a yellow interior, and pungent; and white mustard seeds that are larger, lighter brown with a white interior and less pungent. Mustard has many culinary uses including the seasoning of pickles and meat. Although cultivated in the U.S., nearly 60,000 metric tons of seed were imported from Canada during 1978–79.

Nutmeg and mace are products of the fruit of the 6- to 18-meter tropical tree, *Myristica* (Figure 12–2). The hard, inner seed, the nutmeg, is surrounded by the red, fleshy covering (aril) which is the mace. Both provide flavoring and fragrance. Nutmeg and mace contain the chemical, myristicin, which is toxic in large amounts. Large doses of ground nutmeg produce a narcotic (hallucinogenic) effect. The product, Mace, contains synthetic tear-gas (chloroacetophenone) in an aerosol propellant. Capsaicin, a chemical in cayenne pepper *(Capsicum frutescens)*, can be a strong irritant and has been used in antiattack devices.

*FIGURE 12–1*
From left to right, true cinnamon *(Cinnamomum zeylanicum)*, vanilla *(Vanilla planifolia)* and nutmeg and mace *(Myristica fragrans)* as offered for sale in the market place. *(Courtesy of J. F. Matthews)*

*FIGURE 12–2*
A branch with fruits of the tropical tree, nutmeg *(Myristica)*. The seed of this fruit is the source of the spice, nutmeg, and the red fleshy covering (aril) is the source of mace, another spice. *(Courtesy of R. W. Read, Smithsonian Institution)*

Pepper *(Piper nigrum)*, one of the major spices of the world, produces fruits, actually drupes, from its perennial vine. Harvested unripe fruits are fermented for a few days and then are dried to yield black pepper. The dried seeds of ripe fruits are the source of white pepper. World exports of this spice reached approximately 120,000 metric tons in 1978.

Vanilla *(Vanilla planifolia)* is a vine (Figure 12–3) in the orchid family (Orchidaceae). It produces spice-containing fruits which, because of their appearance, are often called pods or "vanilla beans" but technically are capsules. The fermented pods develop the chemicals which impart the vanilla aroma and flavor. Vanilla flavoring in ice cream, liqueurs, and puddings is enjoyed by many people.

*FIGURE 12–3*
(A) Flower and leaves of vanilla *(Vanilla planifolia)*. (B) Capsule fruits are the source of the spice, vanilla. *(Courtesy of T. L. Mellichamp)*

## Interest in spices and herbs

While millions of people around the world are starving to death, others are planting home gardens to beat rising food prices. In addition, increasing numbers desire food that tastes good and is "good for you." The news media constantly remind us of the advantages of natural foods and the dangers of artificial food additives. For these and other reasons, the interest in spices, and particularly herbs, including growing and harvesting your own, has seldom been greater than it is today. The movement of people to suburban and rural areas, the resultant additional space, the use of these plants in landscaping, earlier retirement, and even the social pressure to have a hobby, have, at least partly, resulted in the current love affair with herbs and their culture.

## Uses of herbs

In addition to providing flavoring, fragrance, and some medicinal needs, herbs can also be used for landscaping, both indoors and outdoors.

Even with lack of outdoor space and in an unfavorable climate, herbs can be grown indoors, some to provide greenery and fragrance during the winter months, and others to serve as tasty additions to foods. In fact, a small windowsill garden can be the source of your favorite herb tea. Since most require maximum light, providing artificial light may be required for successful indoor herb culture.

Even if only limited outdoor space is available, small, but beautiful and productive, herb gardens can become a reality. These need not be large, formal gardens. Indeed, rather than having a separate herb garden, herbs might be used to fill some empty spaces in the vegetable garden and serve as the major components of a rock garden. Since many herbs spread freely, they are often used as ground covers. Some herbs tend to grow slightly taller, and they make ideal borders and edgings. Yet larger herbs are shrublike, and, with the proper training and pruning, can be used as hedges. Container plants can be grown indoors near a sunny window or outdoors on the patio or deck.

An added fringe benefit of growing herbs is that some will control pests, such as insects (see Chapter 8). This is especially useful in food gardens and around outdoor living areas. Around the food preparation area, bay leaves seem to serve as a natural control for cockroaches. If one is social-minded, one might invite friends to view the herb tea garden and sample some herb tea while enjoying the fragrances emanating from the garden. On the other hand, if one tends to be antisocial, then one might consider ringing one's property with a large herb "bee garden."

## Growing herbs

Herbs are inexpensive, relatively easy to grow, and do not require one's constant attention; they are fun to grow.

The propagataion of herbs, with a few exceptions, is not a difficult task. The customary methods include the use of seeds, young plants, cuttings, division, or layering. Seeds and young plants can be purchased from local garden centers, seed companies, or herb specialists (Table 12–2). The seeds and seedlings of some species are more difficult to germinate and transplant than others. One very economical way of increasing the variety of herbs in the garden is to exchange root and stem cuttings with other herb fanciers. Frequency of dividing plants depends on the rate of growth of the particular species. Propagation by means of layering, in which a stem devel-

*TABLE 12–2*
*Herb suppliers in the United States*

1. Caprilands Herb Farm, Silver Street, North Coventry, CT. 06238.
2. Carroll Gardens, P.O. Box 310, Westminster, MD. 21157.
3. Clyde Robin Seed Co., Inc., P.O. Box 2855, Castro Valley, CA. 94546.
4. Fox Hill Farm, P.O. Box 7, Parma, MI. 49269.
5. George W. Park Seed Co., Inc., Greenwood, SC. 29647.
6. Hemlock Hill Herb Farm, Hemlock Hill Road, Litchfield, CT. 06759.
7. Herbst Bros., Seedsmen, Inc., 1000 Main Street, Brewster, NY. 10509.
8. Logee's Greenhouses, 55 North Street, Danielson, CT. 06239.
9. Taylor's Herb Gardens, Inc., 1535 Lone Oak Rd., Vista, CA. 92083.
10. W. Atlee Burpee Company, Warminster, PA. 18991.
11. Well-Sweep Herb Farm, 317 Mt. Bethel Road, Port Murray, NJ. 07865.

ops roots after making contact with the soil, is not as common as other means of propagation (Table 12–3).

Generally, most herbs are sun loving and prefer a slightly fertile, low acidic to low alkaline (pH 6.0 to 7.5) sandy loam soil that is well drained. Maintenance of herbs is not excessive. Pruning or division is required for the more aggressive growers, and protection from harsh winters and occasional pests may be necessary for some.

Even the amateur herb grower should be aware of the numerous public herb gardens across the country (Figure 12–4 and Table 12–4), including the National Herb Gardens at the U.S. National Arboretum in Washington, D.C., and of the existence of the Herb Society of America in Boston, Massachusetts.

*TABLE 12–3*
*Some common herbs: their names and families, parts harvested, uses, and culture*

| *Common Name* | *Scientific Name* | *Family* | *Part* | *Use* | *Annual Biennial Perennial* | *Propagation* | *Indoor Outdoor* |
|---|---|---|---|---|---|---|---|
| Angelica | *Angelica archangelica* | Carrot | Roots, stems, leafstalks, leaves, seeds | Food, perfume | B | Seed | O |
| Anise | *Pimpinella* anisum | Carrot | Leaves, seeds | Food, perfume, medicine | A | Seed | O |
| Bay | *Laurus nobilis* | Laurel | Leaves | Food | P | Cuttings | I/O |
| Basil | *Ocimum basilicum* | Mint | Leaves | Food | A | Seed | I/O |

## TABLE 12–3 Continued

| Common Name | Scientific Name | Family | Part | Use | Annual Biennial Perennial | Propagation | Indoor Outdoor |
|---|---|---|---|---|---|---|---|
| Catnip (Catmint) | *Nepeta cataria* | Mint | Leaves | Food, attract cats | P | Seed, cuttings, division | I/O |
| Chives | *Allium schoeno-prasum* | Lily | Leaves | Food | P | Division, seed | I/O |
| Dill | *Anethum graveolens* | Carrot | Leaves, seeds | Food | A | Seed | I/O |
| Fennel | *Foeniculum vulgare* | Carrot | Leaves, seeds | Food | A/P | Seed | O |
| Hyssop | *Hyssopus officinalis* | Mint | Leaves | Food, perfume, medicine | P | Seed, cuttings, division | O |
| Lavender | *Lavandula officinalis* | Mint | Leaves, flowers | Food, fragrance | P | Seed, cuttings | I/O |
| Marjoram | *Origanum majorana* | Mint | Leaves, seeds | Food, perfume | A/P | Seed, cuttings, layering | I/O |
| Mint | *Mentha* | Mint | Leaves | Food, perfume, fragrance, medicine | P | Cuttings, division, layering | I/O |
| Oregano | *Origanum vulgare* | Mint | Leaves | Food, perfume, medicine | A/P | Seed, cuttings, division | I/O |
| Parsley | *Petroselinum crispum* | Carrot | Leaves | Food | A/B | Seed | I/O |
| Rosemary | *Rosmarinus officinalis* | Mint | Leaves, flowers | Food, fragrance, perfume, medicine | P | Seed, cuttings, division, layering | I/O |
| Sage | *Salvia officinalis* | Mint | Leaves | Food, perfume | P | Seed, cuttings, division, layering | I/O |
| Savory | *Satureia hortensis* | Mint | Leaves | Food | A/P | Seed, cuttings, division, layering | I/O |
| Tarragon | *Artemisia dracun-culus* | Aster | Leaves | Food | A/P | Cuttings, division | I/O |
| Thyme | *Thymus vulgaris* | Mint | Leaves, flowers | Food, perfume, medicine | P | Seed, cuttings, division | I/O |

*FIGURE 12–4*
Cranbrook House Herb Garden located in Bloomfield Hills, Michigan. *(Photograph courtesy of Mrs. Roger Jamison)*

*TABLE 12–4*
*Herb gardens in the United States*

California. Strybing Arboretum, 9th and Lincoln Way, San Francisco, CA.
Colorado. Denver Botanic Garden, Denver, CO.
Connecticut. Caprilands Herb Farm, North Coventry, CT.
Henry Whitfield State Historical Museum, Old Whitfield Street, Guilford.
District of Columbia. National Herb Garden, United States National Arboretum.
Michigan. Cranbrook House Gardens, Bloomfield Hills, MI.
Missouri. Missouri Botanical Gardens, 2315 Tower Grove Avenue, St. Louis, MO.
New Jersey. Well-Sweep Herb Farm, Port Murray, NJ.
New York. Brooklyn Botanical Garden, 1000 Washington Avenue, Brooklyn, NY.
New York Botanical Garden, Bronx Park, New York, NY.
The Cloisters, Fort Tryon Park, New York, NY.
North Carolina. N.C. Botanical Garden, Chapel Hill.
Pennsylvania. John Bartram's Garden, Morris Arboretum, Philadelphia, PA.
Virginia. John Blair Gardens, Williamsburg, VA.
Kitchen Garden, Mount Vernon, VA.

## Drug plants

While spices and herbs add interest and flavorings to our food and life, another group of plants plays a far more practical and essential role in our well-being. These may be designated under the broad umbrella term of *drug plants.* Through the centuries, man's dependence on naturally obtained drugs has been much more extensive; today, the chemist in his laboratory has synthesized many drug compounds and has usurped "Mother Nature" in her production of medicinals from plant sources. The development of drugs from plants is not the first priority of pharmaceutical companies, although approximately 25 percent of the drugs used in the U.S. are derived from plants. In the lesser developed nations of the world, about 75 percent of the people rely on remedies prepared from plants by the local herbalist.

Some of the well-known drugs derived from plants are belladonna, cocaine, digitalis, opium, quinine, and reserpine; and these as well as others are employed in the treatment of a variety of ills ranging from pain to heart disorders.

The medicinal properties of drug plants are based upon active substances such as alkaloids and glycosides present in the plant tissues. An alkaloid is an organic substance, bitter tasting, having alkaline properties and containing nitrogen. Glycosides are also bitter tasting; they contain sugar molecules which are attached to nonsugar molecules.

This discussion will include but a few of these plant-derived medicinals and will conclude with some prospects for the future of such drugs.

Belladonna, also called deadly nightshade *(Atropa belladonna)*, is a toxic, herbaceous, perennial and a member of the nightshade family. It is related to some very important economic plants, such as tobacco, tomato, and potato. The leaves and roots of belladonna contain several alkaloids, chief of which is atropine. Belladonna has many uses, including the stimulation of the sympathetic nervous system, dilation of the pupil of the eye, and the relief of pain.

(A)

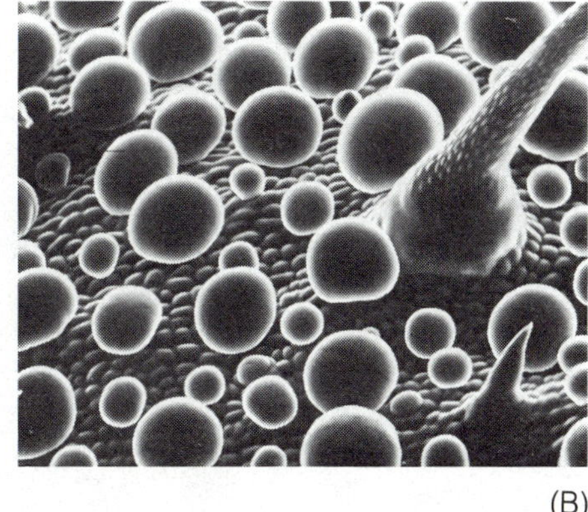

(B)

*FIGURE 12–5*
(A) Young vegetative plant of Indian hemp *(Cannabis sativa)* from which the THC containing drug, marijuana, is derived. *(Courtesy of D. Bay)* (B) Scanning electron micrograph showing THC containing glands of *C. sativa*. × 250. *(Courtesy of P. Dayanandan)*

Cocaine is the product of the leaves of coca *(Erythroxylon coca),* a shrub, growing naturally in the Andes but cultivated in many parts of the world. South American Indians discovered that chewing the toasted leaves of the coca plant enabled them to endure hunger and fatigue and work harder and longer. Medicinal use of cocaine is limited, although it does serve as an anaesthetic.

Colchicine, an alkaloid derived from the corm (underground stem) of the autumn crocus *(Colchicum autumnale),* is used to treat gout. Because it also affects nuclear division in cells, causing cells to become polyploid (having more than one set of chromosomes, that is, chromosome numbers greater than the 2N or diploid level), it is considered to be hazardous to use by humans because it may cause cancer.

Curare is a black, resinous substance, prepared from the bark of a South American tree *(Chondodendron tomentosum)* and used by some Indians to poison the tips of arrows. The alkaloid, d-tubocurarine, is used to bring about muscle relaxation during anaesthesia and to treat spastic conditions, rabies, and tetanus.

Digitalis is a drug obtained from the dried leaves of foxglove *(Digitalis purpurea).* The leaves of this herbaceous perennial contain several glycosides including digitoxin. This medicinal is a major drug used in the treatment of congestive heart failure. An attractive plant with an upright, leafy stem and whitish or purplish flowers, it is a common ornamental in the garden.

Ginseng *(Panax quinquefolium)* is grown in eastern North American forests and is exported to China where it augments the dwindling supply of *P. ginseng* which is indigenous to that country. Ginseng is a "cult drug," easily obtainable in health-food stores in the U.S.; in China, Korea, and Russia, it is widely used as a cure-all. The glycoside, panaquilon, found in this plant, is used as a stimulant and overall tonic.

Marijuana, a well-publicized drug, is obtained from the leaves and inflorescences of Indian hemp *(Cannabis sativa).* It is a native of central Asia but is now distributed in temperate and tropical areas around the world. The most potent substance present in the plant is tetrahydrocannabinol (THC), which is an intoxicant (Figure 12–5). In centuries past, marijuana has been used extensively as a drug; today, it is used legally in the treatment of glaucoma and to relieve the nausea associated with the chemotherapy treatment of cancer patients.

Opium is a constituent of the latex of the green fruits or capsules of the opium poppy *(Papaver somniferum).* It contains several alkaloids, the more important ones being morphine and codeine, both widely used as analgesics and sedatives. Codeine, contained in cough syrup, is used to relieve coughing.

Quinine is derived from the bark of species of *Cinchona,* flowering evergreen trees, native to the Andes, with hard, thick, grayish bark. Large plantations for the commercial production of quinine were established in India and Java, the latter producing the bulk of the drug. Quinine is renowned for the treatment of malaria, the number one killer disease in history.

*FIGURE 12–6*
Madagascar periwinkle *(Catharanthus roseus)*, the source of medically important drugs. *(Courtesy of M. Phillips)*

Reserpine, obtained from the bark of roots of the perennial evergreen shrub, snakeroot *(Rauwolfia serpentina)*, is primarily a product of India. It is used today to treat hypertension and as a tranquilizer to reduce stress and to calm mentally-ill patients (Table 12–5).

## Future prospects for drug plants

Although the primary thrust of the pharmaceutical industry is the development of profitable synthetic medicinals, the interest in learning about and using "nature's gifts" from plant life has never completely disappeared. Home remedies have always had, and always will have, a place in the "medicine cabinets" of the natives of less developed countries as well as those who inhabit the mountains and back valleys of Appalachia and those who dwell in the "canyons" of New York and the greener surroundings of suburbia. Just as the herbalists of an earlier time had an interest in the helpful aspects of plant life on man, so do some botanists and pharmacists of today. Work continues as more and more plants are investigated to unlock any secrets they may contain that will aid mankind. Examples of such potentially new drug plants include the following: (1) a member of the sunflower family, zoapatle *(Montanoa tomentosa)*, which may provide a naturally occurring contraceptive and (2) the discovery that the Madagascar periwinkle *(Catharanthus roseus)* (Figure 12–6) is the source of two drugs that have been used successfully in the treatment of certain kinds of cancer, namely, vinblastine, used in the treatment of Hodgkin's disease, and vincristine, for the treatment of leukemia. Another "gift," from the root of *C. roseus*, is ajmalicine, used in Europe to treat certain heart conditions. Scientists, today, are thus not casting aside the folk remedies that have been handed down for generations but instead are investigating their potential for modern medicine.

*TABLE 12–5*
*Some common drug plants*

| *Drug* | *Scientific Name* | *Family* | *Part* | *Active Substance* | *Use* |
|---|---|---|---|---|---|
| Belladonna | *Atropa belladonna* | Nightshade | All parts | Atropine, hyoscyamine, scopolamine | Stimulant, dilator of pupil, pain relief |
| Cocaine | *Erythroxylon coca* | Coca | Leaves | Cocaine | Anaesthetic |
| Coffee | *Coffea arabica* | Madder | Seeds (beans) | Caffeine | Stimulant |
| Colchicine | *Colchicum autumnale* | Lily | Corms | Colchicine | Treatment of gout |
| Curare | *Chondodendron tomentosum* | Moonseed | Bark | D-tubocurarine | Muscle relaxant |
| Digitalis | *Digitalis purpurea* | Figwort | Leaves | Digitoxin, digoxin, and other glycosides | Heart stimulant |
| Ginseng | *Panax quinquefolium* | Ginseng | Roots | Panaquilon | Stimulant, general tonic |
| Marijuana | *Cannabis sativa* | Mulberry | Leaves, inflorescence | Tetrahydrocannabinols | Glaucoma treatment, relief of nausea in cancer chemotherapy patients |
| Opium | *Papaver somniferum* | Poppy | Fruits | Morphine, codeine, papaverine, and other alkaloids | Pain relief |
| Quinine | *Cinchona officinalis* | Madder | Bark | Quinine, quinidine, and other alkaloids | Antimalarial, cardiac depressant |
| Reserpine | *Rauwolfia serpentina* | Dogbane | Roots | Reserpine | Reduce hypertension, tranquilizer |
| Tea | *Camellia sinensis* | Tea | Leaves | Caffeine | Stimulant |
| Tobacco | *Nicotiana tabacum* | Nightshade | Leaves | Nicotine | Stimulant |

## REFERENCES


Aikman, L. 1974. Nature's gifts to medicine. *National Geographic* 146(3):420–40.

Aikman, L., and Kohl, L. 1983. Herbs for all seasons. *National Geographic* 163(3):386–409.

Ayensu, E. S. 1981. The plants that heal. *Smithsonian* 12(8):86–97.

Baker, H. G. 1978. *Plants and civilization.* Belmont, Calif.: Wadsworth Publishing Co.

Bianchini, F., and Corbetta, F. 1977. *Health plants of the world: atlas of medicinal plants.* New York: Newsweek Books.

Claus, E. P.; Tyler, V. E.; and Brady, L. R. 1970. *Pharmacognosy.* 6th ed. Philadelphia: Lea and Febiger.

*Economic Botany,* quarterly journal. The Society for Economic Botany, New York Botanical Garden, New York. (Publication dealing with plants and their usefulness.)

Emboden, W. A. 1979. *Narcotic plants.* Rev. ed. New York: Macmillan Co. (Deals with drugs derived from plants and contains fine illustrations.)

Gordon, L. 1980. *A country herbal.* New York: Mayflower Books.

Hill, A. F. 1952. *Economic botany.* New York: McGraw-Hill Book Co.

Jacobs, B. 1981. *Growing and using herbs successfully.* Charlotte, Vermont: Garden Way Publishing. (Helpful information concerning herb gardening and practical uses of herbs.)

Janick, J.; Schery, R. W.; Woods, F. W.; and Ruttan, V. W. 1981. *Plant science: an introduction to world crops.* San Francisco: W. H. Freeman and Co.

Krieg, M. B. 1964. *Green medicine: the search for plants that heal.* Chicago: Rand, McNally and Co.

Lathrop, N. J. 1981. *Herbs: how to select, grow, and enjoy.* Tucson: Horticultural Publishing Co. (Popular approach to the world of herbs—growing, harvesting, and utilization.)

Lewis, W. H., and Elvin-Lewis, M. P. F. 1977. *Medical botany: plants affecting man's health.* Somerset, N.J.: Wiley-Interscience. (Reference on plants that injure, heal, and nourish, or alter the conscious mind.)

McGourty, F., Jr., and Nelson, P. K. 1973. *Handbook on herbs.* Plants and Gardens Handbook. Brooklyn: Brooklyn Botanic Gardens. (Well illustrated dictionary of herbs and factual material on use, culture, and harvesting.)

McNair, J. K. 1978. *The world of herbs and spices.* San Francisco: Ortho Books, Chevron Chemical Co. (Complete guide to 240 of the best for gardening, cooking, and family crafts.)

Morton, J. F. 1976. *Herbs and spices.* New York: Golden Press. (Illustrations, descriptions, and uses.)

———. 1981. *Atlas of medicinal plants of middle America.* Springfield, Ill.: Charles C. Thomas.

Parry, J. W. 1972. *Spices.* New York: Chemical Publishing Co.

Pullar, E. 1979. A history of old herbs. *American Horticulturist* 58(5):28–33.

Rosengarten, F., Jr. 1973. *The book of spices.* New York: Jove Publications. (Very useful work containing wealth of information on spices.)

Schery, R. W. 1972. *Plants for man.* Englewood Cliffs, N.J.: Prentice-Hall.

Schultes, R. E. 1976. *Hallucinogenic plants.* New York: Golden Press.

Swain, T., ed. 1972. *Plants in the development of modern medicine.* Cambridge, Mass.: Harvard University Press.

Taylor, N. 1965. *Plant drugs that changed the world.* New York: Dodd, Mead. (Uses of drugs and their plant sources.)

Tippo, O., and Stern, W. L. 1977. *Humanistic botany.* New York: W. W. Norton and Co.

USDA. 1980. *Spices.* Foreign Agriculture Circular. Washington, D.C.: Foreign Agricultural Service.

Woodson, R. E., Jr.; Youngken, H. W.; Schlittler, E.; and Schneider, J. A. 1957. *Rauwolfia: botany, pharmacognosy, chemistry, and pharmacology.* Boston: Little, Brown.

# Plants for crafts and natural dyes 13

One of the most ancient of hobbies is making crafts using plants. It is an activity that provides relaxation and much enjoyment. During the winter months, plant crafts in the house remind us of the summer outdoors. They also help provide focal points of interest in interior decoration. The types of crafts for which dried or preserved plant materials can be used are almost infinite.

Plant materials are also used for dyeing fibers. The science and art of using dyes extracted from plants goes back centuries. Interest in the use of natural plant dyes is currently enjoying a resurgence, after it was largely abandoned for synthetically dyed fibers.

In the first section of this chapter we discuss important ways to prepare plants for crafts. And in the second we consider the extraction of natural plant dyes from the different plant sources and procedures for completing a "dyed-in-the-wool" project.

## Preparing plant materials for crafts

Each of us, at one time or another, has picked up and admired a colorful autumn leaf, an acorn, a dried seed capsule, or a cone. Now that using natural materials is so popular, you might want to collect attractive plant materials for future craft use (Table 13–1 and Figures 13–1, 13–2). While many items are available at the grocery store (such as seeds of peas, beans, and spices) and at hobby-craft stores (such as cones, dried flowers, and seed capsules), many more materials are available in nature for free. For instance,

## *TABLE 13–1*
## *Crafts using plant materials*

- Apple sculpture
- Book marks
- Bottle decorations
- Candle crafts
- Christmas decorations
- Collages
- Cone crafts
- Corn husk crafts
- Dried arrangements
- Egg crafts
- Ink impressions
- Jewelry
- Lamp shade decorations
- Mobiles
- Mosaics
- Note and greeting cards
- Pictures, botanical "prints"
- Placemats
- Plastic crafts (placemats, paperweights, screens, window coverings)
- Pomanders
- Potpourri
- Rubbings
- Sachets
- Sculpture
- Seed crafts
- Shadow boxes
- Spatterpaint prints
- Stationery
- Straw crafts
- Swags, wall hangings
- Wreaths

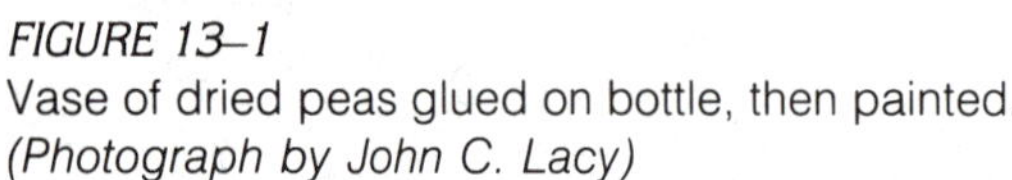

*FIGURE 13–1*
Vase of dried peas glued on bottle, then painted. *(Photograph by John C. Lacy)*

*FIGURE 13–2*
Tree decoration of dried seeds on an egg. *(Photograph by John C. Lacy)*

you can find fasciated stems of wild evening primrose *(Oenothera)*, in which normally round stems are abnormally wide, flattened, often twisted, and topped by many flowers (a fasciated condition) (Figure 13–3). Large collections of one type of plant, such as teasel *(Dipsacus)*, goldenrod *(Solidago)*, and Queen Anne's lace *(Daucus)*, are easily accessible along roadsides and in old fields. While on a collecting trip, common plants may be noticed that are in an unfamiliar form such as the winter seed heads of bee balm *(Monarda)* and the spore cases of sensitive fern *(Onoclea)* (Figure 13–4).

## COLLECTING PLANTS

During plant collection trips keep in mind that owner-permission is necessary if on private property. Generally speaking, collecting along public highways is usually permitted although parking may not be. Areas to be decimated by the

*FIGURE 13–3*
Fasciated and normal stalks of evening primrose *(Oenothera). (Photograph by John C. Lacy)*

*FIGURE 13–4*
Sporecases of sensitive fern *(Onoclea). (Photograph by John C. Lacy)*

construction of roads or buildings are excellent sources of plant materials. Public lands and parks and the plants they contain are protected. These sites should not be used for collection. Also, you should be aware of your state's endangered and threatened plant species. What looks like an obnoxious thistle may be on the list! Collecting in old fields in a city's industrial area or in parking lots is usually appreciated. Growing your own plants to dry is, of course, the ideal method.

Plastic bags of various sizes are ideal containers for all types of plant materials—from acorns to flowering branches. A pair of garden spring-clasp clippers is easier to use than a knife for cutting stems. For delicate items such as mushrooms, lichens, and mosses a woven basket with handle (containing wax paper to keep the materials separated) is easily carried. A pocket notebook is handy for keeping track of good collection locations and the plant materials available there. If traveling some distance by car for press-dry plant materials, a cooler with plastic bottles of frozen water is a good "holding tank." And finally, for an enjoyable collection excursion in the summertime, do not forget to include insect repellent and a hat to keep off flies and spiderwebs.

The time to collect summer plant materials at their freshest is on a dry sunny day, late in the morning after the dew is gone and before the afternoon heat wilts the plants, or during the early evening when they have revived. Choose flowers that are just beginning to open, as drying will open them further still. With cattail *(Typha)* flowers select green flowering shoots before they turn brown; otherwise, flowers will disperse from stems when dry. The green flower heads will turn brown as they dry but will remain intact.

For plants that will be pressed, the refrigerator is a temporary "florist case." Depending on the plant material, place stem ends in a jar of water and cover with a plastic bag. For leaves and flowers, add air to plastic bags and seal with a "twist'em" or rubber band. Aquatic plants, which are time consuming to press, may be placed temporarily in a bucket or pan of water. Fruits to be dried must be taken out of plastic bags and separated on newspaper before fungi become too fond of the moist environment and envelop them in strands of hyphae. Plant materials useful for crafts are listed in Table 13–2.

### *TABLE 13–2*
### *Plant materials for crafts*

| *Foliage*[1] | |
|---|---|
| Angelica *(Angelica)* | Mountain ash *(Sorbus)* |
| Bedstraw *(Galium)* | Oak *(Quercus)* |
| Black gum *(Nyssa)* | Oregano *(Origanum)* |
| Cedar, juniper *(Juniperus)* | Parsley *(Petroselinum)* |

[1]Specific plants have not been designated for air or press drying so as not to limit creativity. List includes a mixture of wild and tamed (horticultural) specimens.

Corn *(Zea)*
Cycas *(Cycas)*
Eucalyptus *(Eucalyptus)*
Ferns (Polypodiaceae)
Geranium *(Pelargonium)*
Gladiola *(Gladiolus)*
Grape *(Vitis)*
Heartsease *(Viola)*
Holly *(Ilex)*
Iris *(Iris)*
Lady's mantle *(Alchemilla)*
Lavender *(Lavandula)*
Locust *(Robinia)*
Maple *(Acer)*
Marjoram *(Majorana)*
Poplar *(Populus)*
Rosemary *(Rosmarinus)*
Rue *(Ruta)*
Sage *(Salvia)*
Sassafras *(Sassafras)*
Savory *(Satureja)*
Seedbox *(Ludwigia)*
Silverking *(Artemisia)*
Silverrod *(Solidago)*
Sweet Cicely *(Osmorhiza, Myrrhis)*
Sweet fern *(Myrica)*
Sweet gum *(Liquidambar)*
Staghorn sumac *(Rhus)*
Thyme *(Thymus)*
White poplar *(Populus alba)*

*Flowers*

Artichoke *(Cynara)*
Baby's breath *(Gypsophila)*
Bachelor button *(Centaurea)*
Beard-tongue *(Penstemon)*
Bells-of-Ireland *(Molucella)*
Bergamot, bee balm *(Monarda)*
Blue-eyed grass *(Sisyrinchium)*
Butter & eggs, toadflax *(Linaria)*
Butterflyweed *(Asclepias tuberosa)*
Cattail *(Typha)*
Chives *(Allium)*
Clover *(Trifolium)*
Cockscomb, celosia *(Celosia)*
Coral bells *(Heuchera)*
Coneflower, brown-eyed Susan *(Rudbeckia)*
Daisy *(Chrysanthemum leucanthemum)*
Everlasting *(Helipterum)*
False dragonhead *(Physostegia)*
False goat's-beard *(Astilbe)*
Forget-me-not *(Myosotis)*
Feverfew *(Chrysanthemum parthenium)*
Globe thistle *(Echinops)*
Honesty, money plant *(Lunaria)*
Hops *(Humulus)*
Joe-Pye weed, boneset *(Eupatorium)*
Johnny-jump-up, pansy *(Viola)*
Lady's thumb *(Polygonum)*
Larkspur *(Delphinium)*
Lavender *(Lavandula)*
Liatris *(Liatris)*
Loosestrife *(Lythrum)*
Love-lies-bleeding *(Amaranthus caudatus)*
Marigold *(Tagetes)*
Meadow rue *(Thalictrum)*
Mullein *(Verbascum)*
Pearly everlasting *(Anaphalis)*
Pink, carnation *(Dianthus)*
Prince's feather *(Polygonum orientale)*
Pussywillow *(Salix)*
Pussy toes *(Antennaria)*
Queen Anne's lace, wild carrot *(Daucus)*
Ragwort *(Senecio)*
Salvia, sage *(Salvia)*
Sea lavender *(Limonium)*
Statice, thrift, sea pink *(Armeria)*
Steeplebush, meadow-sweet *(Spiraea)*
St. John's wort *(Hypericum)*
Strawflower *(Helichrysum)*
Tansy *(Tanacetum)*
Tassel flower *(Emilia sagittata)*
Vervain *(Verbena)*
Violet *(Viola)*
Virgin's bower *(Clematis)*
Winged everlasting *(Ammobium alatum)*
Yarrow *(Achillea)*
Zinnia *(Zinnia)*

*(Continued)*

*TABLE 13–2* (continued)

*Fruit (see Fruit Types)*

- Alder *(Alnus)*
- Almond *(Amygdalus)*
- Ash *(Fraxinus)*
- Beech *(Fagus)*
- Bittersweet *(Celastrus)*
- Bladdernut *(Staphylea)*
- Black walnut *(Juglans)*
- Bur-reed *(Sparganium)*
- Corn *(Zea mays)*
- Dock *(Rumex)*
- Dogbane *(Apocynum)*
- Eucalyptus *(Eucalyptus)*
- Grasses (Poaceae) such as:
  - Bentgrass *(Agrostis)*
  - Foxtail grass *(Setaria)*
  - Panic grass *(Panicum)*
  - Rye *(Secale)*
  - Sea oats *(Uniola)*
  - Squirreltail, foxtail barley *(Hordeum)*
  - Wheat *(Triticum)*
- Ground cherry, Chinese lantern *(Physalis)*
- Hazelnut *(Corylus)*
- Hickory *(Carya)*
- Horse chestnut, buckeye *(Aesculus)*
- Lotus *(Lotus)*
- Love-in-a-mist *(Nigella)*
- Magnolia *(Magnolia)*
- Maple *(Acer)*
- Milkweed *(Asclepias)*
- Mullein *(Verbascum)*
- Nandina *(Nandina)*
- Oak *(Quercus)*
- Okra *(Hibiscus esculentus)*
- Osage orange *(Maclura)*
- Ostrich fern *(Matteuccia)*
- Pecan *(Carya)*
- Pepper grass *(Lepidium)*
- Pine *(Pinus)*
- Poppy *(Papaver)*
- Porcelain vine *(Ampelopsis)*
- Prickly cucumber *(Echinocystis)*
- Queen Anne's lace, wild carrot *(Daucus)*
- Redwood *(Sequoia)*
- Rose *(Rosa)*
- Sedges *(Cyperaceae)* such as:
  - Bulrush *(Scirpus)*
  - Cotton grass *(Eriophorum)*
  - Foxtail sedge *(Carex vulpinoidea)*
  - Nut rush *(Scleria)*
- Sensitive fern *(Onoclea)*
- Spruce *(Picea)*
- Sweet gum *(Liquidambar)*
- Unicorn plant *(Proboscidea)*
- Velvet leaf *(Abutilon)*
- Vervain *(Verbena)*
- Wild yam *(Dioscorea)*
- Wood rose *(Ipomoea tuberosa)*
- Yarrow *(Achillea)*

*Seeds*

- Allspice *(Pimenta)*
- Apple *(Malus)*
- Caraway *(Carum)*
- Cardamom *(Elettaria)*
- Celery *(Apium)*
- Chick-pea *(Cicer)*
- Coffee *(Coffea)*
- Corn, popcorn *(Zea)*
- Dill *(Anethum)*
- Fennel *(Foeniculum)*
- Flax *(Linum)*
- Juniper *(Juniperus)*
- Kidney bean *(Phaseolus vulgaris)*
- Lima bean *(Phaseolus lunatus)*
- Mace *(Myristica)*
- Millet *(Pennisetum)*
- Mung bean *(Phaseolus aureus)*
- Mustard *(Brassica)*
- Orange, lemon, lime *(Citrus)*
- Pea *(Pisum)*
- Peach *(Prunus)*
- Peppercorn *(Piper)*
- Pinto bean *(Phaseolus vulgaris* var. *humilis)*
- Poppy *(Papaver)*
- Pumpkin, squash *(Cucurbita)*

| Lentil *(Lens)* | Rice *(Oryza sativa)*<br>Sesame *(Sesamum)* |
|---|---|
| ***Vines*** | ***Other*** |
| Clematis *(Clematis)*<br>Grape *(Vitis)*<br>Honeysuckle *(Lonicera)*<br>Wisteria *(Wisteria)* | Galls<br>Lichens<br>Mosses (Bryophyta)<br>Mushrooms (Fungi)<br>Spike mosses (Lycopodiophyta) |

## AIR DRYING

The principle involved in air drying plant materials is rapid removal of moisture to retain color, texture, and shape. Determine whether plant material is to be air dried or press dried.

*SUMMER AND FALL COLLECTIONS.* Plants collected for dried arrangements in vases need to be separated into small bunches of four to five stems (Figure 13–5). With flowering shoots, remove leaves so the stems dry faster. Tightly wrap stem ends with a rubber band as shrinkage will occur. Hang upside down to prevent flower heads from drooping as they dry. An easy way to hang stem bunches is to secure a string or wire across a room, then attach them with paper clips through the rubber band on the stems, allowing space between for air circulation. Wire coat hangers instead of string can be

*FIGURE 13–5*
Mixed arrangement of strawflowers *(Helichrysum)*, pearly everlasting *(Anaphalis)*, goldenrod *(Solidago)*, cattail *(Typha)*, milkweed *(Asclepias incarnata)*, lily *(Lilium)*, bee balm *(Monarda)*, spore cases of sensitive fern *(Onoclea)* and ostrich fern *(Matteuccia)*. *(Photograph by John C. Lacy)*

used for a few bunches. Some stems air dry in attractive arches if set upright in a dry container instead of hung upside down. Damp basements and garage areas will cause rotting; so an airy, dry area is essential. Exposure to sunlight will fade plant colors. After drying, strip off any remaining stem leaves. Dock *(Rumex)* in all stages—from green, green-red, red, to brown—will retain the color stage in which it was collected.

*WINTER COLLECTIONS.* Usually plant materials collected during the winter are stiff and dry within and just need a short drying time if they were exposed to rain and snow. To remove dirt accumulation, rinse under water, then hang or set upright to dry.

*FRUITS.* Fruits such as black walnut, oak acorns, and conifer cones need to be dried to prevent fungal growth (Figure 13–6). Air dry or oven dry at 66° C (150° F) until dry. Closed cones may be opened and dried by placing on an aluminum foil-covered cookie sheet which is then set in a 93° C (200° F) oven until the cones open and resins begin to run. Stems with small fleshy fruits such as bittersweet *(Celastrus)* and Chinese lantern *(Physalis)* can be air dried. Large fleshy fruits to be used in crafts, such as osage orange *(Maclura)*, can be wired before drying. Force wire completely through the fruit so it will dry with a "handle" already attached for future use. An alternative is to drill a puncture hole or to saw into slices after drying. Gourds (Cucurbitaceae) may be left to air dry until the seeds rattle. For the large dandelionlike seed heads of salsify or goat's-beard *(Tragopogon)*, use hair spray before cutting from the plant to prevent seed dispersal.

*FIGURE 13–6*
Cone and fruit wreath. *(Photograph by John C. Lacy)*

*FLOWERS.* For drying individual flowers with short or limp peduncles, such as strawflowers *(Helichrysum)* or marigolds *(Tagetes)*, push a wire "stem" into the back of the flower. As the flower head dries its wire stem will become permanent (Figure 13–7). Immersion in sand, borax, or silica gel has been much publicized as an ideal way to preserve large, whole flowers. However, upon removal, air humidity will cause the brittle, fragile flowers to crumble or wilt. So unless the humidity is very low, these methods are not recommended.

*FUNGI, LICHENS, AND GALLS.* For an interesting variety small mushrooms, shelf (bracket) and earthstar fungi, reindeer moss lichen, and insect-caused galls on plants can be dried. While mushrooms are fresh, carefully push a threaded needle through the base of the stalk. Allow several inches of thread for each mushroom, to provide a fastener in craft use. Place on aluminum foil in a 66° C (150° F) oven about 15 minutes or until dry, or air dry.

## PRESS DRYING

To flat dry plant materials a "press" is necessary. A telephone book for small items or a stack of newspapers for larger materials can be used as a press. The principles of press drying are absorption of moisture, pressure, dry heat, and air circulation.

*FIGURE 13–7*
Wired strawflowers *(Helichrysum)* with baby's breath *(Gypsophila)*. *(Photograph by John C. Lacy)*

*PRESSING PROCEDURE.* For each specimen such as a flower, a leaf, or a branch of leaves and/or flowers, place in a single folded sheet of newspaper not to be removed until the specimen is completely dry and slides easily. To blot moisture out quickly, thereby retaining color and texture, sandwich the specimen in its paper between two newspaper sections (of several sheets each). The next specimen in its paper sheet is placed on top, then another section is added, then the next specimen in its paper, then another section and so on, in an even-edged stack. More effective drying is possible when corrugated cardboard pieces are placed between blotter-newspaper sections. To be able to locate specimens in the stack, arrange plant paper folded edge to right, newspaper section folded edge to left. When the stack is not much over 30 centimeters (12 inches) high, place two flat 30-centimeter by 38-centimeter (12-inch by 15-inch) boards, such as plywood, one on the top and the other on the bottom of the stack. Wrap two rope pieces, such as clothesline, over boards and half-hitch tie the stack as tightly as possible (Figure 13–8). It is helpful if you stand on the stack to compress the papers while someone else secures the ropes. Canvas straps with toothed buckles can be purchased to substitute for ropes. Every day the wet blotter sections of newspapers must be replaced by dry sections to achieve fast drying. If only a few specimens are to be pressed, instead of compressing with boards and ropes, a thick cardboard and heavy weight, such as a pile of books, a concrete block, or bricks, can be placed on top. While the paper blotters absorb moisture, ideally dry heat is also available from, for example, a nearby heater vent, a radiator, a clothes dryer, or a refrigerator motor. A closed car, setting in the sun, is a good daytime oven for press drying. Presses can also be tied on top of a car to accelerate the drying of specimens while driving. For small leaves and flowers a telephone book with a weight on top can be used. Space specimens several pages apart.

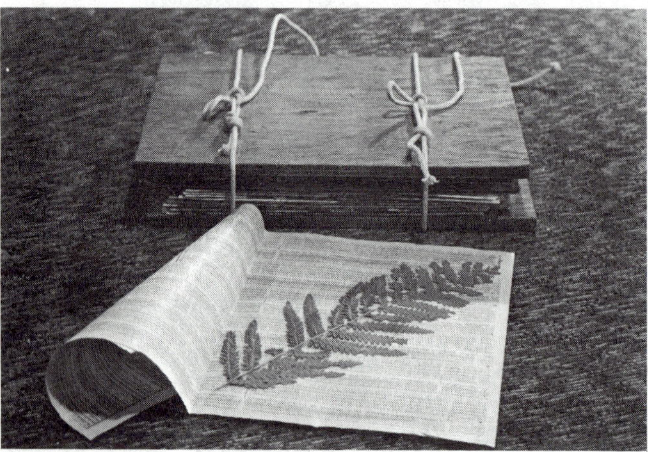

*FIGURE 13–8*
Dried specimen in single newspaper, press. *(Photograph by John C. Lacy)*

*PLANT PREPARATION.* Flowers tend to flatten well if placed face-down on the pressing paper. Long flowers can be laid for a side view. To hold flowers in position, cover with small pieces of wet newspaper or tissue paper. Leaves can be arranged as desired with margins unfolded. Fold stems to fit paper size or cut them off. If attached to a stem, a few leaf lower sides turned up add a pleasing variety. Separate plant parts so they do not overlap. Separate leaves of corn husks before pressing. A water weed which appears as a hopeless soggy mass when out of water can be pressed into a beautiful specimen (Figure 13–9). Cut art paper or bond paper the size of a half-sheet of newspaper (30 cm by 37 cm or 11½ in by 14½ in). Place the plant in a deep tray or pan of water to spread naturally. Slide the paper under the plant and slowly lift out with the plant on top. Drain excess water off one end, blot bottom of paper on newspaper, then place plant and paper in a single folded newspaper sheet. Stack with blotter sections as for land plants. Plant materials (not flowers) that have become wrinkled during collection or temporary storage can be revived by soaking in water before pressing. For example, fern fronds which begin to curl around the edges will flatten in water. Besides ferns there are other plants with compound leaves (of many small leaflets) which can be pressed to use in air-dried flower arrangements such as locust *(Robinia)*, mountain ash *(Sorbus)*, and sumac *(Rhus)*.

*FIGURE 13–9*
Press-dried water plant. *(Photograph by John C. Lacy)*

### *Pressing problems*

1. *Wrinkles* If the stems are too thick to press flat, the raised space allows leaves and flowers to wrinkle as they dry. If drying time was too long, or if not enough pressure was applied even though the stems were flat, the leaves will wrinkle.
2. *Brown discoloration* The specimen has started to rot because the blotter sections needed to be changed more often or the specimen was not pressed flat to allow moisture to be absorbed by blotter papers.
3. *Sticking* When a plant sticks to its paper, it is not dry yet.
4. *Fading* Drying fades the colors. But the faster the drying, the less the fading. Exposure of the dry, pressed plant to the sun or strong light accelerates fading.
5. *Uneven flatness* If a stem and leaves are mostly flat, except where fruits are attached, remove the fruit to dry separately. After drying, glue fruit back in place.
6. *Unattractive* Perhaps the plant material was not in very good condition before pressing. It will look worse after pressing.

## PLIABLE PREPARATION

To obtain pliable plant materials, such as stems, leaves, and flowers, infiltrate with a glycerin solution. Make up a solution of one part glycerin to two parts water by boiling water, mixing in the glycerin, and cooling to room temperature. Glycerin is available in drug stores or use car radiator antifreeze (ethylene glycol). Green or other food coloring may be added to the solution because the process darkens the materials. Some plants such as oak leaves turn a rich bronze color without food dyes. Dunk stem leaves (not flowers) into the solution, then place the freshly cut stem ends into the solution. With woody stems, first smash ends with a hammer to increase absorption area. Weigh down floating stems with pebbles. It will take about two weeks to complete infiltration depending on the plant. When the materials are ready, wipe off excess solution with paper towels. If the stems are too pliable for intended craft use, wires can be inserted or wrapped around. The glycerin solution can be stored and reused. This method works well with leaves such as oak, magnolia, maple, fern, and ivy stems with leaves and cedar branches.

For wreaths and swags, plant materials that are to be bent can be used fresh and worked immediately to dry in place or be stored dry until use. To rewet such materials as wheat and rye stems for straw crafts or corn husks, soak materials in warm water, weighed down to prevent floating.

## COPIES FOR CRAFTS

For needle-work, paint stencils, and paper collage, plant materials can be used as patterns. Leaves, flowers, and stems used fresh, air or press dried,

*FIGURE 13–10*
Machine copy of press-dried plant on left, ink-traced pattern on right. *(Photograph by John C. Lacy)*

can be placed on a commercial copy machine for paper reproductions. Then, the reproductions can be pencil transferred using tracing or carbon paper, or cut out (Figure 13–10).

## STORAGE OF PLANT MATERIALS

With flat-pressed plant materials, retain in original single newspaper sheets and stack in paper bags or boxes until used. Some plant materials are very susceptible to insect damage. To prevent loss until the materials are used for crafts, store in paper (not plastic) bags or boxes with moth crystals (paradichlorobenzene). Plastic prevents air circulation, retains humidity, and disintegrates with moth crystals. The vapors from moth crystals are cumulatively carcinogenic so *avoid inhalation* and store materials out of the way in *closed containers.*

# Natural plant dyes

For thousands of years, and up until about a century ago, the color of our fabrics came chiefly from plant extracts. The use of the indigo plant *(Indigofera tinctoria)* for dyeing fabrics blue dates beyond 3000 B.C. Probably a thousand years later, people knew how to combine other substances in the dyeing process to make colors "fast." This later discovery is known as *mordanting.* An English chemist, Sir Henry Perkin, discovered the first aniline dye, mauve, in 1856; this led to the establishment of the English synthetic dye-stuff industry. After the advent of synthetic dyes, particularly the aniline dyes in France in the late 1800s, use of plant dyes in Europe and the United States nearly died out. Today there is a resurgence of interest in the use of natural dyes, especially coincident with attempts to revive some of the crafts used by our early pioneers in the United States.

## PLANTS FOR NATURAL DYES

Natural plant dyes come from many parts of plants, including roots, leaves, bark, flowers, fruits, and bulb scales. Lichens (apparent individual organisms composed of a fungus and an alga in combination) are also used. The dye plants are either cultivated in the garden or are collected in the wild.[1] Most people rely on the local flora for their dye plants. The listing in Table 13–3 gives you an idea of some of the most common plants, the plant parts used to obtain different dye colors, mordants used, and the colors produced. Two

*TABLE 13–3*
*Natural plant dyes*

| Plant | Plant Part | Mordant | Color |
|---|---|---|---|
| Bloodroot *(Sanguinaria canadensis)* | Roots | Alum | Red |
| Ladies bedstraw *(Galium verum)* | Roots | Alum | Light red |
| Dyer's broom *(Genista tinctoria)* | Flower tops | Alum | Yellow |
| Scotch broom *(Cytisus scoparius)* | Flowering branches | Alum | Yellow |
| Saffron *(Crocus sativus)* | Stigmas | Alum | Yellow |
| Safflower, dyer's thistle *(Carthamus tinctorius)* | Flowers | Alum | Yellow |
| Agrimony *(Agrimonia eupatoria)* | Leaves and stalk | Alum | Yellow |
| Goldenrod *(Solidago)* | Flowers | Alum | Yellow tan |
| Smartweed *(Polygonum hydropiper)* | All parts except the roots | Alum | Yellow |
| Privet *(Ligustrum vulgare)* | Branch tips (clippings) | Alum | Yellow |
| Golden marguerite *(Anthemis tinctoria)* | Flower heads | Alum<br>Alum + 2nd dyebath | Yellow<br>Khaki |

[1]One of the best listings of dye plants is given in *Dye Plants and Dyeing* (Brooklyn Botanic Garden, 1964).

| | | | |
|---|---|---|---|
| Dock *(Rumex obtusifolius)* | Roots | Alum | Dark yellow |
| Marigold *(Tagetes)* | Flower heads | Alum | Yellow, buff, old gold |
| St-John's wort *(Hypericum perforatum)* | Plant tops | Alum | Yellow |
| Queen of the meadow *(Filipendula ulmaria)* | Whole plant (when in bloom) | Alum | Greenish yellow |
| Broom sedge *(Andropogon virginicus)* | Stalks and leaves | Alum | Greenish yellow |
| Nettle (wear gloves when collecting) *(Urtica dioica)* | Everything except roots | Alum | Greenish yellow |
| Bracken *(Pteridium aquilinum)* | Young shoots (spring) | Alum | Yellowish green |
| Bayberry *(Myrica pensylvanica)* | Leaves (summer) | Alum | Gray green |
| Sumac *(Rhus glabra* or *Rhus typhina)* | Berries | Alum and iron | Yellowish tan, gray |
| *(Rhus glabra* or *Rhus typhina)* | Leaves and young stalks | None | Tan to dark brown (depends on simmering time) |
| Osage orange *(Maclura pomifera)* | Bark extract | Alum | Yellowish tan |
| Lichen *(Parmella conspersa)* | Whole plant | Alum | Tan |
| Butternut hulls *(Juglans cinerea)* | Nut hulls (still green) | None | Tan |
| Black walnut *(Juglans nigra)* | Walnut hulls (green or dried) | None | Dark brown |
| Lombardy poplar *(Populus nigra, P. italica)* | Leaves | Alum | Lime yellow |
| Black oak *(Quercus velutina)* | Inner bark (dried) | Alum | Buff |
| Coreopsis, calliopsis *(Coreopsis tinctoria)* | Flower heads | Tin and cream of tartar | Bright yellow |
| Onion *(Allium cepa)* | Brown papery skin | Alum | Burnt orange |
| Blackberry *(Rubus)* | Young shoots | Alum | Light gray |
| Indigo *(Indigofera tinctoria)* | Available in supply houses | Caustic soda | Blue |

articles made of wool fibers that were dyed with natural plant dyes are illustrated (Figures 13–11 and 13–12). One is a wall hanging and the other is a very long winter hat!

Regarding dye plants, most can be dried and frozen and thus be stored for future dyeing operations. Bark, flowers, leaves, roots, and some berries are air dried in a warm, dry place, then stored in jars. Other plants such as fleshy berries, leaves, and flowers can be placed in plastic bags and stored in a freezer. Recycled coffee cans, jars, and plastic containers serve as excellent receptacles for dried plants. Dry onion bulb scales are easily obtained (with permission!) at markets in the bottom of onion bins. Dye sources, such as indigo, cochineal, saffron, and madder, are often available at supply houses. We should mention that cochineal, the source of a carmine red dye, is not obtained from a plant; rather, it comes from the dried cochineal bugs that inhabit the spine clusters on the pads of prickly pear cactus *(Opuntia)* in the southwestern United States.

## MATERIALS FOR DYEING FIBERS

Most of the materials needed for natural plant dyeing are readily available either in the home or from commercial sources. The basic equipment includes the following:

1. Scales for weighing chemicals.
2. Pyrex measuring cup and liter or quart jar.
3. Plastic spoons or metric kitchen scale for mordants.
4. Strainer and cheesecloth for straining plant material.
5. Mild soap.
6. Porcelain kettles for mordanting and dyeing, three glass bowls for rinsing.
7. Wooden spoons and thermometer.
8. Kitchen timer and stove.
9. Rubber gloves.
10. Glass rods to hang dried wool.
11. Mordants (see following paragraph).
12. Tags and "Sharpie" pen to indicate different mordants used.

Mordants are chemicals that bind the dye to the fiber and usually change the color in the process. Some of the common mordants and how they are prepared are listed in Table 13–4. There are several methods of mordanting: *before* dyeing, *during* dyeing, and *after* dyeing. Many people prefer to *premordant* since they can keep several samples of yarn premordanted with several mordants on hand to test with the same dye whenever convenient.

*FIGURE 13–11*
Wool wall hanging. The wool in this wall hanging was dyed with extracts from onion skins and rabbit-brush *(Chrysothamnus)*, a desert shrub in the southwestern United States. This wall hanging was made on a home-built loom by Linda M. Ryan in Colorado. *(Photograph by David Bay)*

*FIGURE 13–12*
A long wool hat. The wool fibers were dyed with extracts from walnut hulls (black walnut, *Juglans nigra)*, goldenrod inflorescences *(Solidago* sp.*)*, staghorn sumac infructescences *(Rhus typhina)*, marigold flowers *(Tagetes)*, onion skins *(Allium cepa)*, and cochineal bugs that live on pads of the prickly pear cactus *(Opuntia* spp.*)*. The hat was made by Anne Weismann, a former botany undergraduate student at the University of Michigan. *(Photograph by David Bay of Spence Bay)*

*TABLE 13–4*
*Common mordants and their properties*

| *Common Name* | *Chemical Name* | *As a Premordant in 0.95 liter/1 quart water (Amounts per 28 gm/1 oz medium weight 2-ply natural wool)* | *As An Additive* |
|---|---|---|---|
| Alum plus cream of tartar | Aluminum potassium sulfate plus potassium bitartrate | 2 gm/3/4 tsp alum plus 0.65 gm/1/4 tsp cream of tartar | May be used along with wool and dyestuff |
| Iron | Ferrous sulfate | Primarily used as an additive to darken or "sadden" a dyebath | Use a pinch |
| Tin | Stannous chloride | More commonly used as an additive as it can make wool brittle | To lighten or brighten a dyebath, use a pinch (well dissolved in water) before adding to bath |
| Copper sulfate | Cuprous sulfate | Primarily used as an additive, gives wool a light blue or blue green color | 1/4 tsp in water |
| Vinegar | Acetic acid | 89 ml/1/3 of a cup | Frequently used to heighten the color of dyebath, esp. in the red color range |
| Ammonia | Ammonium hydroxide | Frequently used to draw color out of dye materials especially grasses and lichens | Whether a premordant or an additive, the amount varies with the different dye material |

*Note:* Chrome as a mordant is not included in this list. It is no longer recommended since it is carcinogenic and not worth the risk.
*Source:* The chart is modified from that of Mollie Harker Rodriguez in an article by Palmy Weigle, "Natural Plant Dyeing," in *Natural Plant Dyeing: A Handbook* (Brooklyn Botanic Garden, 1973), p. 5.

## PROCEDURES FOR DYEING WOOL

*PREPARING WOOL.* Wool can be used before it is spun or after. In either case, it must be washed thoroughly to remove lanolin before it will accept the dye. Either natural untreated yarn or bleached white yarn may be used.

*MORDANTING.* For the novice, wool is recommended. Later, try mordanting silk and cotton, which takes longer and is more complicated. For our example we shall use alum and cream of tartar. For 454 grams (1 pound) of dry wool, use 85 to 113 grams (3 to 4 ounces) of alum and 28 grams (1 ounce) of cream of tartar, in 11 to 15 liters (3 to 4 gallons) of water.

When working with smaller amounts of wool, reduce the weight of mordants proportionately. Extreme care should be taken in handling some of the mordants because they are toxic and harmful to the skin (such as caustic soda); the alum and cream of tartar used here are relatively mild, however. In any event use rubber gloves when handling wool. Also use a porcelain kettle to mordant the wool. Metal containers can release metal ions such as aluminum ($Al^{3+}$), iron ($Fe^{2+}$ and $Fe^{3+}$), or tin ($Sn^{2+}$), depending on the container, which can interfere with the mordant reaction and ultimate dye color.

Soak the washed, clean yarn in lukewarm water for a few minutes. Meanwhile, dissolve the alum and cream of tartar in a small quantity of boiling water and add to 15 liters (4 gallons) of cold water, using a porcelain kettle. Put in yarn and slowly bring to a boil and simmer. Simmer one hour longer. Let yarn cool in pot, then rinse and dry.

*PREPARING THE DYEBATH.* Our next procedure, after mordanting the wool, is to prepare the dyebath. Take dried or frozen plant parts and chop them finely. Soak in water one hour to overnight, depending on the plant. Boil plant parts in the water approximately one hour, then strain and cool.

*DYEING WOOL.* Heat the dyebath and premordanted wool to just *below* boiling point, stirring with a wooden spoon. Simmer according to the recipe. Since the volume of the dyebath water decreases in this operation, add *hot* water to correct the level *after* removing the wool. When wool is dyed, it must be rinsed in progressively cooler water. If subjected to drastic changes in water temperature, wool will mat and/or shrink. For the "cooling operation," set up three glass bowls of rinse water, each successively cooler. With a wooden spoon pass the dyed wool through each successive rinse until the rinse water stays clear. Now squeeze out the excess water, roll in a towel, shake out, and hang to dry in a cool, shady place. Now you have completed your "dyed-in-the-wool" operation!

## REFERENCES

### *Preparing plant materials for crafts*

Bager, Bertel. 1976. *Nature as designer.* New York: Van Nostrand Reinhold. (A book of excellent black and white photographs of dried plant parts.)

Bailey, Ralph, ed. 1972. *Good Housekeeping illustrated encyclopedia of gardening.* 16 vols. New York: Book Division, Hearst Magazine. (This series has only a small section on dried flowers for arrangements, but is useful for finding information on horticulturally important plants by alphabetical order of both Latin and common names.)

Bragdon, Allen Davenport, ed. 1974. *Family creative workshop.* 24 vols. New York: Plenary Publications International. (An excellent series of books on crafts of all kinds including many, large sections on how to preserve and use plant materials in an array of crafts.)

Smith, Helen V. 1973. *Winter wildflowers.* Michigan Botanical Club Special Publication No. 2. Pp. 1–64. Michigan Botanical Club, Herbarium, University of Michigan, Ann Arbor, Michigan, 48109. (A booklet of black and white photographs shows mainly wild plant fruiting stalks as they appear in winter, dried by nature, and ready for collecting.)

Taylor, Norman, ed. 1948. *Taylor's encyclopedia of gardening.* Boston: Houghton Mifflin Co. (Newer, enlarged editions of this perennial gardening book provide a wealth of facts on plants for gardens.)

### *Natural plant dyes*

Brooklyn Botanic Garden. 1964. *Dyeplants and dyeing.* Plants and Gardens 20 (3): 1–100. (This booklet tells about the history of dyeing and basic steps in dyeing. It has 35 recipes for various dye plants used by people around the world and a color plate showing results.)

Brooklyn Botanic Garden. 1973. *Natural plant dyeing.* Plants and Gardens 29 (2): 1–65. (An excellent handbook, giving recipes for how to prepare dyes from plants and how to dye wool and other fibers. Indicates many different plant sources for dyes.)

LaRue, Jane. 1977. Natural dyeing with plants in Michigan. *Michigan Botanist* 16 (1): 3–14. (An article by an experienced user of dye plants. Explains step-by-step methods and lists book references and suppliers of dye materials, mordants, wool, and spinning supplies. Includes a a *Suppliers Directory* which lists firms specializing in supplies and equipment for natural dyeing, spinning, and weaving. Available for $2 from Handweavers Guild of America Publications, 998 Farmington Avenue, West Hartford, Conn. 06107.)

# Practical plant photography 14

Photographing plants as subjects of artistic and scientific interest can be a very rewarding experience. There is a great variety of equipment, techniques, and literature available, and we will not be able to describe all of it here. Instead, we present a simple system which beginners might employ and leave it up to them to pursue advanced techniques as they become more familiar with what they are doing. We discuss certain general aspects of photography that will help beginners improve their technique, but it is up to them to *practice* and learn by doing and studying.

Nature photography may be viewed as photographing natural objects or scenery outdoors in the environment, as opposed to laboratory or studio set-ups. It can be a rewarding end in itself as you get outside to hike through field and forest in search of favorite subjects to photograph. It is as good an excuse as any to go for a hike; but it can be more than that. One of the great rewards of nature photography is that it teaches us to be more observant, to learn about the plants which we enjoy, where they grow, how they live, and the problems they encounter in their struggle for survival. In order to photograph a certain wildflower that you have seen in a book, you first must learn its habitat, where to find it, and the season when it will be in bloom. That is as much as part of the nature photographer's technique and ability as the knowledge of the use of the camera equipment. One nice thing about photographing flowers is that once you have found one, it does not run and hide, or fly away, as you approach!

## The SLR camera

The best way to begin is to buy, borrow, or rent a good but inexpensive single-lens reflex (SLR) camera. The SLR is preferred over other types of cameras because it is lightweight but sturdy; provides maximum versatility through interchangeable lenses of different focal lengths (wide angle, normal, telephoto, macro); and allows you to actually view the subject you are focusing on through the same lens that will take the picture. This last feature allows sharply focused close-ups to be obtained. It is also recommended that you start by taking slides instead of prints. Slides are less expensive, and since you are practicing and experimenting, you should take lots of pictures to perfect your technique. You can always have prints made from slides.

### FILM

Kodachrome or Ektachrome film is recommended as being the most consistent in color quality for flowers, plants, and scenery. Most common SLR cameras utilize the 35-mm film format, with image size proportions of a 1″ × 1½″ rectangle. This provides the option of a horizontal or a vertical orientation and can affect composition (Figures 14–1 and 14–2). Thus, you might start out with a 35-mm SLR using Kodachrome 64 film, which has a film speed, or ASA, of 64. The film speed indicates the sensitivity of the film to light; the higher the ASA number, the less light you need to take a picture. So, ASA 400 would be a good choice of film for situations of low light.

*FIGURE 14–1*
**Composition and orientation.** *Iris cristata,* dwarf iris, horizontal format. (North Carolina, April)

*FIGURE 14–2*
*Iris cristata,* vertical format. Note that here the frame is better filled and that the subject is larger due to the natural vertical growth of the plant. (North Carolina, April)

## CAMERA CONTROLS

The next thing to understand about SLR cameras is the adjustment of the two main controls: aperture and shutter speed. The aperture is the degree of openness of the iris diaphragm in the lens which lets in light. It works like the iris of the eye, opening and closing as you turn the aperture ring on the lens barrel. On this ring are marked f/stops. Each standard f/stop number (2.8, 4, 5.6, 8, 11, 16, 22, 32) indicates a standard setting for the iris diaphragm: going up one f/stop number higher makes the "light hole" smaller and lets in one-half the amount of light as the next lower number (f/11 lets in one-half the amount of light as f/8).

Behind the iris diaphragm, in the camera, is a shutter that opens and closes at variable speeds to allow in the amount of light determined by the iris diaphragm. The shutter speed setting determines the exposure time of the film to the light coming in (from 1 second up to 1/1000 second on most cameras).

Thus, a correctly exposed picture is determined by the correct choice of f/stop (amount of light) and shutter speed (exposure time). How do you determine the right combination of light (aperture or f/stop) and exposure time? You use a light meter. Another good feature of the SLR camera is the built-in light meter that most of them have. It reads the incoming light through the lens and indicates, or helps you determine, what the correct exposure should be to get a good picture. So, for example, a typical exposure on a bright sunny day of an evenly lighted tree would be 1/125 sec at f/11 using ASA 64 film. Only practice will help you understand the relationship of shutter speed to f/stop. If you go, at the same light level, from f/8 to f/11, you must change the shutter speed from 1/250 sec to 1/125 sec. As you cut the amount of light in half, you must double the exposure time to keep the balance the same.

## CAMERA LENSES

Learning to use the SLR camera is a snap! Find a nice subject, preferably a plant, and practice correct combinations of exposure (f/stops) and shutter speeds as indicated by the automatic light meter in the camera. Be careful to hold the camera very still because it vibrates as the shutter opens and closes. You are now ready for the next step.

For taking normal pictures of people and scenery, you use a normal lens with a focal length of around 50–55 mm. For taking pictures of distant objects, you use a telephoto lens, something like 135 mm, 200 mm, or 300 mm, to bring the subject in closer. These lenses "squeeze" the subject into the picture, reducing apparent depth. To get more of a panoramic view of scenery into the picture, you use a wide-angle lens, like 28 mm or 35 mm (Figure 14–3). These lenses "stretch" the picture, adding depth. The versatility of the SLR camera allows you to change from one lens to another at any time by merely unscrewing one lens and screwing on another. Of course, extra lenses cost money and you do not need them to start as a beginner.

In order to take close-ups of flowers, you have to overcome the problem of getting close to your subject (Figure 14–4). Most lenses allow you to get no closer than three feet to your subject, giving you a reproduction ratio of about 1:20; that is, the image size on the film is 1 (or 1 inch) and the size of the object you are photographing is 20 inches. For close-up photography it is necessary to work in the range of reproduction ratios of 1:10 down to 1:1 (read "one-to-ten" down to "one-to-one"). Life size is 1:1 (when the image size on film is the same as the actual size of the object) (Figure 14–5). Taking close-up pictures can be accomplished in three ways.

1. Extend the lens away from the camera using bellows or extension tubes.
2. Use special, inexpensive, close-up lenses which screw onto the front of the normal lens, so as to magnify the image and produce a close-up picture.
3. Use of a special lens called a *macro* lens.

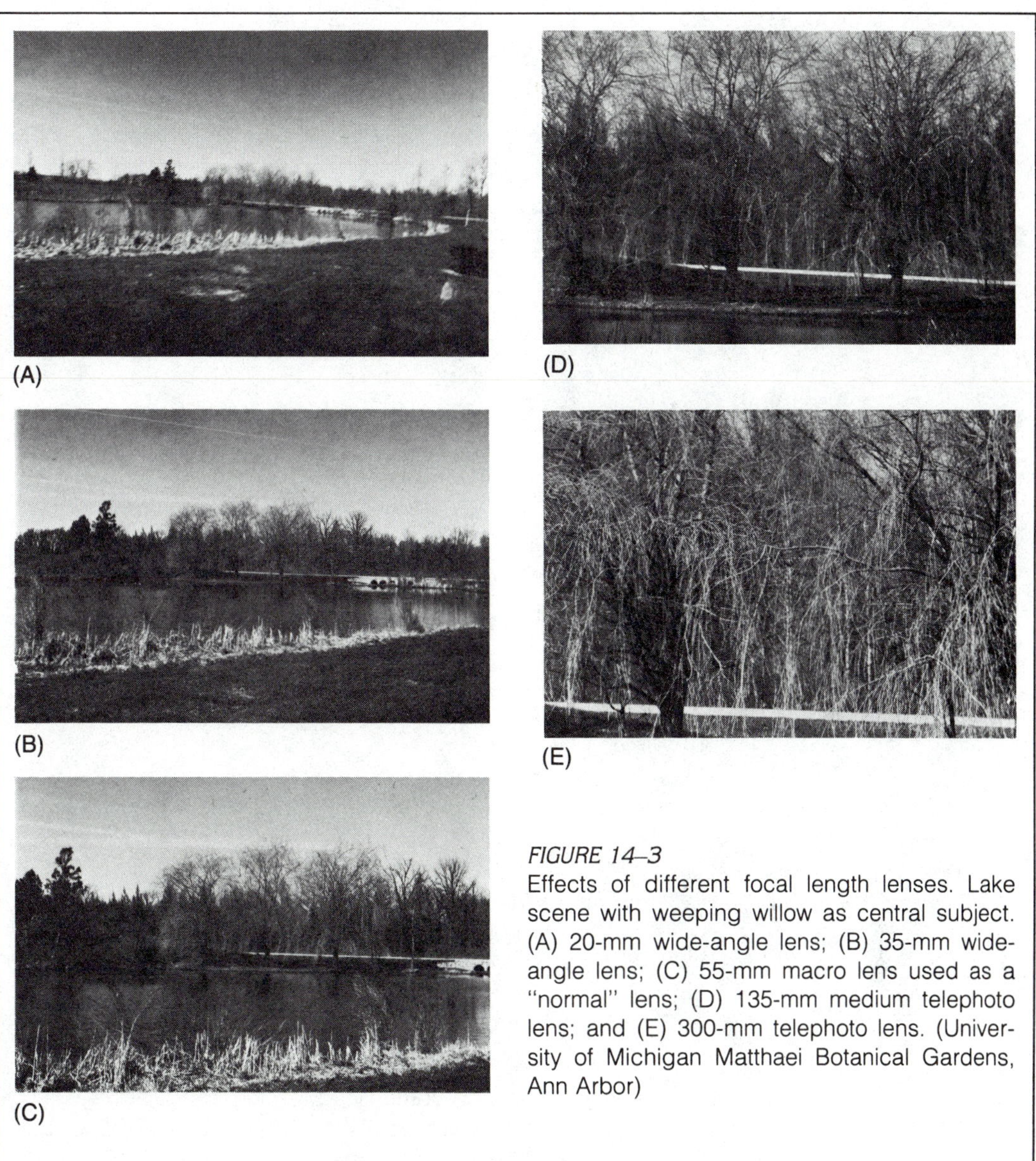

*FIGURE 14–3*
Effects of different focal length lenses. Lake scene with weeping willow as central subject. (A) 20-mm wide-angle lens; (B) 35-mm wide-angle lens; (C) 55-mm macro lens used as a "normal" lens; (D) 135-mm medium telephoto lens; and (E) 300-mm telephoto lens. (University of Michigan Matthaei Botanical Gardens, Ann Arbor)

The third way is the ideal way, because a macro lens is specially designed to give you maximum optical quality at close-up distances. The remarkable thing is that a macro lens can also be used as a normal lens to take "normal" pictures. We are fortunate to have a wide variety of macro lenses on the market these days for taking close-up pictures. Using a macro lens is the easiest way to begin nature photography because you need only one lens for many jobs. They are a little more expensive than normal lenses, but are worth it if you plan to do a lot of close-up work.

*FIGURE 14–4*
Henbit *(Lamium amplexicaule),* a small early spring lawn weed. In this photograph, the subject is too far away to fill the frame, and the background is too conspicuous and distracting.

*FIGURE 14–5*
Henbit, close-up. Reproduction ratio is now 1:1 (or life-size on a 35-mm slide). At this distance, the background is no longer a problem since it is out of focus.

## Plant photography pointers

Nature photography can be an educational, rewarding, enjoyable, even lucrative endeavor. Once you have mastered the equipment, you should approach your task from a personal viewpoint and look through the lens of the camera to find scenes in nature that please you. You may be after crisp, clean, well-proportioned portraits of wildflowers for identification or teaching purposes. Or you may be after those soft, moody, artistic shots that capture fleeting impressions of nature. Someone once said "pictures in nature are found, not made." Whether you go out to *make* pictures or to *look for* pictures, you are the creator with the photographic tools, and you have many variables to consider in performing your task. Trial and error will help make you a better photographer, but here are a few hints of things to consider and watch out for as you begin your adventure.

## EXPOSURE RECORDS

Initially, keep records of your subjects and the exposure data (f/stop and shutter speed) for each picture you take. This will help you in criticizing your own work and in trying to figure out what went wrong, or right. Make notes on lighting conditions, wind movement, actual colors, amount of shade, shadows, and so forth so you can remember what you liked about each scene as you compare the developed photographs. You will soon be able to judge whether or not you want to attempt a picture by the conditions of the situation.

## BACKGROUND

Watch the background in each picture. More often than not the eye concentrates on the subject and general composition, and it may conveniently overlook distractions in the background (Figure 14–6). Only after the pictures come back do you notice a stray twig, a bright object, or objectionable clutter in the background. Train yourself to observe all aspects of the picture before you snap the shutter. You may want to use a blue cloth or paper as a natural looking (sky) artificial background for some close-up shots (Figure 14–7).

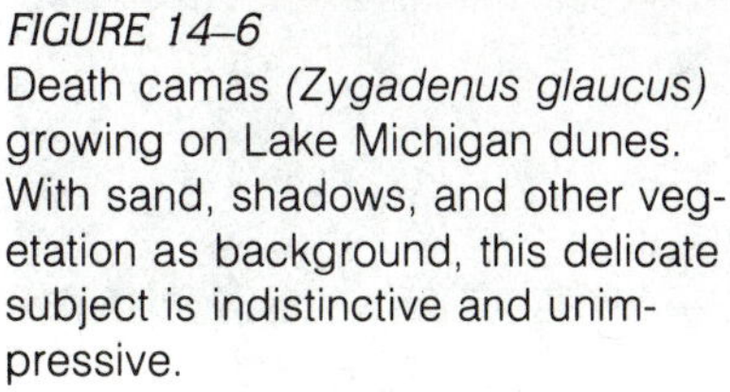

*FIGURE 14–6*
Death camas *(Zygadenus glaucus)* growing on Lake Michigan dunes. With sand, shadows, and other vegetation as background, this delicate subject is indistinctive and unimpressive.

*FIGURE 14–7*
By taking a lower angle shot of Death camas, the sky becomes the background and the subject stands out as distinctive and striking while preserving the impression of the habitat and surrounding vegetation.

## DEPTH OF FIELD

This is the amount of a scene that is in focus at a given aperture setting. The higher the f/stop number, the smaller the aperture, and the greater the depth of field in the picture. That is, more of the scene will be in acceptable focus as you look behind the main subject that you have sharply focused on. Most lenses have a scale on them that approximates how much in front and behind will be focus. And most cameras have a depth of field preview button that allows you to "stop down" the lens to see what will actually be in focus before you shoot. This is because most cameras have a mechanism whereby the lens is held "wide open," until you press the shutter button, to allow in the maximum light for critical focusing. Then, when you press the shutter, the lens stops down to the f/stop set according to the light meter reading.

Contrary to a beginner's belief, it is not always desirable to have the

*FIGURE 14–8*
**Depth of field.** White trillium *(Trillium grandiflorum).* Photo was taken at f/22 at 1/60 second. It shows maximum depth of field under these conditions. Note that even in the far background the flowers are recognizable and in focus.

greatest depth of field (unless you are working at high magnifications close to 1:1 or 1:2) (Figures 14–8 and 14–14). You might want to make more of the background a soft blur to allow the main subject to stand out (Figure 14–9). You accomplish this by opening the aperture wider (a lower f/stop number), and then compensating by increasing the shutter speed. For example, f/16 at 1/64 sec would be a typical setting for a fully lighted subject on a sunny day, and f/16 would provide a lot of depth of field. If you want less, stop down to f/8 and increase the shutter speed to 1/250 sec (or f/11 and 1/125 sec). This will be the *same* exposure, since you went down two f/stops (four times more light), you go up two shutter speeds faster (one-quarter less exposure time); the needle on the light meter in the camera viewfinder should still indicate an acceptable exposure level. Consider this alternative when taking portrait pictures so that your subject stands out sharply from the background.

*FIGURE 14–9*
White trillium under same lighting conditions as in Figure 14–8. This photograph was taken at f/5.6 at 1/500 second. Note the lack of focus in background flowers. The flowers in the foreground stand out more as the main subjects. It is not always desirable to have maximum depth of field in a picture. (Ann Arbor, Michigan in May)

## LIGHTING

There are more creative lighting techniques than just normal lighting. Normal lighting is "front lighting," that is, with the sunlight coming over your shoulder, hitting the subject face-on, and bouncing back into the camera (Figure 14–10). Of course, you can take some excellent pictures this way, but it is worthwhile to experiment with back lighting and side lighting to get a different representation of moods, forms, and especially textures (Figure 14–11). For example, hairs on stems and the translucent nature of delicate flowers come through better with back lighting. Side lighting gives you a greater sense of depth, and three-dimensional shape. As a general rule for back-lighted shots, you should give two extra stops of exposure from the normal front-lighted meter reading; and one to one and a half stops more exposure for side-lighted subjects. For example, if the meter reading for a front-lighted subject is f/11, use an exposure of f/5.6 to take a back-lighted picture of the same subject from behind at the original f/11 shutter speed. Experiment!

*FIGURE 14–10*
**Front lighting.** Morel mushroom *(Morchella hybrida).* Typically, a front-lighted subject shows the surface features of the specimen well.

*FIGURE 14–11*
**Back lighting.** Here the subject is back lighted (by taking the picture from the other side of the mushroom with light coming directly into the lens) and shows the delicate translucent texture of the mushroom body. (Waterloo Recreation Area, Washtenaw County, Michigan, May 18, 1972)

While nothing beats natural, available light for nature photography, you may reach a time when you want to experiment with the creative use of artificial light from an electronic flash unit. There are many models on the market from the simple box flash to the sophisticated flash units that automatically determine the correct amount of light to put out, based on their calculation of the distance from the subject to the camera. These can be adapted for use in close-up photography. You may find it rewarding to use flash for fill-in lighting, that is, as supplemental to natural sunlight for providing soft highlights on the subject and reducing the harshness of shadows (Figures 14–12 and 14–13).

*FIGURE 14–12*
**Lighting.** Wake-robin trillium *(Trillium erectum).* Photo taken front lighted with existing light on a sunny day. Note the harsh shadows on the flower. (North Carolina mountains)

*FIGURE 14–13*
**Lighting.** Wake-robin trillium. Flower taken with existing light and with electronic flash providing fill-in light to eliminate shadows. Background is still apparent and unevenly lighted. (Southeastern Michigan in May)

For sharp pictures of subjects for critical scientific examination, you may find close-up flash photography a must. The use of a ring-light flash, a circular flash unit that fits on over the front of the macro lens, gives an even cone of illumination for close-up photography. It eliminates all shadows, but it also eliminates most three-dimensional depth impression in the picture. Still, it may have a use for you by providing a well-lighted subject on an evenly lit (or even dark) background (Figure 14–14).

## TIME OF DAY

Time of day and lighting conditions can make a difference. Early morning or late afternoons shots are "warm" in appearance from the abundance of red light when the sun is low in the sky. These can produce dramatic color effects. Taking pictures at high noon on a bright day can be disappointing because of harsh shadows from overhead lighting. The ideal times for normal lighting conditions are between "two hours after sun-up until 11 A.M." and

*FIGURE 14–14*
Extreme close-up photo of twigs of black ash (*Fraxinus nigra,* left) and white ash *(Fraxinus americana,* right) taken with a ring-light flash. Flat, uniform, shadowless lighting is achieved on the subject and background material. ×4. Reproduction ratio on the actual slide is 1:1.

"1 P.M. until two hours before sundown;" in other words, midmorning and mid-afternoon. Flowers are usually fresher in the morning, and I prefer it for photography.

Dust and haze can also affect light quality which will in turn affect the color of your pictures. Some photographers use a clear haze filter over their lens to correct for slight color changes, and a polarizing filter can reduce glare from bright objects. I have also become fond of slightly overcast days, when there is just enough translucent cloud cover to create only very faint shadows (or no shadows at all). This diffused light (from a totally overcast sky, not just a passing cloud that covers the sun) provides conditions for maximum color saturation and absence of harsh shadows. It does, however, change the colors very slightly from that of bright sunlight. Let personal preference be your guide.

## HIGH CONTRAST CONDITIONS

Avoid contrasty lighting conditions, that is, scenes or subjects which are too dark or too light. Your light meter in the camera will give you a reading that is set to render average light intensity as if it were a neutral gray (18 percent gray to be technical). This means that the meter is designed to optimize exposure; that is, give you the best exposure under average conditions of front lighting on objects of *average* brightness. If you want to do back lighting or photograph unusually bright or dark objects, you will have to override the light meter readings and make your own settings. For these reasons, I prefer not to use a fully automatic camera, but one whose meter gives me a reading that I can take and adjust if necessary.

Experience helps you determine when you are dealing with average or contrasty situations. For example, if you are photographing a white cat in the snow on a bright day and you take the picture at what the light meter indicates, the whole scene will come out appearing much too dark. In fact, it is perfectly exposed from the meter's point of view; the meter rendered it as an average neutral gray. To compensate, you would take a meter reading (say f/16 at 1/125 sec) and increase exposure two stops, from f/16 to f/8 at 1/125 sec. In effect, you are overexposing the picture; but that is the only way to get it "bright white." Likewise, for dark objects you would underexpose to get the true darkness rather than average tones.

If your scene has some parts bright and some parts dark, as when dark shadows are present, you can either expose for the most important parts and let the other areas go darker or lighter as they will, or use fill-in flash to lighten the shadows and soften the harsh tones. Stick to slightly overcast days and you will avoid most instances of contrasty lighting. Good results may also be obtained by taking a meter reading off of an average lighted subject (such as blue sky or green grass, or a neutral gray card) before photographing your contrasty subject (Figure 14–15).

*FIGURE 14–15*
Taking photos under contrasty situations requires careful use of the camera's light meter. In this setting of a log cabin in a snow-covered woodland on a bright sunny day, the most accurate meter reading was taken directly off (at very close range) the girl's blue coat, the green tree branches, or the light brown logs of the cabin. The photographer then moved back to take the picture. A reading including any of the bright, highly reflective snow would have resulted in an underexposed, gray picture. (University of North Carolina at Charlotte, Van Landingham Rhododendron Glen)

## USING A TRIPOD

Use a tripod if you are unsteady with a camera or if you want to insure the sharpest pictures. Usually, when you are taking pictures with an SLR camera, you can hand-hold the camera at a shutter speed of 1/64 sec or 1/125 sec with no problems. As you get down to 1/30 or 1/15 sec (when the light is low) you begin to be unable to obtain a sharp hand-held photo. Often, however, conditions outdoors are slightly windy such that low shutter speeds are unsuitable, even with a tripod. A tripod is an extra piece of equipment that can slow you down in the field, but it may be a useful tool under certain circumstances (Figure 14–16).

## INDICATING SCALE

Unless you are taking extreme close-ups (1:1 or 1:2) or purposely abstract pictures, try and have some subtle indication of scale in your pictures of plants. A leaf, pine cone, or twig of recognizable dimension can be perfectly natural in a picture and help give the viewer some feeling for the size of your subject. This is most important in educational or scientific photos where an actual ruler may be used or you may show a hand holding the object (Fig-

*FIGURE 14–16*
A good, sturdy tripod can be an asset to some aspects of nature photography. On a calm day, it can help you take a longer exposure of a close-up subject with increased sharpness. Be sure to get a tripod with a central column that is invertible for close-up work near the ground, as is shown here.

*FIGURE 14–17*
**Scale and size.** *Draba verna,* a close-up of a tiny plant with fingernail for scale and palm of the hand for background. ×2. Reproduction ratio on the actual slide is 1:2.

ure 14–17). One interesting solution is to get pictures of flowers with insects visiting them; it is a natural phenomenon and shows relative size.

## Concluding photographic tips

In selecting camera equipment for purchase, consult a respectable dealer, ask him or her to show you several different cameras and lenses in the price range you are prepared to pay (good SLRs with macro lenses cost about $250 to $300 or more). Ask experienced photographers to show you their equipment and explain the advantages and disadvantages. Seek up-to-date comparison literature found periodically in Consumer Reports or various photographic magazines. Do not buy more expensive equipment than you need until you determine how you want to use it. It is best to begin by borrowing or renting an SLR camera to try out the feel of it, and then you will have a better basis for comparison shopping.

In conclusion, if you have the interest and motivation to learn nature photography, obtain some simple equipment, learn the basics, and then practice! Film is no longer cheap, but the rewards of taking pictures may be worth it as you interact with nature and learn to appreciate the hidden beauty that only the trained eye can reveal. Attend slide lectures, peruse books that contain photographs of plants and scenery (state wildflower guides are excellent), take courses on photography, talk to experienced photographers, and do all that you can to observe how others have approached nature photography. Then adapt techniques that you like, develop your own style, and perfect a formula that consistently satisfies you. This is the essence of nature photography.

## REFERENCES

Angel, Heather. 1972. *Nature photography: its art and techniques.* Kings Langley, England: Fountain Press. (One of the best idea and technique book available.)

Blaker, Alfred A. 1976. *Field photography.* San Francisco: W. H. Freeman and Co. (An excellent, general photography manual with details on advanced techniques.)

Blatter, Dave. 1974. When macro equals standard. *Peterson's Photographic Magazine* 3(4):64–68. (A discussion of macro techniques.)

Davis, Phil. 1979. *Photography.* 3d. ed. Dubuque, Iowa: W. C. Brown Co. (An excellent general photographic manual, especially for beginners.)

Eastman Kodak. *Close-up photography.* Publ. Number N-12A. Rochester, N.Y.: Eastman Kodak Co. (Good introduction to close-up technique.)

______. *Photomacrography.* Publ. Number N-12B. Rochester, N.Y.: Eastman Kodak Co. (Good for photomacrography technique.)

Fell, Derek. 1980. *How to photograph: flowers, plants, and landscapes.* Tucson: H. P. Books. (An excellent book for ideas and techniques.)

Greham, F., and Rickett, H. W. 1964. *The Odyssey book of American wildflowers.* New York: Odyssey Press. (Artistic, impressionistic, moody photographs of wildflowers.)

Justice, W. S., and Bell, C. R. 1968. *Wildflowers of North Carolina.* Chapel Hill: University of North Carolina Press. (Source of some of the best photographs of wildflowers in a small book for identification.)

Rickett, H. 1967. *Wildflowers of the United States.* Vol. 1 Northeastern States. Vol. 2 Southeastern States. New York: McGraw-Hill Book Co. (A large book of outstanding wildflower photographs for identification.)

Schmid, Rudolf. 1971. *Close-up photography of flowers with (ringlight) flash.* American Orchid Society Bulletin 40:232–43 (Technical "how to" for flash close-ups.)

Wells, James R., Case, F. W.; and Mellichamp, T. L. 1976. Perfecting wildflower photography. *Michigan Botanist* 15:183–90. (General principles of photography.)

# 15 New uses of plants

Today's plant science research, whether in the field or laboratory, is dependent, to some extent, upon the findings of the past. Some of these discoveries were made long ago, in distant lands, by plant hunters who endangered life and limb, often overcoming great obstacles, to discover plants new to them and rediscover old ones. Their discoveries ranged from ornamentals to medicinals and food plants, which were eventually introduced to the world.

One of these men was America's first plant explorer, David Fairchild, after whom the Fairchild Tropical Garden in Miami, Florida, is named. His expeditions and years of plant hunting around the world took him to Africa, Europe, Asia, the East Indies, and South and Central America. His collections and introductions of exotic plants, fruits, vegetables, grains, vines, trees, and flowers, included zoysia grass, Japanese flowering cherry trees, soybeans, dates, and improved varieties of rice, wheat, and cotton.

Just as the objective of some early plant hunters was to locate economically beneficial plants, that same mission is true of plant explorers today. They also strive to preserve rare and endangered species. Although twentieth-century chemical research has produced a multitude of synthetic drugs, the search for medicinal plants in Appalachia and other areas as well as other useful plants around the globe continues. In an attempt to meet some of the modern challenges confronting mankind—world hunger, the energy crisis, pollution, disease problems—plant and chemical scientists are investigating the genetic and chemical potential of a variety of plant species the world over and are reporting some surprising new uses of plants.

## Alleviating hunger

One such high-protein, new crop plant that, unlike soybean, grows well in minimum and low-protein areas of the world, is the tropical and subtropical winged bean *(Psophocarpus tetragonolobus).* This herbaceous, perennial legume, which is also called four-angled bean, asparagus bean, Goa bean, and Manila bean, is cultivated over a wide geographic area including Southeast Asia and Papua, New Guinea. Another attractive feature of the plant is that practically all of it is edible; parts that are usually cooked—roots, leaves, flowers, pods, and seeds—can be and are consumed (Figure 15–1). The tuberous roots contain at least 20 percent protein, substantially higher than cassava, potatoes, sweet potatoes, and yams. The protein and oil content of seeds is also especially high, up to 42 percent and 19 percent, respectively. In addition, parts such as roots, pods, and seed contain the nutrients iron, calcium, and phosphorus, while vitamin E is present in the seed and A in the leaves. Here, then, is a plant that could help alleviate hunger in developing countries.

As well as being an excellent source of protein-rich food, the winged bean may prove to be a boon to a hungry world in another way. As do other legumes, it possesses the ability to biologically fix nitrogen; however, perhaps because its roots develop a greater number of larger nodules than other legumes, it has the capacity to fix above average amounts of nitrogen. Since this nitrogen is obtained by the plant from the air, it is available without limit. The fact that the plant can produce some of its own fertilizer is especially important in the tropics where soil is often deficient in nitrogen.

*FIGURE 15–1*
Winged Bean *(Psophocarpus tetragonolobus).* (Drawing published in Horticulture, June 1982, used with permission of Robin Rothman, St. Johnsbury, Vermont.)

The significance of the capability of winged bean and some other plants, including the mesquite tree *(Prosopis)* and honey locust *(Gleditsia triacanthos)*, to biologically fix nitrogen in a world in which fertilizer is in short supply and therefore high priced, and, at the same time, supply protein, is noteworthy indeed.

For centuries a type of "green manure" has been used for fertilizing rice paddies in Asian countries. This green manure, composed of the small, water fern *Azolla* in symbiotic relationship with the blue-green alga, *Anabaena azollae*, also fixes substantial amounts of nitrogen from the air and serves to naturally fertilize the rice paddies. This results in a higher yield of rice grains.

Leucaena, an extremely, fast-growing treelike legume, which thrives under poor growing conditions in Southeast Asia and Latin America, also is capable of fixing nitrogen biologically, thus enriching the soil. This invaluable plant, in addition, produces high-quality firewood and high-protein animal feed.

Another interesting leguminous plant is *Lupinus mutabilis*, called chocho or tarwi. This native of the South American Andes Mountains produces protein and oil rich seeds that are, unfortunately, unpalatable, until treated, due to the presence of a bitter alkaloid. Genetic research and the development of pleasant tasting seeds could make this an important food source.

Another new crop species is a hybrid, triticale (also called *Triticosecale*). Hybridization of *Triticum aestivum* (bread wheat) or *T. turgidum* (durum wheat) with *Secale cereale* (rye) produces this high-yield grain. Besides gigantic increases in yield per plant as well as per hectare, the grain is higher in protein than wheat; and, because the plants are shorter, lodging (falling over due to the action of wind and/or rain) is greatly reduced. Triticale, which is used in bread and for livestock feed, is grown as a crop in lesser developed countries including India and Pakistan. In 1979, approximately 800,000 hectares of triticale were grown world wide, 10 percent of which was grown in the United States. Further research on this cereal grain continues and could ultimately result in its becoming a major crop and thereby increasing world production of food.

Just as triticale was produced through the efforts of plant geneticists utilizing strains of older, well-established food grains, so too have plant breeders at Purdue University developed a high lysine and tryptophan type of corn known as opaque-2, as well as a high lysine variety of sorghum (lysine and tryptophan are essential amino acids, necessary for human growth).

Rice, a dietary staple among Asian peoples, has undergone a series of changes due to the efforts of plant breeders and geneticists who crossed strains of older types of rice, then continued working with these newer strains until, in 1973, IR-26 was developed. This semidwarf strain gives high yield, shows resistance to disease and lodging and, along with other new IR strains (from IRRI, International Rice Research Institute, Los Baños, Philippines), is being introduced into the rice paddies of Asia.

Although the emphasis is on food production at the U.S. Department of Agriculture's Agricultural Research Center at Beltsville, Maryland, crop plants, including wheat, corn, alfalfa, and soybeans, are being grown for use in satellite monitoring. Such space-age technology could serve to diagnose the need for water, fertilizer, control of pests, the approach and arrival of harvest, and even some indication of the impending crop yield. Meanwhile, research still goes on at the Center to develop new, high-yield strains of such grains as millet, barley, and oats.

## Echoes from the past

While some scientists have been developing new, improved varieties of food grains, others have been reexamining the value of plants that were once widely used but have been neglected during recent times. These plants can be valuable food sources and can provide natural materials for industrial use that will become increasingly important to our mechanized society as our more traditional raw materials are less and less available.

Although there are many examples of these plants, three southwestern, old Indian crops worthy of note are: amaranth, guayule, and jojoba. Amaranth *(Amaranthus hypochondriacus)*, as a member of the pigweed family, is related to many common noxious weeds such as pigweed and tumbleweed. It grows natively in Arizona and Mexico, produces a large congested inflorescence and thousands of seeds, which are protein rich and high in the essential amino acid, lysine, as well as leaves that are tasty.

The shrub guayule *(Parthenium argentatum)*, related by family (Asteraceae or Compositae) to dandelion, goldenrod, ragweed, and sunflower, grows on both sides of the Texas-Mexico border, produces latex and, long ago, was used for rubber production (Figure 15–2). For a variety of reasons, including a dwindling supply and the cost of importing Asian rubber *(Hevea brasiliensis)*, it was recently recommended by the National Academy of Sciences that the guayule be cultivated as an alternative source of natural as well as synthetic rubber. Perhaps, in the future, we will realize that it was fortunate that the U.S. Department of Agriculture agreed to plant, and will plant more, guayule in the southwestern United States. Protoplast cloning (done through protoplast fusion techniques) is being used to improve yields of this rubber-bearing shrub. The threat of conflict in natural rubber-producing countries of the world as well as a shortage of petroleum, which is required for the production of synthetic rubber, could bring transportation to a standstill.

Still another economically promising plant is jojoba *(Simmondsia chinensis)*, a shrub growing in the Sonoran Desert of Arizona, California, and Mexico. The seed of this plant is especially valuable since it contains up to 60 percent liquid wax which is very similar to sperm whale oil. Substitution of jojoba wax for sperm oil, which has many important industrial and home uses, could result in the survival of the sperm whale, already an endangered species. Cultivation of the crop in the southwest has resulted in some successes—look on your local drugstore shelf for jojoba shampoo.

*FIGURE 15–2*
Guayule *(Parthenium argentatum). (Courtesy N. Vietmeyer, National Academy of Sciences, Washington, D.C.)*

## Easing the energy crisis

Without too much risk, one can state confidently that the primary objective of early plant hunters was to discover new plants which would serve as sources of food and, perhaps, drugs. Most assuredly, it is unlikely that they ever conceived of wild or cultivated plants as producers of hydrocarbons that could be turned into motor fuel and thereby function as sources of energy. That once undreamed of plant use has now become a reality.

The Office of Technology Assessment (OTA) recently reported that wood, plants, and other biomass (living matter, plant and animal, in any form) could supply as much as 20 percent of the U.S. annual energy requirements by the year 2000. In order for this undertaking to be beneficial, it must be wisely managed so that cropland shortages do not develop and an advantageous energy balance is maintained. At no time can we permit farming for fuel to compete with farming for food. The OTA further stated that nontraditional land-based biomass crops may have more potential than the conventional crops which are presently serving as energy sources. Some of these rather unusual crops include various species of hardwoods, such as red alder and hybrid poplar; grasses, including Bermuda grass, big bluestem, kenaf, and Sudan grass; as well as vegetable oil and hydrocarbon plants. At the present time, however, conventional sources such as corn, sugarbeets, sugarcane and sweet sorghum, sugar and starch crops, are used to produce ethanol.

Melvin Calvin of the University of California at Berkeley has reported, based on his research, the possibility of the production of oil from species of spurge *(Euphorbia)* and *Copaifera,* both hydrocarbon-producing plants. *Euphorbia* is a vast genus of about 2,000 species growing in subtropical and warm temperate regions. According to Calvin's data, *Euphorbia lathyris* (gopher plant) a small bush that grows wild in northern California, could give a yield of between 10 and 20 barrels of oil per 0.4 hecatre per year. Utilizing plant tissue-culture methods and better agronomic practices, he projected a yield of up to 50 barrels of oil per 0.4 hectare per year; and because such plants could satisfy at least some of our chemical and energy needs for hydrocarbons, he recommended the development of petroleum plantations for fuel and other materials.

There is now substantial evidence around the United States that we are planting as well as drilling for oil. Milkweed, laden with latex which contains 30 percent hydrocarbons, has been planted in Alabama. It has been estimated that the plant can produce an annual yield of 160 barrels of oil per 0.4 hectare for approximately $18.50 per barrel. *Euphorbia* plantings in California as well as *Euphorbia* and milkweed in Utah and Chinese tallow *(Sapium sebiferum)* trees in Texas attest to the interest of researchers in the future of oil plants. Investigators at the University of Minnesota are currently studying the common cattail as a potential energy source. Fuel, in the form of methanol, derived from *Eucalyptus* growing in Florida and Brazil, may in the future help to fill the void created by the dwindling supplies of oil from the Middle East. Many of the plant species that are being investigated possess attractive characteristics that make them especially desirable. They can often be grown on marginal land that is not satisfactory for growing food crops, require little rainfall and, of course, they are a renewable resource.

More traditional plants, including grains, sugar crops, potatoes, and other starchy plants are used to produce ethanol (ethyl alcohol), an ingredient in gasohol, which in the U.S. is a mixture of 10 percent alcohol and 90 percent unleaded gasoline and is therefore a gasoline extender. The synfuel, gasohol, first appeared at gas stations in the mid-1970s and experienced a meteoric rise. A yield of approximately 950 liters of alcohol is obtained from 0.4 hectare of corn, which is the major source of gasohol in the U.S., and with sweet sorghum, sugarcane, and sugarbeets, the yield is even higher. Furthermore, following distillation, there remains the mash or dried grain, which contains protein and can be used as livestock feed. The prediction for the early 1980s was that around 760–950 million liters of alcohol would be distilled and that by the mid-1980s, gasohol would be in widespread use. The Energy Security Act of 1980 set a target of 38 billion liters of fuel alcohol by 1990, or 10 percent of the nation's gasoline consumption. It is no longer a novelty, and presently gasohol is available in many states at well over 2,000 gas stations. In the state of Michigan, more than 200 stations sell the fuel. State sales of gasohol rose from less than 380,000 liters in September, 1979, to 16.3 million liters in March, 1981. Gasohol now makes up approximately 1 percent of all the automotive fuel sold in Michigan, and its popularity is even greater in Corn Belt states such as Iowa where approximately 30 percent of the unleaded fuel is gasohol. For a variety of reasons, however, the interest of the oil companies in gasohol and the consumption of this so-called "fuel for the future," have fallen off suddenly and dramatically. Energy-conservation measures that have paid off, new fuel finds, and the current recession have resulted in an oil glut, and this surplus has reduced oil prices and therefore the price of gasoline. Sky-rocketing heating oil and fuel prices in the last decade demanded that consumers conserve fuel, and automobile manufacturers, both foreign and domestic, have produced more fuel-efficient vehicles. Oil demand in the U.S. decreased 6.3 percent in 1981 and even larger reductions occurred in Great Britain, Japan, and France. Greatly expanded worldwide oil exploration and production by the U.S. and other countries such as Mexico, Egypt, and China as well as a widespread and crippling recession in the U.S. has resulted in a surplus of world oil. Furthermore, the cost of producing gasohol is greater than the production cost of gasoline particularly when the corn crop harvest is poor. Such costs are passed on to the consumer, and the price differential between gasohol and gasoline ceases to exist. However, in spite of some temporary relief from our energy woes, experts caution that long-term energy problems are one of man's challenges and heavy dependence on gasohol could easily become a reality in the future. In Brazil, which committed itself to the alcohol program in 1975, hectares of sugarcane grow alongside fuel tanks. Gasohol containing 20 percent ethanol was their initial answer to the energy crisis and, ultimately, vehicles powered with alcohol.

## Pollution abatement and landscaping

Although forms of pollution—water, air, solid waste, and noise—have been a part of our environment for centuries, the use of plants to control pollution is relatively new.

An aquatic floating weed called water hyacinth *(Eichhornia crassipes)*, long considered a nuisance, produces roots which are now known to naturally absorb toxic pollutants such as lead, mercury, and strontium 90, as well as some organic compounds. It has also been discovered that some aquatic plants, including cattail *(Typha angustifolia)*, pondweed *(Potamogeton pectin-*

*atus)*, and an alga *(Cladophora)* can be instrumental in removing pollutants such as heavy metals from water.

The National Institute for Environmental Studies of Japan has reported that sunflowers and poplars rid the air of auto exhaust pollution. Sunflowers are capable of absorbing both sulfur dioxide and nitrogen dioxide, while enzymes produced by poplars detoxify sulfur dioxide. As an added benefit, attractive birds eat sunflower seeds and roost in poplars. Again, at the Beltsville Agricultural Research Center, the pursuit and investigation of plants that might be resistant to pollution continue.

The male maidenhair tree *(Ginkgo biloba)* has now become a more common addition to the landscape because of its apparent ability to resist air pollutants and many pests.

While many of the uses of plants in landscaping remain the same, changes in our environment and our attitudes have altered some of the ways in which we now utilize plants in landscaping. There are many factors responsible for this new usage: population growth and density of population in both cities and suburbs; increasing concern about our environment; a need for privacy; increased food and energy costs; everyman's inate desire for a "touch of green"; and a resurgence of our love affair with plants.

As our cities have become more densely populated, it is necessary for people to live in ascending columns above the streets. Rooftop areas provide a place, sometimes limited, for a few potted plants, a small garden, a miniature farm, or even a few small trees. Though transpiration rate may be higher than normal, providing adequate light is not a problem. Suburbanites have other problems. They are not as concerned about a lack of space as they are about a lack of privacy around their patios and outdoor pools; plantings around such areas usually ease their anxiety. Wherever one lives, sharply rising food prices have convinced millions to "grow their own." Plants can also function as energy savers as well as sources of food. Indoors or outdoors, they generally improve the appearance of an area. Indoor landscaping has become fashionable and, one might add, excessive at times. Plantings along busy freeways certainly serve to relieve the monotony.

Whether it is home or industrial landscaping, there are some relatively new functional uses of plants—to conserve energy, to control pollution, erosion, and even to control pests. Nondeciduous evergreens, particularly on the northwest side of a building, can serve as effective windbreaks. Deciduous, shade trees on the southwest will provide shade in the summer, and along with vines, help reduce air-conditioning costs. Once these trees shed their leaves in the fall, the available solar radiation will increase and help reduce heating costs in the winter. Some plants are capable of removing gases and particulate matter, an especially important consideration on heavily traveled streets and in parking areas adjacent to residences and office buildings. Noise pollution, about which we are slowly, but fortunately becoming more concerned, can be reduced when dense evergreens, such as pine, spruce, arborvitae, and Douglas fir are utilized. To the manufacturer of the first noiseless power mower, this author would like to propose the awarding of the Nobel PEACE Prize. Erosion can reach serious proportions in bulldozed suburban areas but can be controlled, too, with plantings of ground covers, grass, shrubs, and trees. Though they are not new plants, the organic gardener heartily recommends the use of companion plants such as marigolds, catnip, chives, garlic, and horseradish to control pests.

## Problems of disease

Chemicals present in a tropical tree called *Virola surinamensis* have been reported as giving protection from schistosomiasis, a parasitic disease of man, resulting from an infestation of larvae of blood flukes or schistosomas, which are flatworms *(Schistosoma)*. Control of the disease, which is common in tropical America, Africa, and Asia, is also believed possible because of the presence of toxic compounds present in a plant related to ragweed; these compounds kill the larvae-containing snails, which are intermediate hosts of the parasite.

Alkaloids such as vincristine, vinblastine, and ajmalicine, present in a small, tropical shrub called periwinkle *(Catharanthus roseus)*, suggest additional medical value in plant species. Some success has been achieved using vincristine to prolong the life of leukemia patients, vinblastine in the treatment of Hodgkin's disease, and ajmalicine used for some heart diseases.

At the present time, 32 states in the U.S. have legalized the use of the drug marijuana *(Cannabis sativa)* for the treatment of particular medical disorders, and public health officials have opened legalized marijuana working programs for persons seeking medical attention in California, Georgia, Michigan, New Mexico, and Washington. The drug is known to relieve the very painful eye pressure in the blinding eye disease, glaucoma, and minimizes the nausea experienced by cancer patients who have undergone chemotherapy treatment. For a variety of reasons, the majority of patients and doctors in Michigan, at least, have not been overly enthusiastic about the use of the drug for glaucoma. It is likely that the long-standing, scientific-legal controversy concerning marijuana's medical value as opposed to its negative aspects will continue until, at least research unequivocally proves its worth.

## Genetic engineering

The "Green Revolution" of the 1960s and 1970s which was looked upon as the ultimate answer to the world hunger problem has proven in the 1980s not to be the panacea it was expected to be. Although greatly increased crop yields have been attained, the measures necessary to obtain these yields—heavy applications of fertilizer, the use of pesticides, mechanization of the farms, and irrigation—are much too costly for a world that is yearly becoming poorer and poorer in its supply of fossil fuels.

Now the farmer must look to the laboratory, again, for help. Perhaps genetic engineering can provide some of the answers.

Primitive varieties of crops, such as *Zea diploperennis*, a perennial, early relative of corn that will cross-breed with corn, must be sought out, brought back to laboratories for study, and their genetic contents preserved. From the use of such gene banks, plant breeders may develop new types of crop plants which exhibit characteristics that will enable them to thrive in the world to come. Gene banks will also insure the survival of plants that are in danger of becoming extinct.

Protoplast fusion is a process whereby the cell walls of two cells from different plants, such as potato and tomato, are dissolved by enzymatic action, and the resulting naked protoplasts merge, divide, and form a callus. This callus will grow and divide further and, eventually, produce a hybrid showing traits of both parents. The protoplast fusion technique has resulted in a number of plant improvements. Hybrid tobacco plants resistant to commercial crop diseases have been developed following the union of protoplasts of commercial and wild strains. This same tissue culture technique is also used to produce nursery plants and to study the development and survival of plants growing under adverse environmental conditions. Cloning by this means is also the approach used in forest tree improvement studies, but is in its infancy, and much work must be done before an entire tree results from protoplast fusion.

Work is also being done on the transfer of genes responsible for such traits as nitrogen fixation and disease resistance, from one cell to another. For example, if the nitrogen fixation gene found in legumes could be transferred to corn, it might result in a corn plant that would not require such large amounts of fertilizer to attain good yields.

The more conventional methods of plant breeding along with the new techniques of genetic engineering can be used in tandem to produce new crop plants and modify old ones so that man may be the beneficiary.

## REFERENCES

Baker, H. G. 1978. *Plants and civilization.* Belmont, Calif.: Wadsworth Publishing Co.

Barton, K. A., and Brill, W. J. 1983. Prospects in plant genetic engineering. *Science* 219:671–676.

Blair, J. G. 1982. Test-tube gardens. *Science 82* 3(1):70–76.

Borlaug, N. E. 1983. Contributions of conventional plant breeding to food production. *Science* 219:689–693.

Calvin, M. 1979. Petroleum plantations for fuel and materials. *BioScience* 29(9):533–38.

______. 1983. New sources for fuel and materials. *Science* 219:24–26.

Chaleff, R. S. 1983. Isolation of agronomically useful mutants from plant cell cultures. *Science* 219:676–682.

Chrispeels, M. J., and Sadava, D. 1977. *Plants, food, and people.* San Francisco: W. H. Freeman and Co.

Clark, W. 1980. China's green manure revolution. *Science 80* 1(6):68–73.

Delwiche, C. C. 1978. Legumes: past, present, and future. *BioScience* 28(9):565–70.

Erdman, M. D., and Erdman, B. A. 1981. *Calotropis procera* as a source of plant hydrocarbons. *Economic Botany* 35(4):467–72.

Farney, D. 1980. Agriculture Department's plant hunters risked lives to bring us new trees, shrubs, flowers, food crops. *Smithsonian* 11(3):128–40.

Farnum, P.; Timmis, R.; and Kulp, J. L. 1983. Biotechnology of forest yield. *Science* 219:694–702.

Feldman, M., and Sears, E. R. 1981. The wild gene resources of wheat. *Scientific American* 244(1):102–12.

Felker, P., and Bandurski, R. S. 1979. Uses and potential uses of leguminous trees for minimal energy input agriculture. *Economic Botany* 33(2):172–84.

Garmon, L. 1981. Getting guayule into the rubber market. *Science News* 119: 365–66.

Hammond, A. L. 1977. Alcohol: a Brazilian answer to the energy crisis. *Science* 195:564–66.

Heiser, C. B. 1981. *Seed to civilization.* San Francisco: W. H. Freeman and Co.

Karnosky, D. F. 1981. Potential for forest tree improvement via tissue culture. *BioScience* 31(2):114–20.

Maugh, T. H. 1977. Guayule and jojoba: agriculture in semiarid regions. *Science* 196:1189–90.

National Academy of Sciences (NAS). 1975. The winged bean, a high-protein crop for the tropics. Washington, D.C.: NAS. (An in-depth examination of the plant including its agronomy, composition, and nutritive value.)

———. 1975. Products from jojoba: a promising new crop for arid lands. Washington, D.C.: NAS.

———. 1976. Making aquatic weeds useful: some perspectives for developing countries. Washington, D.C.: NAS. (Examines methods for controlling aquatic weeds and using them to best advantage, especially those methods that show promise for less-developed countries.)

———. 1977. Jojoba feasibility for cultivation on Indian reservations in the Sonoran Desert Region. Washington, D.C.: NAS.

———. 1977. Guayule: an alternative source of natural rubber. Washington, D.C.: NAS. (An account of this rubber-producing shrub.)

Olmert, M. 1982. Uncle Sam, he had a farm. *Smithsonian* 12(12):54–63.

Parfit, M. 1981. Corn-belt enthusiasts, practicing what they preach, see our energy salvation in alcohol as a fuel. *Smithsonian* 11(12):44–53.

Pimentel, D. et al. 1981. Biomass energy from crop and forest residues. *Science* 212:1110–15.

Rodale, R. 1981. Amaranth: super nutrition, tasty leaves, new varieties ahead. *Organic Gardening* 28(7): 26–31.

Shepard, J. F. 1982. The regeneration of potato plants from leaf-cell protoplasts. *Scientific American* 246(5):154–66.

Shepard, J. F.; Bidney, D.; Barsby, T.; and Kemble, R. 1983. Genetic transfer in plants through interspecific protoplast fusion. *Science* 219:683–688.

Sprague, G. F.; Alexander, D. E.; and Dudley, J. W. 1980. Plant breeding and genetic engineering: a perspective. *BioScience* 30(1):17–21.

Venne, R. V., ed. 1982. Genetic engineering of plants. *California Agriculture* 36(8): 1–36.

Vietmeyer, N. 1979. The greening of the future. *Quest* 3(6):25–32.

______. 1979. A wild relative may give corn perennial genes. *Smithsonian* 10(9): 68–76.

______. 1981. America's forgotten crops. *National Geographic* 159(5):702–12.

Wells, J. R.; Kaufman, P. B.; and Jones, J. D. 1980. Heavy metal contents in some macrophytes from Saginaw Bay. *Aquatic Botany* 9:185–93.

# Metric system and conversion tables APPENDIX I

**A COMPARISON OF THE INTERNATIONAL METRIC SYSTEM AND THE ENGLISH SYSTEM OF MEASUREMENT[1]**

| 1 Centimeter | = .3937 inch | 1 Kilometer | = 1000 meters | 1 Gallon | = 3.785 liters |
|---|---|---|---|---|---|
| 1 Inch | = 2.54 centimeters | 1 Kilometer | = .62137 mile | 1 Gram | = 15.43 grains |
| 1 Foot | = 30.48 centimeters | 1 Sq. Centimeter | = .155 sq. inch | 1 Ounce | = 28.35 grams |
| 1 Meter | = 39.37 inches | 1 Sq. Decimeter | = 100 sq. centimeters | 1 Kilogram | = 1000 grams |
| 1 Meter | = 100 centimeters | 1 Cu. Centimeter | = .061 cu. inch | 1 Kilogram | = 2.205 pounds |
| 1 Meter | = 1.094 yards | 1 Cu. Decimeter | = 1000 cu. centimeters | 1 Pound | = 7000 grains |
| 1 Meter | = 1000 millimeters | 1 Cu. Meter | = 1000 liters | 1 Pound | = 4536 kilograms |
| 1 Yard | = .001 meter | 1 Fluid Ounce | = 29.54 milliliters | | |
| 1 Mile | = 1609.344 meters | 1 Liter | = 1000 cu. centimeters | | |
| | | 1 Liter | = 1.057 quarts | | |

[1]From Hummert Seed Co. Catalog, St. Louis, Mo.

## CONVERSION TABLE[1]

| MULTIPLY | TO OBTAIN | MULTIPLY | TO OBTAIN | MULTIPLY | TO OBTAIN |
|---|---|---|---|---|---|
| Bushels by .8 | Cu. Feet | Cu. Yards by 764600 | Cu. Centimeters | Liters by 1.057 | Quarts (Liq.) |
| Bushels by 4 | Pecks | Cu. Yards by 22 | Bushels | Meters by 100 | Centimeters |
| Bushels by .04545 | Cu. Yards | Cu. Yards by 27 | Cu. Feet | Meters by 3.281 | Feet |
| Centimeters by 0.3937 | Inches | Cu. Yards by 46.656 | Cu. Inches | Meters by 39.37 | Inches |
| Centimeters by 0.01 | Meters | Cu. Yards by 0.7646 | Cu. Meters | Meters .001 | Kilometers |
| Centimeters by 10 | Millimeters | Cu. Yards by 202.0 | Gallons | Meters by 1000 | Millimeters |
| Cu. Centimeters by .000001 | Cu. Feet | Cu. Yards by 764.6 | Liters | Meters by 1.094 | Yards |
| Cu. Centimeters by .06102 | Cu. Inches | Cu. Yards by 1616 | Pints (Liq.) | Microns by .000001 | Meters |
| Cu. Centimeters by .000001 | Cu. Meters | Cu. Yards by 807.9 | Quarts (Liq.) | Miles by 160900 | Centimeters |
| Cu. Centimeters by .000001308 | Cu. Yards | Fathoms by 6 | Feet | Miles by 5280 | Feet |
| Cu. Centimeters by .0002642 | Gallons | Feet by 30.48 | Centimeters | Miles by 1,609 | Kilometers |
| Cu. Centimeters by .001 | Liters | Feet by 12 | Inches | Miles by 1760 | Yards |
| Cu. Centimeters by .002113 | Pints (Liq.) | Feet by 0.3048 | Meters | Miles per Hr. by 44.70 | Centimeters per Sec. |
| Cu. Centimeters by .001057 | Quarts (Liq.) | Feet by 1/3 | Yards | Miles per Hr. by 88 | Feet per Min. |
| Cu. Feet by 28,317 | Cu. Centimeters | Gallons by 3785 | Cu. Centimeters | Miles per Hr. by 1.467 | Feet per Sec. |
| Cu. Feet by 1728 | Cu. Inches | Gallons by 0.1337 | Cu. Feet | Miles per Hr. by 1.609 | Kilometers per Hr. |
| Cu. Feet by 0.02832 | Cu. Meters | Gallons by 231 | Cu. Inches | Miles per Hr. by 0.8684 | Knots |
| Cu. Feet by 0.03704 | Cu. Yards | Gallons by .003785 | Cu. Meters | Miles per Hr. by 26.82 | Meters per Min. |
| Cu. Feet by 7.48052 | Gallons | Gallons by .004951 | Cu. Yards | Millimeters by 0.03937 | Inches |
| Cu. Feet by 28.32 | Liters | Gallons by 3.785 | Liters | Ounces by 2 | Tablespoons (liq.) |
| Cu. Feet by 59.84 | Pints (Liq.) | Gallons by 8 | Pints (Liq.) | Ounces by 6 | Teaspoons (liq.) |
| Cu. Feet by 29.92 | Quarts (Liq.) | Gallons by 4 | Quarts (Liq.) | Ounces by 3 | Tablespoons (dry) |
| Cu. Feet by 1.25 | Bushels | Gallons Water by 8.3453 | Lbs. of Water | Ounces by 9 | Teaspoons (dry) |
| Cu. Inches by 16.39 | Cu. Centimeters | Grams by 0.03527 | Ounces | Ounces by 28.349527 | Grams |
| Cu. Inches by .0005787 | Cu. Feet | Grams by 0.03215 | Ounces (Troy) | Ounces by 0.9115 | Ounces (Troy) |
| Cu. Inches by .00001639 | Cu. Meters | Grams by .002205 | Pounds | Ounces (Fluid) by 1.805 | Cu. Inches |
| Cu. Inches by .00002143 | Cu. Yards | Inches by 2.540 | Centimeters | Pounds by 16 | Ounces |
| Cu. Inches by .004329 | Gallons | Kilometers by 100000 | Centimeters | Pounds of Water by 0.01602 | Cu. Feet |
| Cu. Inches by .01639 | Liters | Kilometers by 3281 | Feet | Pounds of Water by 27.68 | Cu. Inches |
| Cu. Inches by 0.03463 | Pints (Liq.) | Kilometers by 1000 | Meters | Pounds of Water by 0.1198 | Gallons |
| Cu. Inches by 0.01732 | Quarts (Liq.) | Kilometers by 0.6214 | Miles | Tablespoons (liq.) by 0.5 | Ounces |
| Cu. Meters by 1,000,000 | Cu. Centimeters | Kilometers by 1094 | Yards | Tablespoons (dry) by 0.3333 | Ounces |
| Cu. Meters by 35.31 | Cu. Feet | Liters by 1000 | Cu. Centimeters | Tablespoons by 3 | Teaspoons |
| Cu. Meters by 61.023 | Cu. Inches | Liters by 0.03531 | Cu. Feet | Teaspoons (liq.) by 0.1666 | Ounces |
| Cu. Meters by 1.308 | Cu. Yards | Liters by 61.02 | Cu. Inches | Teaspoons (dry) by 0.1111 | Ounces |
| Cu. Meters by 264.2 | Gallons | Liters by .001 | Cu. Meters | Teaspoons by 0.3333 | Tablespoons |
| Cu. Meters by 1000 | Liters | Liters by .001308 | Cu. Yards | Temp (C) + 17.78 by 1.8 | Temp (F) |
| Cu. Meters by 2113 | Pints (Liq.) | Liters by 1.2642 | Gallons | Temp (F) − 32 by 5/9 | Temp (C) |
| Cu. Meters by 1057 | Quarts (Liq.) | Liters by 2.113 | Pints (Liq.) | Tons (Long) by 2240 | Pounds |

[1]From Hummert Seed Co. Catalog, St. Louis, Mo.

## RATE OF APPLICATION[1] EQUIVALENT TABLE

| Liquid Materials | | | Dry Materials | | |
|---|---|---|---|---|---|
| *Rate per Acre* | *Rate per 1000 Sq. Ft.* | *Rate per 100 Sq. Ft.* | *Rate per Acre* | *Rate per 1000 Sq. Ft.* | *Rate per 100 Sq. Ft.* |
| 1 pt. | ¾ T | ¼ t | 1 lb. | 2½ T | ¼ t |
| 1 qt. | 1½ T | ½ t | 3 lbs. | 2¼ T | ¾ t |
| 1 gal. | 6 T | 2 t | 4 lbs. | 3 T | 1 t |
| 25 gals. | 4½ pts. | ½ pt. | 5 lbs. | 4 T | 1¼ t |
| 50 gals. | 4½ qts. | 1 pt. | 10 lbs. | ½ cup | 2 t |
| 100 gals. | 8 qts. | 1 pt. | 100 lbs. | 2¼ lbs. | ¼ lb. |
| 200 gals | 4½ qts. | 2 qts. | 200 lbs. | 4½ lbs. | ½ lb. |
| 300 gals. | 6¾ gals. | 3 qts. | 300 lbs. | 6¾ lb. | ¾ lb. |
| 400 gals. | 9 gals. | 1 gal. | 400 lbs. | 9 lbs. | 1 lb. |
| 500 gals. | 11¼ gals. | 1¼ gals. | 500 lbs. | 11¼ lbs. | 1¼ lbs. |

## DILUTION OF LIQUID PESTICIDES[1] AT VARIOUS CONCENTRATIONS

| *Dilution* | *1 Gal.* | *3 Gal.* | *5 Gal.* | *15 Gal.* |
|---|---|---|---|---|
| 1– 100 | 2 tbs. + 2 tsp. | ½ cup | ¾ cup + 5 tsp. | 1 cup + 3 tbs. |
| 1– 200 | 4 tsp. | ¼ cup | 6½ tbs. | ½ cup + 2 tbs. |
| 1– 400 | 2 tsp. | 2 tbs. | 3 tbs. | 4 tbs. + 2½ tsp. |
| 1– 800 | 1 tsp. | 1 tbs. | 1 tbs. + 2 tsp. | 3 tbs. + 2½ tsp. |
| 1–1000 | ¾ tsp. | 2¼ tsp. | 1 tbs. + 1 tsp. | 1 pt. + ½ cup |

## LIQUID EQUIVALENT TABLE

| *100 Gal.* | *25 Gal.* | *12½ Gal.* | *5 Gal.* | *1 Gal.* |
|---|---|---|---|---|
| 2 gal. | 2 qt. | 1 qt. | 12⅞ oz. | 2½ oz. |
| 1 gal. | 1 qt. | 1 pt. | 6½ oz. | 1¼ oz. |
| 2 qts. | 1 pt. | 8 oz. | 3¼ oz. | ⅝ oz. |
| 1 qt. | ½ pt. | 4 oz. | 1 9/16 oz. | 5/16 oz. |
| 1½ pt. | 6 oz. | 3 oz. | 1¼ oz. | ¼ oz. |
| 1 pt. | 4 oz. | 2 oz. | ⅞ oz. | 3/16 oz. |
| 8 oz. | 2 oz. | 1 oz. | 7/16 oz. | ½ tsp. |
| 4 oz. | 1 oz. | ½ oz. | ¼ oz. | ¼ tsp. |

## SOLID EQUIVALENT TABLE

| *100 Gal.* | *25 Gal.* | *12½ Gal.* | *5 Gal.* | *1 Gal.* |
|---|---|---|---|---|
| 4 oz. | 1 oz. | ½ oz. | 3/16 oz. | ¼ tsp. |
| 8 oz. | 2 oz. | 1 oz. | ⅜ oz. | ½ tsp. |
| 1 lb. | 4 oz. | 2 oz. | ⅞ oz. | 1 tsp. |
| 2 lb. | 8 oz. | 4 oz. | 1¾ oz. | 2 tsp. |
| 3 lb. | 12 oz. | 6 oz. | 2⅜ oz. | 2⅖ tsp. |
| 4 lb. | 1 lb. | 8 oz. | 3¼ oz. | 3⅘ tsp. |
| 5 lb. | 1¼ lb. | 12 oz. | 4 oz. | 4⅘ tsp. |

[1]From Hummert Seed Co. Catalog, St. Louis, Mo.

### LIQUID CONVERSIONS[2]

1 teaspoon = 1/3 tablespoon = 1/6 oz.
1 tablespoon = 3 teaspoons = 1/2 oz.
1 fluid oz. = 2 tablespoons = 6 teaspoons
1 pt./100 gal. = 1 teaspoon/gallon
1 qt./100 gal. = 2 tablespoons/gallon

### LIQUID EQUIVALENT TABLE[2]

| *Concentrate* | *Water = Concentrate* | *Water* |
|---|---|---|
| 2 gallons | 100 gal. = 2½ oz. | 1 gallon |
| 1 gallon | 100 gal. = 1¼ oz. | 1 gallon |
| 2 quarts | 100 gal. = 5/8 oz. | 1 gallon |
| 1 quart | 100 gal. = 5/16 oz. | 1 gallon |
| 1½ pints | 100 gal. = 1/4 oz. | 1 gallon |
| 1 pint | 100 gal. = 3/16 oz. | 1 gallon |
| 8 oz. | 100 gal. = 1/2 teaspoon | 1 gallon |
| 4 oz. | 100 gal. = 1/4 teaspoon | 1 gallon |

### SOLID EQUIVALENT TABLE[2]

| *Concentrate* | *Water = Concentrate* | *Water* |
|---|---|---|
| 1 pound | 100 gal. = 1 tablespoon | 1 gallon |
| 8 ounces | 100 gal. = 1/2 teaspoon | 1 gallon |
| 4 ounces | 100 gal. = 1/4 teaspoon | 1 gallon |

### TABLE OF DILUTIONS FOR LIQUIDS AND DUSTS[2]

*Equivalent Quantities of Liquid Materials When Mixed by Parts*

| *Water* | *1-400* | *1-800* | *1-1600* |
|---|---|---|---|
| 100 gal. | 1 qt. | 1 pt. | 1 cup |
| 50 gal. | 1 pt. | 1 cup | 1/2 cup |
| 5 gal. | 3 tbsp. | 5 tsp. | 2½ tsp. |
| 1 gal. | 2 tsp. | 1 tsp. | 1/2 tsp. |

### WEIGHTS[2]

28.35 grams = 1 ounce
16 ounces = 1 pound = 453.6 grams
1 kilogram = 1000 grams = 2.205 pounds
1 gallon water = 8.34 pounds = 3.8 kilograms
1 cubic foot water = 62.4 pounds = 28.3 kilograms
1 kilogram water = 33.81 ounces
1 gallon No. 2 fuel oil = 7 pounds
1 gallon kerosene = 6.7 pounds
1 ton = 2000 pounds = 907 kilograms
1 metric ton = 1000 kilograms = 2205 pounds

[2]From Brawley Seed Co. Catalog, Mooresville, NC.

## VOLUME AND LIQUID MEASURE[2]

3 teaspoons = 1 tablespoon = 14.8 cc.
2 tablespoons = 1 fluid ounce = 29.6 cc.
8 fluid ounces = 16 tablespoons = 1 cup = 236.6 cc.
2 cups = 32 tablespoons = 1 pint = 473.1 cc.
2 pints = 64 tablespoons = 1 quart = 946.2 cc.
1 liter = 1000 cc. = 0.2642 gallon = 33.81 ounces
4 quarts = 256 tablespoons = 1 gallon = 3785 cc.
1 gallon = 128 fluid ounces = 231 cubic inches = 3785 cc.
1 bushel = 2.244 cubic feet = 9.309 liquid gallons = 35.2 liters
1 cc = 1 ml.

## EQUIVALENT QUANTITIES OF DRY MATERIALS (WETTABLE POWDERS) FOR VARIOUS QUANTITIES OF WATER[2]

| *Water* | *Quantity of Material* | | | | | |
|---|---|---|---|---|---|---|
| 100 gal. | 1 lb. | 2 lb. | 3 lb. | 4 lb. | 5 lb. | 6 lb. |
| 50 gal. | 8 oz. | 1 lb. | 1½ lb. | 2 lb. | 2½ lb. | 3 lb. |
| 5 gal. | 3 tbsp. | 1½ oz. | 2½ oz. | 3¼ oz. | 4 oz. | 5 oz. |
| 1 gal. | 2 tsp. | 1 tbsp. | 1½ tbsp. | 2 tbsp. | 3 tbsp. | 4 tbs. |

## EQUIVALENT QUANTITIES OF LIQUID MATERIALS (EMULSION CONCENTRATES, ETC.) FOR VARIOUS QUANTITIES OF WATER[2]

| *Water* | *Quantity of Material* | | | | | |
|---|---|---|---|---|---|---|
| 100 gal. | ½ pt. | 1 pt. | 2 pt. | 3 pt. | 4 pt. | 5 pt. |
| 50 gal. | 4 fl. oz. | 8 fl. oz. | 1 pt. | 24 fl. oz. | 1 qt. | 2½ pt. |
| 5 gal. | 1 tbsp. | 1 fl. oz. | 1½ fl. oz. | 2¼ fl. oz. | 3 fl. oz. | 4 fl. oz. |
| 1 gal. | ½ tsp. | 1 tsp. | 2 tsp. | 3 tsp. | 4 tsp. | 5 tsp. |

[2]From Brawley Seed Co. Catalog, Mooresville, NC.

# A directory of selected plant societies APPENDIX II

**AFRICAN VIOLET Society of America, Inc.,** PO Box 1326, Knoxville, TN 37901. $6.00 a year. Founded 1946; 30,000 members. Organized to stimulate interest in propagation and culture of African violets, to foster distribution of varieties and to gather and publish reliable information on African violets. Publishes the *African Violet Magazine* five times a year. Brochure on African violet culture and fact sheet on the Society free on request.

**American BEGONIA Society,** Kathy Brown, Membership Secretary, 10692 Bolsa St. #14, Garden Grove, CA 92643. $6.00 a year. Founded 1932. Organized to stimulate and promote interest in *begonias* and other shade-loving plants; to encourage the introduction and development of new types of these plants; to standardize the nomenclature of *begonias;* and to gather and publish information in regard to kinds, propagation and culture of *begonias* and companion plants. Publishes a monthly magazine, *The Begonian.*

**BONSAI Clubs International,** PO Box 2098, Sunnyvale, CA 94087. $13.00 a year, $24.00 for two years. Outside U.S., add one dollar. Founded 1959; 3000 members. Organized to exchange and publish educational materials and promote activities to increase appreciation and understanding of bonsai, and to improve the quality of bonsai growing. B.C.I. is not a club, but a federation of clubs. Publishes *Bonsai Magazine* ten times a year. Brochure free on request.

**The BROMELIAD Society, Inc.,** PO Box 41261, Los Angeles, CA 90041. $10.00 a year. Founded 1950; 3500 members. Organized to promote and maintain public and scientific interest and research in bromeliads throughout the world. Publishes the *Journal of the Bromeliad Society* six times a year.

**CACTUS AND SUCCULENT Society of America, Inc.,** PO Box 3010, Santa Barbara, CA 93105. $20.00 a year; 6000 members. Publishes the *Cactus and Succulent Journal* six times a year; publication is now in its fifty-second year. Send 20¢ stamp for a catalog of cactus and succulent books and information about the Society.

**CARNIVOROUS PLANT Newsletter,** Fullerton Arboretum, Fullerton, CA 92634. $10.00 a year. Founded 1972; 800 members. The *Newsletter* is published quarterly and includes articles on taxonomy, history and culture for amateurs as well as professional botanists. A seed bank is available to subscribers. Fact sheet free on request.

**American FERN Society, Inc.,** Dr. Judith E. Skog, Membership Secretary, Dept. of Biology, George Mason University, Fairfax, VA 22030. Membership fee of $5.00 a year includes newsletter; $8.00 a year includes newsletter and journal. Founded 1893; 800 members. Organized to foster a scholarly interest in ferns and fern allies and to stimulate interest in native, exotic and cultivated ferns. Publishes the scientific quarterly *American Fern Journal* and the bimonthly *Fiddlehead Forum* for amateur botanists and horticulturists.

**American GLOXINIA and GESNERIAD Society,** PO Box 312, Ayer, MA 01432. $7.00 a year. Founded 1951; 3000 members. Organized to stimulate interest in and publish reliable information about the identification, nomenclature, culture and propagation of *gloxinias* and other gesneriads and to encourage the origination and introduction of new cultivars. Publishes a bimonthly magazine, *The Gloxinian,* and a manual *How to Know And Grow Gesneriads.* Send self-addressed stamped envelope for illustrated brochure.

**GESNERIAD Society International,** PO Box 549, Knoxville, TN 37901. $6.00 a year. Organized to stimulate interest in the gesneriads, which include *Episcia, Columnea* and *Streptocarpus. Gesneriad Saintpaulia News,* published six times a year, is the official publication of this and an affiliated group, Saintpaulia International. Brochure on care and culture of gesneriads free on request.

**International GERANIUM Society,** 6501 Yosemite Dr., Dept. HPPG, Buena Park, CA 90650. $5.00 a year. Founded 1953; 1500 members. Organized to promote interest in geraniums and *pelargoniums.* Publishes a quarterly, *Geraniums Around The World.* Seed exchange for members. Sample copies of the magazine available for $1.00 a copy.

**Hobby GREENHOUSE Association,** Mrs. Joyce Smith, Membership Chairman, 1607 N. 49th, Omaha, NE 68104. $5.00 a year. Founded 1975; 3200 members. Organized to bring greenhouse and plant buffs together to exchange information. Presently interested in solar energy also. Publishes a bimonthly, *The Planter.*

**American HORTICULTURAL Society,** Mount Vernon, VA 22121. $15.00 a year. Founded in 1922; 25,000 members. Organized to promote and expand interest in the horticultural sciences. Continues today with that purpose, as well as to promote interest in leisure pursuit of gardening throughout the country. Publishes a bimonthly magazine, *American Horticulturist,* and a bimonthly newsletter, *News & Views.*

The AHS headquarters is situated on 25 acres of land along the Potomac River that once was a part of George Washington's Mount Vernon estate. The property contains hundreds of different plants and several experimental gardens. The grounds are open to the public Monday to Friday, from 9 a.m. to 5 p.m., and from 8:30 a.m. to noon Saturdays.

**HYDROPONIC Society of America,** Carlene Anderson, Secretary, PO Box 516, Brentwood, CA 94513. $30.00 a year. 1000 members. Organized to provide information about techniques of growing plants by soilless culture. Provides books, pamphlets, university documents and a hydroponics bibliography. Publishes a quarterly journal and newsletters. Free literature on request.

**The HOYA Society International,** Mrs. Danny Greenhaw, Membership Secretary, Rt. 2, Box 222-B, Kaufman, TX 75142. $12.50 individual, $17.50 commercial. Founded 1979; 200 members. Organized to bring *Hoya* lovers together to share knowledge and plant material, and to bring order to the present confused state of *Hoya* nomenclature. Publishes a quarterly, *The Hoyan.*

**Indoor LIGHT GARDENING Society of America, Inc.,** Membership Secretary, c/o New York Horticultural Society, 128 W. 58 St., New York, NY 10019. $8.00 a year. Founded 1966; 3000 members. Organized to disseminate information on all aspects of growing plants under artificial light. Publishes a bimonthly magazine, *Light Garden,* and various cultural guides.

**American ORCHID Society, Inc.,** Botanical Museum of Harvard University, Cambridge, MA 02138. $20.00 a year. Founded 1932; 23,000 members. Dedicated to the furtherance of horticultural, botanical and scientific interest in orchids. More than 400 affiliated orchid societies in over 70 countries are guided by this organization. Publishes a monthly, 96-page *American Orchid Society Bulletin* and an *Awards Quarterly,* a record of all A.O.S.-awarded orchids. Free brochure on request.

**American PEONY Society,** Greta M. Kessenich, Secretary, 250 Interlachen Rd., Hopkins, MN 55343. $7.50 a year. Founded 1904; 700 members. Organized to increase general interest in herbaceous and tree peonies and their uses, cultivation and care. Publishes a quarterly *American Peony Society Bulletin.* Free peony seeds are given to anyone interested in hybridizing the plants. The A.P.S. has published a new book, *The Best of 75 Years,* available for $15.00.

**The PEPEROMIA Society,** Betty L. Davenport, Membership Secretary, 2013 Road 44, Pasco, WA 99301. $3.00 a year (individual), $4.50 a year (family). Founded 1977; 100 members. Organized to encourage interest in and culture of peperomias and serve as a clearing-house of information and proper nomenclature. Publishes a quarterly, *Peperomia Gazette.* Send self-addressed stamped envelope for brochure.

**New England WILDFLOWER Society, Inc.,** Membership Secretary, Garden In The Woods, Hemenway Rd., Framingham, MA 01701. $10.00 a year, $12.50 for families, $25.00 sustaining (includes an autographed copy of *Newcomb's Wildflower Guide*). Founded 1922; 1800 members. Six sanctuaries in New England are owned and maintained by the Society. The most important of these is Garden In The Woods, a 45-acre botanical garden which contains 2500 species and varieties of plants, including the largest living collection of native plants in the Northeast. The Society supports an active education program for both children and adults.

**SAINTPAULIA International,** PO Box 549, Knoxville, TN 37901. $6.00 a year. Organized in 1963 for the purpose of bringing together people interested in growing and showing African violets. *Gesneriad Saintpaulia News,* published six times a year, is the official publication of this and an affiliated group, the Gesneriad Society International. Brochure on care and culture of African violets free on request.

**The TERRARIUM Association,** Robert C. Baur, 57 Wolfpit Ave., Norwalk, CT 06851. $8.00 a year. Founded 1974. Organized to encourage continuing interest in gardening in glass containers and to provide information to garden groups and individuals. Membership includes four issues of the quarterly *Terrarium Topics* and all publications on their list. Send self-addressed stamped envelope for information.

# A listing of indoor gardening catalogs[1] APPENDIX III

## HOUSE PLANTS

*A.B.C. Herb Nursery:* Dept. HP05, Rt. #1 Box 313, Lecoma, MO 65540. African Violets and herbs. Brochure 25¢.

*The Alestaker:* Elkwood, VA 22718. Ivies. Catalog $1.00.

*Amaryllis Inc.:* PO Box 318, Baton Rouge, LA 70821. Catalog free.

*Antonelli Bros:* 2545-HPG Capitola Rd., Santa Cruz, CA 95062. Begonias. Catalog free.

*Applewood Seed Co:* 833 Parfet St., Lakewood, CO 80215. Indoor and outdoor seeds. Catalog free.

*Brant's Exotic Greenhouses:* Rt. 3, Box 72-HPG, Bessemer, AL 35020. Catalog $1.50.

*The Banana Tree:* 245-HP N. Ninth St., Allentown, PA 18102. Bananas, other tropicals. Catalog 50¢.

*Bumer Inc:* Dept. HP, 9900 N. Michigan Rd., Carmel, IN 46032. Greenhouse cuttings. Catalog free.

*Martin Beckerman:* 454 Fort Washington Ave., Apt. 50B, New York, NY 10033. Marantaceae. Send stamp for list.

*Begonia Paradise Nursery:* 9471 Dana Rd., Cutler Ridge, FL 33157. List 35¢.

*Bishop Nursery:* PO Box 303, Galesburg, MI 49053. Catalog with instruction book free.

*Black Copper Kits:* 266HP Kipp St., Hackensack, NJ 07601. Carnivorous plants. Catalog 25¢.

*Bombadil's Plants:* 2126A E. Locust, Milwaukee, WI 53211. Begonias, gesneriads. Catalog 50¢.

*The Bonsai Farm:* Rt. 1, Box 156-B1, Adkins, TX 78101. Catalog $1.00.

*Bonsai Information Guild Inc:* Box 6140HP, Shirlington Station, Arlington, VA 22206. Information free.

*John Brudy's Rare Plant House:* PO Box 1348 PG, Cocoa Beach, FL 32931. Seeds. Catalog $1.00.

*Brussel's Bonsai Nursery:* 308 Colonial Rd., Memphis, TN 38177.

*Burgess Seed & Plant Co:* Dept. HPG, Galesburg, MI 49053. Catalog with instruction book free.

*Carobil Farms:* Dept. B, Church Rd., Brunswick, ME 04011. Geraniums. Catalog 35¢.

*Carolina Exotic Gardens:* PO Box 1492, Greenville, NC 27834. Price list 25¢.

*Chung's Mountain Studio:* Box 47-K, PO Box 1216, RR #1, Gold Beach, OR 97444. Bonsai. Write for information.

*Country Hills Greenhouse:* Dept. HPG, RR #1, Corning, OH 43730. Catalog $1.00.

*Dow Seeds:* PO Box 30144, Honolulu, HI 96820. Palms. Catalog free.

[1]From *House Plants and Porch Gardens*, October, 1978.

*Endangered Species:* 842 Walnut Ave., Carpinteria, CA 93013. Rare plants. Catalog $1.00.

*Exotic Plants Hawaii:* PO Box 23-HP32, Hilo, HI 96720. Anthuriums.

*Exotica Seed Co:* 820 S. Lorraine Blvd., Los Angeles, CA 90005. Catalog $1.00.

*Exotics Hawaii Ltd:* 1344 Hoakoa Pl., Honolulu, HI 96821. Send self-addressed stamped envelope for brochure.

*Flower Bulbs Inc:* PO Box 344, Edison, NJ 08817. Amaryllis. Catalog free.

*Jim Fobel:* Dept. HP, 598 Kipuka Pl., Kailua, HI 96734. Catalog and 3 packets of seed $1.00.

*Four Winds Growers:* PO Box 3538, Fremont, CA 94538. Dwarf Citrus.

*Herb Fuller Dignitas:* Box 308-HP, 801 W. Vine, Wopakoneta, OH 45895. Ivies.

*Fuchsiarama:* Dept. KB, PO Box 1153, Fort Bragg, CA 95437. Catalog $1.00.

*Fuku-Bonsai:* Box 178G, Homestead Rd., Kurtistown, HI 96760. Send SASE for listings.

*The Garden Spot:* 4032 Rosewood Dr., Columbia, SC 29205. Ivies. Send self-addressed stamped envelope for list.

*The Gift Horse Greenhouse:* Dept. HP-319, Rt. #1, New Johnsonville, TN 37134. Begonias. List 50¢.

*Green Plant Research:* PO Box 735P, Kanawa, HI 96730. Catalog $1.00.

*Greenland Flower Shop:* Dept. HPG-2, RD #1, (Stormstown), Port Matilda, PA 16870. Exotic house plants. Catalog 30¢.

*Greenlife Gardens Greenhouses:* Box HP, 164 Meadovista Rd., Griffin, GA 30223. Catalog 50¢.

*Robert B. Hamm:* Dept. HP, 2951 Elliott, Wichita Falls, TX 76308. Begonias. Quarterly lists $1.50.

Hana Gardenland: Dept. HPP-10, PO Box 177, Hana, HI 96713. Brochure free.

*Heatherbloom Nursery:* Box 230-100, Rt. 1, Lakeville, CT 06039. New Guinea impatiens. Catalog $1.00.

*Heavenly Flowers Hawaii:* Dept. KC-3, PO Box 4040, Honolulu, HI 96813.

*Dorothy L. Henkie:* PO Box 2514, Capistrano Beach, CA 92624. Aroids. Write for information.

*Herb Shop:* Dept. HPP-10, Box 362, Fairfield, CT 06480. Catalog 50¢.

*The Herb Shop:* Dept. HP, 1942½ Cerrillos Rd., Santa Fe, NM 87501. Catalog 35¢.

*Herbs 'N Honey Nursery:* Rt. 2, Box 205, Monmouth, OR 97361. Information free. Botanical catalog $2.00.

*Hidden Springs Nursery:* Dept. HP, Rt. 1, Box 186-1A, Cookeville, TN 38501. Fuchsias. List 40¢.

*Hilltop Herb Farm:* PO Box 1734, Cleveland, TX 77327. Also house plants. Mini-catalog 50¢.

*Jerry Horne:* Dept. HB, 10195 S.W. 70 St., Miami, FL 33173. Send stamp for list.

*Hortica Gardens:* PO Box 308-HP, Placerville, CA 95667. Bonsai. Catalog 25¢.

*Horticulture Associates:* PO Box 1008, Indian Hills, CO 80450. Catalog 50¢.

*House of Wesley:* Dept. 1949-39, 2200 E. Oakland Ave., Bloomington, IL 61701. Catalog free.

*Howe Hill Herbs:* Dept. HP, Howe Hill Rd., Camden, ME 04843. Also scented geraniums. Catalog 35¢.

*Hurov's Tropical Seeds:* Dept. HP, PO Box 10387, Honolulu, HI 96816. Rare seeds. List free.

*International Growers Exchange Inc:* Box 397, HH, Farmington, MI 48024. Rare bulbs and plants. Catalog $2.00.

*J.L. Hudson:* PO Box 1058K, Redwood City, CA 94064. Seeds. Catalog 50¢.

*Japan Bonsai:* 135-HP W. 28th St., New York, NY 10001. Catalog free.

*Kuaola Farms, Ltd:* PO Box 4038, Hilo, HI 96720. Anthuriums. Price list free.

*Lehua Anthurium Nursery:* Dept. HP, 80 Kokea St., Hilo, HI 96720. Catalog free.

*Leilani Nursery:* 69 HP, Kapualani St., Hilo, HI 96720. List free.

*Le Jardin Du Gourmet:* Box 404-HP, West Danville, VT 05873. Herbs. Catalog 25¢.

*Robert Lester's Bamboo:* 280 W. 4th St., New York, NY 10014. List free.

*Logee's Greenhouses:* Dept. HPP-10, 55 North St., Danielson, CT 06239. Begonias, other house plants. Catalog $2.00.

*Loyce's Flowers:* Dept. HP, Rt. 2, Box 11, Granbury, TX 76048. Bougainvillea and hoyas. Catalog 50¢.

*Maile's Anthuriums:* 41-1019 Kakaina St., Waimanalo, HI 96795. Send SASE for list.

*McDaniel's Miniature Roses:* Dept. PG, 7523 Zemco St., Lemon Grove, CA 92045. Catalog free.

*Merry Gardens:* PO Box 595-PG, Camden, ME 04843. Catalog 50¢.

*Michigan Bulb:* Dept. HP-ED, 1950 Waldorf, Grand Rapids, MI 49550. Catalog free.

*Miniature Plant Kingdom:* Dept. HP, 4125 Harrison Grade Rd., Sebastopol, CA 95472. Catalog free.

*Mini-Roses:* Sta. A, Box 4255-B, Dallas, TX 75208. Catalog free.

*Mobile Gardens:* Dept. 078, RR #1, Box 250, Salisbury Turnpike Rd., Rhinebeck, NY 12572. Tropical foliage. Catalog 50¢.

*Nor'East Miniature Roses Inc:* Dept. HPPP, 58 Hammond St., Rowley, MA 01969. Catalog free.

*Osceola Gardens:* Dept. ED, PO Box 1229, St. Cloud, FL 32769. Write for list.

*Palm Coast Nursery Corp:* 15321 One Mile Rd., Boca Raton, FL 33444. Staghorn ferns. List free.

*Park Gardens:* 1435 Huntington Tpke, Trumbull, CT 06611. List free.

*Geo. W. Park Seed Co:* Dept. 91, PO Box 34, Greenwood, SC 29647. Catalog free.

*Peter Pauls:* Rt. 21, Canandaigua, NY 14424. Carnivorous plants. Catalog 25¢.

*Pixie Treasures:* Dept. HP, 4121 Prospect Ave., Yorba Linda, CA 92686. Miniature roses. Catalog 25¢.

*Plant Gallery:* 1491 Vaughn Rd., Mt. Vernon, WA 98273. Send 75¢ and SASE for list.

*Plantmaster:* PO Box 2486. Allentown, PA 18105. Catalog 50¢.

*Plant Oddities:* Dept. HP8, Box 94, Kennebunk, ME 04043. Carnivorous plants. Catalog 25¢.

*The Plant Room:* 6373 Trafalgar Rd., Hornby, Ontario, Canada LOP 1E0. Catalog $1.00.

*Plants from Morrison/Heffner:* Dept. HPP-10, 5305 S.W. Hamilton St., Portland, OR 97221. Begonias and gesneriads. List 35¢.

*The Plant Shop Botanical Garden:* 18007 Topham St., Reseda, CA 91335. Begonias. Catalog $1.00.

*Polk Miniature Roses:* Dept. HZ, PO Box 1051, 1200 Hill St., Hannibal, MO 63401. Catalog free.

*The Pot Shop:* Dept. HP-108, The Buttons, 5070 Ivy Rd. NE, Atlanta, GA 30342. Ceropegias, hoyas. List 25¢.

*The Redwood Planter:* Dept. HP, R#2, Halfway, MO 65663. House plants. Booklet free.

*Redwood City Seed Co.:* Box 361-OD, Redwood City, CA 94064. Catalog 25¢.

*Rex Bulbs:* Box 774, Port Townsend, WA 98368. Catalog 10¢.

*Routh's Greenhouse:* Dept. HP, Louisburg, MO 65685. Begonias. Catalog 25¢.

*Rutland of Kentucky:* 3 Bon Haven, #1, Maysville, KY 41056. Herbs. Catalog $1.50.

*Shadow Lawn Nursery:* Dept. BC, 637 Holly Lane, Plantation, FL 33317. Seeds. Catalog 50¢.

*Shady Hill Gardens:* 821 Walnut St., Batavia, IL 60510. Geraniums.

*Small World Miniature Roses:* PO Box 562B, Rogue River, OR 97537. Catalog free.

*The Specialty Plant Nursery:* Dept. 12-HPPG, 6728 18th NW, Seattle, WA 98117. Miniature roses and miniature African violets. Catalog 25¢.

*Sun Dew:* PO Box 11-H, Denver, NY 12421. Carnivorous plants. Catalog 50¢.

*Sunnybrook Farms Nursery:* Dept. HPS, PO Box 6, Chesterland, OH 44026. Catalog 50¢.

*Sunrise Plants:* Dept. HPP-1, Box 481, Robbinsville, NJ 08691. Carnivorous and succulent plants. Catalog 25¢.

*Sunset Nurseries:* Dept. HP, 4007 Elrod Ave., Tampa, FL 33616. Bonsai. Send self-addressed stamped envelope for catalog.

*Taylor's Herb Garden, Inc.:* Dept. HPP-10, 1535 Lone Oak Rd., Vista, CA 92083. Catalog 50¢.

*Tecumseh Greenhouse:* Mohawk Rd., Tecumseh, MI 49286. Catalog 50¢.

*Millie Thompson:* 310A Hill St., PO Drawer PP, Southampton, Long Island, NY 11968. Begonias. List 35¢.

*Unilab Products:* PO Box 84-PG, Redlands, CA 92373. Palms. Write for information.

*Van Bourgondien:* 245 Farmingdale Rd., Box A-HP-1, Babylon, NY 11702. Catalog free.

*Western Arboretum:* Dept. HPP-10, PO Box 2827, Pasadena, CA 91105. Bonsai. Catalog $1.00.

*Wilson Bros.:* Dept. HP, Roachdale, IN 46172. African violets, geraniums. Catalog free.

*World Insectivorous Plants:* Catalog Dept., Rt. 3, Box 338-S, Arroyo Grande, CA 93420. Catalog 50¢.

*Yankee Peddler Herb Farm:* Dept. HP-2, Rt. 4, Hwy. 36 N, Brenham, TX 77833. Also scented geraniums. Catalog $1.00.

## ORCHIDS AND BROMELIADS

*Alberts & Merkel Bros., Inc.:* HPP-78, 2210 S. Federal Hwy., Boynton Beach, FL 33435. Orchid list $1.00. Foliage list and bromeliad brochure $1.50.

*Armacost & Royston:* 3376 Foothill Rd., PO Box 385, Carpinteria, CA 93013. Orchids. Catalog $2.00.

*Bates Orchids:* 7911 US 301, Ellenton, FL 33532. Catalog free.

*Beach Garden Nursery:* 2131-PG Portola Dr., Santa Cruz, CA 95062. Bromeliads.

*Black River Orchids, Inc.:* Dept. HP, PO Box 110, So. Haven, MI 49090. List free.

*Blossom World Bromeliads:* Dept. HP, Rt. 2, Box 479-B, Sanford, FL 32771. Regular list, send self-addressed stamped envelope. Extensive list 50¢.

*Cornelison's Bromeliads:* 225 San Bernardino St., N. Fort Myers, FL 33903. Send 20¢ stamp for list.

*Down To Earth:* Dept. HHP, PO Box 1972, 1611 Jackson St., Hollywood, FL 33020. Send self-addressed stamped envelope for catalog.

*Edelweiss Gardens:* Box 66-78E, Robbinsville, NJ 08691. Orchids. Catalog 50¢.

*Everglades Orchids:* Dept. OD, 1101 Tabit Rd., Belle Glade, FL 33430. Free list.

*Exoticus:* 108-HP Blossom Rd., Westport, MA 02790. Bromeliads and orchids. Catalog free.

*John Ewing Orchids, Inc.:* PO Box 384, Aptus, CA 95003. Catalog $2.00.

*Exotic World:* 219½ Sepulveda, Manhattan Beach, CA 90266. Bromeliads.

*Fennel Orchid Co.:* 26715 SW 157 Ave., Homestead, FL 33030. Catalog free.

*Finck Floral Co.:* 9849-P Kimker Ln., St. Louis, MO 63127. Orchids. Price list free.

*F & K Nursery:* Dept. HPP-10, 4361 Freeman Rd., Marietta, GA 30062. Bromeliads. Send self-addressed stamped envelope for price list.

*Fox Orchids, Inc.:* 6615 W. Markham, Little Rock, AK 72205. Send 40¢ in coin or stamp for catalog.

*Arthur Freed Orchids:* Dept. HP, 5731 S. Bonsall, Malibu, CA 90265. Catalog free.

*Genus: Tillandsia:* 3964 Grandview Blvd., Mar Vista, CA 90066. Send self-addressed stamped envelope for list.

*Great Lakes Orchids:* 28805 Pennsylvania, Romulus, MI 48174, Monroe, MI 48161. List free.

*Greenbrier Orchids:* 4711 Palm Beach Blvd., Myers, FL 33905. Send stamp for list.

*Hardy & Fouquette Orchids:* 9943-HP E. Heaney Circle, Santee, CA 92071.

*Hoernig's Bromeliads:* 3228 Gerle Ave., Placerville, CA 95667. Price list 50¢.

*Jones & Scully, Inc.:* 2200 N.W. 33rd Ave., Drawer HP, Miami, FL 33142. Catalog and reference guide $4.00.

*Jungle Gems:* Dept. HPX, PO Box 95, Bel Air, MD 21014. Orchid information free.

*Kensington Orchids, Inc.:* 3301-HP Plyers Mill Rd., Kensington, MD 20795. List free.

*Kent's Bromeliad Nursery:* 703 Pomelo Dr., Vista, CA 92083. Catalog $1.00.

*Keraseva Ltd.:* Dept. HPP-10, Consaul Rd., RD #5, Amsterdam, NY 12010. Orchids. List 25¢.

*Lager & Hurrell:* 805 Garner Rd., Lilburn, GA 30247. Orchids Cultural handbook and listing $2.50.

*Ann Mann's Orchids:* Rt. 3, Box 202-HP, Orlando, FL 32811. Catalog 50¢.

*Marilynn's Garden:* 13421 Sussex Pl., Santa Ana, CA 92705. Bromeliads. Send self-addressed stamped envelope for list.

*Marz Bromeliads:* 10782 Citrus Dr., Moorpark, CA 93021. Catalog 50¢.

*McCauley Plants:* Dept. HPP, 1426 Eastern Heights Mesquite, TX 75149. Bromeliads. Send stamp for list.

*Rod McLellan Co.:* 1450-HP El Camino Real, S. San Francisco, CA 94080. Orchids. Catalog free.

*Orchid Imports:* 11802-HP Huston St., N. Hollywood, CA 91607. List free.

*Orchids by Hausermann:* PO Box 363-HPP, Elmhurst, IL 60126. Catalog $1.00.

*S&G Exotic Plant Co., Inc.:* Box A. 22 Goldsmith Ave., Beverly, MA 01915. Orchids. Catalog $1.00.

*Seaborn Del Dios Nursery:* Rt. 3, Box 455-HP, Escondido, CA 92025. Bromeliads. Catalog $1.00.

*Sea Breeze Orchids, Inc.:* PO Box 1416, Bayville, NY 11709. Price list free.

*Shelldance Bromeliads:* 2000-HP Cabrillo Hwy., Pacifica, CA 94044. Send self-addressed stamped envelope for catalog.

*Stewart's Orchids:* Dept. HP78, 1212 E. Las Tunas, San Gabriel, CA 91778. Catalog $1.00.

*Tropiflora Nursery:* Dept. HPP-10, 5439 3rd St. E, Bradenton, FL 33507. Bromeliads. Catalog 50¢.

*Walther's Exotic:* RD-3, Box 30-GO, Catskill, NY 12414. Bromeliads. Catalog $1.00.

## CACTI AND SUCCULENTS

*Abbey Garden:* Dept. HP, 176 Toro Canyon Rd., Carpinteria, CA 93010. List $1.00.

*Alpine Gardens:* 280H SE Fir Villa Rd., Dallas, OR 97338. Catalog 50¢.

*Altman Specilty Plants:* Dept. HP-F10, 26963 Sea Vista Dr., Malibu, CA 90265. Catalog 50¢.

*Arizona Cactus Farms:* PO Box 249, Superior, AZ 85273. List 25¢.

*Barnett Cactus Garden:* 1104 Meadowview, Bossier City, LA 71111. List 30¢.

*Beahm Epiphyllum Gardens:* Dept. HP78, 2686 Paloma St., Pasadena, CA 91107. Catalog $1.00.

*B & G Greenhouse:* Dept. HPP-1 4950 Howell St., Golden, CO 80401. Catalog 50¢.

*Cactus by Mueller:* 10411 Rosedale Hwy., Bakersfield, CA 93308. Catalog 30¢.

*The Cactus Corner:* 1027 Prescott, E. Lansing, MI 48823. Plant list 25¢.

*Cactus Gardens:* Box 2654-HP, Vero Beach, FL 32960.

*The Cactus Lady:* Dept. HPP, RD 1, Box 347, Woodvine, NJ 08270. Catalog 50¢.

*California Epi Center:* PO Box 2474-HPP, Van Nuys, CA 91404. Epiphyllums. Catalog $1.00.

*Collector's Succulents:* PO Box 1998-HP, Rancho Sante Fe, CA 92067. Send self-addressed stamped envelope for catalog.

*Desert Dan's:* Summer Ave., Minotola, NJ 08341. Catalog $1.00.

*Fernwood Plants:* PO Box 268, Topanga, CA 90290. Catalog 50¢.

*Linda Goodman's Sun Plants:* PO Box 20014-HPX, Riverside, CA 92516. Catalog 50¢.

*Grigsby Cactus Gardens:* Dept. HP, 2354 Bella Vista, Vista, CA 92083. Catalog $1.00.

*Grotes Cactus Gardens:* 13555 S. Leland Rd., Oregon City, OR 97045. Price list 25¢.

*Hahn's Cactus Nursery:* Dept. H-2, 2663 Loomis Dr., San Jose, CA 95121. Brochure 30¢ in stamps.

*Hawks Nursery:* 2508-HP, E. Vista Way, Vista, CA 92083. Epiphyllums. Catalog 40¢.

*Henrietta's Nursery:* 1345 N. Brawley Ave., Fresno, CA 93711.

*Hurst Nursery:* Dept. HPP, 9751 East Valley Blvd., Rosemead, CA 91770. Epiphyllums. Catalog $1.00.

*Kelly's Epiphyllum Collection:* Dept. HPP-10, 141 Quail Dr., Encinitas, CA 92024. Catalog $1.00.

*Kimura International, Inc.:* Dept. P, PO Box 327, Aromas, CA 95004. Price list 25¢.

*Kirkpatrick's Nursery:* 27785 De Anza, Barstow, CA 92311. Catalog 25¢.

*K&L Cactus Nursery:* 12712 Stockton Blvd., Galt, CA 95632. Catalog 75¢.

*Loehman's Cactus Patch:* PO Box 871-G, 8014 Howe St., Paramount, CA 90723. Catalog 50¢.

*Meristem Cactus Co.:* 968 Central St., East Bridgewater, MA 02333. Catalog 50¢.

*Mesa Garden:* 905 Impala Dr., Belen, NM 87002. Send 20¢ stamp for list.

*Nature's Curiosity Shop:* 2560 Ridgeway Dr., National City, CA 92050. Catalog 75¢.

*New Mexico Cactus Research:* PO Box 787-HP, Belen, NM 87002. Seeds only. Send self-addressed stamped envelope for list.

*Nichols Cactus Nursery:* 570 South Hughes, Fresno, CA 93706. Send self-addressed stamped envelope for list.

*Oakhill Gardens:* Rt. 3, Box 87A, Dallas, OR 97338. Catalog 25¢.

*Pantree:* Dept. HPP, 1150 Beverly Dr., Vista, CA 92083. Catalog 50¢.

*Rainbow Gardens Nursery:* Dept. HPP-10, PO Box 721, La Habra, CA 90631. Catalog $1.00.

*Scotts Valley Cactus:* Dept. P, PO Box 66302, Scotts Valley, CA 95066. Catalog 50¢.

*Singers' Growing Things:* 6385 Enfield Ave., Reseda, CA 91355. Catalog $1.00.

*Southwest Desert:* 218 Tennessee NE, Albuquerque, NM 87108. Seeds only. Send 20¢ stamp for list.

*Spring Hollow Plants:* Dept. D, Rt. 1, Box 287, Sparta, MO 65753. Catalog free.

*Ed Storms, Inc.:* Dept. 3, PO Box 775, Azle, TX 76020. Lithops. Catalog 50¢.

*Sturtevant's Cactus Nursery:* Rt. 2, Box 7010, Richland, WA 99352. Catalog 30¢.

*Sweetheart Botanical Gardens:* Dept. V, PO Box 9084, Long Beach, CA 90810. Newsletter 50¢.

*Howard Wise:* Dept. EW-06, 3710 June St., San Bernardino, CA 92405. Catalog. 25¢.

# A listing of outdoor gardening catalogs[1] APPENDIX IV

## ORNAMENTAL TREES AND SHRUBS

*Andrews Nursery:* 200 Andrews Lane, Faribault, MN 55021. Catalog free.

*Bachman's:* 6010 Lyndale Ave. South, Minneapolis, MN 55423. Catalog free.

*Vernon Barnes & Sons: Nursery:* PO Box 250-L, McMinnville, TN 37110. Catalog free.

*Dutch Mountain Nursery:* Dept. PGM, 7984 No. 48th St., Augusta, MI. Berries for birds. Catalog 25¢.

*Earl Ferris Nursery:* Dept. PG, Hampton, IA 50441. Catalog free.

*Gurney's Seed & Nursery:* Yankton, SD 57078. Catalog free.

*Harvest Farms:* PO Box 278, Morriston, TN 37357. Catalog free.

*Hortica Gardens:* PO Box 308, Placerville, CA 95667. Catalog 25¢.

*Kelly Bros. Nurseries:* Dansville, NY 14437. Catalog free.

*Lee's Nursery:* Rt. 2, Box 184A-6F, McMinnville, TN 37110. Catalog free.

*McConnell Nurseries:* RFD 1, Port Burwell, Ontario. Catalog free.

*McMinnville Tree Farm:* Box 712, McMinnville, TN 37110. Catalog free.

*Silver Falls Nursery:* 19542 Jack Lane S.E., Silverton, OR 97381. Catalog free.

*Spruce Brook Nursery:* Rt. 118, Litchfield, CT 06759. Catalog $2.00.

*Stark Bros. Nursery:* Box A 24496, Louisiana, MS 63353. Catalog free.

*Stern's Nurseries:* Dept. PG, Geneva, NY 14456. Catalog free.

*Tingle Nursery:* Pittsville, MD 21850. Catalog free.

*Valley Nursery:* Box 4845, Helena, MT 59601. Catalog free.

*Wayside Gardens:* 100 Garden Lane, Hodges, SC 29695. Catalog $1.00.

[1]From *House Plants and Porch Gardens,* January 1978.

## GARDENING SUPPLIES

*American Standard Company:* PO Box 325, One West St., Plantsville, CT 06479. Pruning and trimming tools. Catalog free.

*B.F. Baker:* PO Box 123, Dept. HP-178, Monsey, NY 10952. Horticultural tools. Catalog $1.00.

*Ball Superior:* 1155 Birchview Dr., Mississauga, Ontario L5H 3E1. *Burgess Seed* and *Carefree Garden Products* in Canada. Catalogs free.

*Bonide Chemical Co. Inc.:* II Wurz Ave. HP, Yorkville, NY 13495. Pest and disease control. Catalog free.

*Easy Heat—Wirekraft:* U.S. 20 East, New Carlisle, IN 46552. Electric tools. Catalog free.

*Gardening Goodies:* Box 5081-H, Beverly Hills, CA 90210. Gardening accessories. Catalog free.

*Gardening Naturally:* Rt. 102, Stockbridge, MA 01262. Composters. Catalog free.

*Heald Inc.:* Dept. HPN, PO Box 1148, Benton Harbor, MI 49022. Tractors and tillers. Catalog free.

*Hermitage Gardens:* Canastota, NY 13032. Decorative garden pools, rock and fiberglass. Catalog 50¢.

*Lethelin Products Co., Inc.:* 15 MacQueen Pkwy. S., Mt. Vernon, NY 10550.

*Mellinger's Inc.:* 2310 W. South Range, North Lima, OH 44452. 1000 horticultural items. Catalog free.

*Miracle Earth Corp.:* 180L Penrod Ct., Glen Burnie, MD 21061. Seed starter cubes. Catalog free.

*Walter F. Nicke:* Box 667HP, Hudson, NY 12534. Gardening accessories. Catalog 25¢.

*Rotocrop:* 38 Buttonwood St., New Hope, PA 18938. Compost bins. Catalog free.

*Smith and Hawken Tool Co.:* 68 Homer, Palo Alto, CA 94301. Hard to find tools and accessories. Catalog free.

*Troy-Bilt Tillers:* c/o Garden Way Mfg. Co., Dept. 84421, Troy, NY 12180. Power tiller and composter, two models. Catalog free.

*Wheel Horse Products:* 515 W. Ireland Rd., South Bend, IN 46614. Lawn and garden tractors. Catalog free.

## FLOWER SPECIALISTS

*Armstrong Nurseries:* Dept. 145, Box 4060, Ontario, CA 91761. Roses, fruit trees. Catalog free.

*Blackthorne Gardens:* 48 Quincy St., Holbrook, MA 02343. Lilies, other bulbs. Catalog $1.00.

*Heatherbloom Farm:* White Hollow Rd., Lakeview, CT 06039. *Coleus* and *impatiens.* Catalog free.

*Hildebrandt's Iris Garden:* Star Rt., Box 4, Lexington, NE 68850.

*Honeywood Lilies:* A. J. Porter, PO Box 63, Parkside, Saskatchewan SOJ 2AO. Catalog free.

*Imperial Flower Gardens:* Cornell, IL 61319. Irises. Send stamp for catalog.

*Jackson & Perkins:* Medford, OR 97501. Roses. Also fruits. Catalog free.

*King's Mums:* 3723 E. Castro Valley Blvd., Castro Valley, CA 94546. *Chrysanthemums.* Catalog free.

*Klehm Nursery:* 2 E. Algonquin Rd., Arlington Hts., IL 60005. Peonies, other perennials. Catalog free.

*McDaniel's Miniature Roses:* 7523 Zemco St., Lemon Grove, CA 92045. Catalog free.

*Melrose Gardens:* 309 Best Rd., South, Stockton, CA 95206. Irises. Catalog $1.00.

*Miniature Plant Kingdom:* 4125 Harrison Grade Rd., Dept. HP, Sebastapol, CA 95472. Miniature roses. Catalog free.

*Mini-Roses:* PO Box 4255 Station A, Dallas, TX 75208. Catalog free.

*Nor'East Miniature Roses:* 58 Hammond St., Rowley, MA 01969. Catalog free.

*Noweta Gardens:* 900 Whitewater Ave., St. Charles, MN 55972. *Gladiolus.* Catalog free.

*Pixie Treasures Miniature Roses:* 4121 Prospect, Dept. PG, Yorba Linda, CA 92686. Catalog 25¢.

*Polk Miniature Roses:* 1200 Hill, Hannibal, MO 63401. Catalog free.

*Sequoia Nursery:* 2519 E. Noble Ave., Visalia, CA 93277. Miniature Roses. Catalog free.

*Small World Miniature Roses:* 6383 E. Evans Creek Rd., Rogue River, OR 97537. Catalog free.

*Star Roses:* Dept. HP, West Grove, PA 19390. Catalog free.

*Stocking Rose Nursery:* 785 N. Capitol Ave., San Jose, CA 95133. Catalog free.

*Sunnyslope Gardens:* 8638 Huntington Dr., San Gabriel, CA 91775. *Chrysanthemums.* Catalog free.

*Tate Nursery:* Rt. 3, Box 307HP, Tyler, TX 75705. Roses. Catalog free.

## GREENHOUSES AND SUPPLIES

*Aluminum Greenhouses, Inc.:* 14605 Lorain Ave., Dept. PG, Cleveland, OH 44111. Catalog free.

*Archway Greenhouses:* PO Box 246-HP, Duck Hill, MS 38925. Brochure free.

*Barrington Industries:* PO Box 133-HP, Barrington, IL 60010. Catalog $1.00.

*Bramen Co., Inc.:* PO Box 70H, Salem, MA 01970. Solar ventilator.

*Burpee Seed Co.:* Dept. GH, 1068 Burpee Bldg., Warminster, PA 18974 or Clinton, IA 52732 or Riverside, CA 92502 (mail to nearest address). Catalog free.

*Charley's Greenhouse Supplies:* 12442 N.E. 124th St., Kirkland, WA 98033. Catalog $1.00.

*Dalen Products Inc.:* 201-HP Sherlake Dr., Knoxville, TN 37922. Coldframes, solar ventilators. Catalog free.

*D&M Enterprises:* PO Box 739-PG, Wildomar, CA 92395. Catalog $1.00.

*Edward Owen Engineering:* Dept. B, Snow Shoe, PA 16874. Color brochure 25¢.

*Environmental Dynamics:* 3010 V7G Vine, Riverside, CA 92507. Catalog free.

*Four Seasons Greenhouse:* Dept. HP32, 17 Avenue of the Americas, New York, NY 10013. Catalog free.

*Garden Year:* Dept. HP1, PO Box 17936, Tampa, FL 33682. Catalog free.

*Gothic Arch Greenhouse:* PO Box 1564, Mobile, AL 36601. Catalog free.

*Greenhouse Specialties Co.:* 9849-R Kimker Lane, St. Louis, MO 63127. Catalog $1.00.

*Horticulture Associates:* 4318 S. Eagle Circle, Denver, CO 80232. Catalog. $1.50.

*Instant Greenhouses:* 333 E. Airy St., Norristown, PA 19401. Brochure free.

*J. A. Nearing Co., Inc.:* 9390-HP Davis Ave., Laurel, MD 20810. Catalog free.

*Lord & Burnham:* Division Burnham Corp., Dept. PPG, Irvington, NY 10533. Catalog free.

*McGregor's Greenhouses:* 1195 Thompson Ave., Dept. K3, Santa Cruz, CA 95063. Catalog free.

*Mellinger's Inc.:* 2308 Range Rd., North Lima, OH 44452. Catalog free.

*National Greenhouses:* PO Box 100 Pana, IL 62557. Literature free.

*Pacific Aquaculture:* 3A Gate 5 Rd., Sausalito, CA 94965. Hydroponics. Catalog 50¢.

*Pacific Coast Greenhouse Mfg. Co.:* Dept. HP2, 430 Hurlingame Ave., Redwood City, CA 94063. Catalog free.

*Peter Reimuller:* Dept. 73-CP, PO Box 2666, 980 17th Ave., Santa Cruz, CA 95062. Catalog free.

*Ripa Industries:* 112 N. College Rd., Wilmington, NC 28401. Catalog free.

*Santa Barbara Greenhouses:* 390 Dawson Dr., Dept. H, Camarillo, CA 93010. Catalog free.

*Semispheres:* PO Box 26273, Richmond, VA 23260. Catalog free.

*Shelter Dynamics:* PO Box 616, Round Rock, TX 78664.

*Slater Supply Co.:* 143 Allen Blvd., Farmingdale, NY 11735. Catalog $2.00.

*Sol-Air:* 7858-HP Barlow Rd., Mobile, AL 36608. Catalog free.

*Sturdi-Built Mfg. Co.:* 11304 SW Boones Ferry Rd., Dept. K, Portland, OR 97219. Catalog free.

*Sun-Glo:* 3714 S. Hudson, Seattle, WA 98118. Catalog free.

*Terrella Energies, Inc.:* PO Box 2842, Santa Fe, NM 87501. Catalog free.

*Texas Greenhouses:* 2719 St. Louis Ave., Forth Worth, TX 76110. Catalog free.

*Turner Greenhouses:* PO Box 1260-HP, Goldsboro, NC 27530.

*Vegetable Factory, Inc.:* PO 2235, Dept. HPE-1, Grand Central Station, New York, NY 10017. Catalog free.

*Victoria Greenhouses, Ltd.:* Box 947B, Southampton, PA 18966. Catalog free.

*Water Works Gardenhouses:* Box 905, El Cerrito, CA 94530. Hydroponics, too. Catalog free.

## RHODODENDRONS AND AZALEAS

*Baldsieffen Nursery:* Box 88A, Bellvale, NY 10912. Catalog $2.00.

*The Bovees Nursery:* 1737 S.W. Coronado St., Portland, OR 97219. Catalog 50¢.

*Carlson's Gardens:* Box 305HP, South Salem, NY 10590. Catalog $1.00.

*E. B. Nauman:* 170 Hill Hollow Rd., Watchung, NJ 07060. Catalog free.

*Orinda Nursery:* PO Box 217, Bridgeville, DE 19933. *Camellias*, too. Catalog 50¢.

*Van Veen Nursery:* 4201 S.E. Frankline, Portland, OR 97206. Catalog free.

## FRUIT TREES

*W.F. Allen:* 577 Plant St., Salisbury, MD 21801. Strawberries, raspberries, blackberries, blueberries, grape and asparagus plants. Catalog free.

*Alexander's Nursery:* PO Box 309H, Middleboro, MA 02346. Blueberries and lilacs. Send self-addressed stamped envelope for catalog.

*Armstrong Nurseries:* Dept. 145, Box 4060, Ontario, CA 91761. Fruit trees. Catalog free.

*Bountiful Ridge:* 268 Nursery Lane, Princess Anne, MD 21853. Deciduous fruits, dwarf patio trees. Other ornamentals. Catalog free.

*Brittingham Plant Farms:* Ocean City Rd., Salisbury, MD 21801. Strawberries, many other berries, grapes. Catalog free.

*Dean Foster Nurseries:* Hartford, MI 49057. Strawberries, fruit trees, ornamental trees and flowers. Catalog free.

*Four Winds Growers:* 42186 Palm Ave., Box 3538 Mission San Jose Dist., Fremont, CA 94538. True dwarf indoor and outdoor citrus. Catalog free.

*Raynor Bros., Inc.:* Dept. HP1, Salisbury, MD 21801. Berry plants, nut trees, rhubarb. Catalog free.

*Stark Bros.:* Box A 24496, Louisiana, MO 63353. Fruit trees, nut trees, berry plants. Catalog free.

*Waynesboro Nurseries:* Waynesboro, VA 22980. Fruit trees and ornamentals. Catalog free.

*Dave Wilson:* 4306 Santa Fe Ave., Hughson, CA 95326. Dwarf fruit trees. Information free.

## BULBS

*Breck's Bulbs:* Dept. HP, 6523 N. Galena Rd, Peoria, IL 61632. Holland bulbs. Catalog $1.00.

*De Jager Bulbs:* S. Hamilton, MA 01982. Catalog free.

*French's Bulb Imports:* Box 87, Center Rutland, VT 05736. Catalog free.

*Gladside Gardens:* 61 Main St., Northfield, MA 01360. Gladiolus, dahlias, mums and other bulbs. Catalog free. Stamp appreciated.

*International Growers Exchange:* Box 397-H, Farmington, MI 48024. Orchids, bulbs, gesneriads and rare species. Catalog $3.00.

*Messelaar Bulb:* P.O. Box 269, Ipswich, MA 01938. Catalog free.

*Van Bourgondien's:* 245 Farmingdale Rd., Box A-HP 1, Babylon, NY 11702. Bulbs and plants. Catalog free.

## PERENNIALS, ROCK GARDENS

*Conley's Garden Center:* 145 Townsend Ave., Boothbay Harbor, ME 04538. Wildflowers and ferns. Catalog 35¢.

*Far North Gardens:* 15621 Auburndale St., Dept. HP, Livonia, MI 48154. Primrose plants. Catalog 50¢.

*Fleming's Flower Fields:* PO Box 4607, 3100 Leighton Ave., Lincoln, NE 68504. *Chrysanthemums* and perennials. Catalog free.

*Intermountain Cactus:* 1478 N. 705 East, Kaysville, UT 84037. Winter-hardy cacti. Send stamp for catalog.

*Island Gardens:* Box 101, Grosse Ile, MI 48138. Lilies, ferns, ground cover. Send stamp for catalog.

*Lamb Nurseries:* E. 101 Sharp, Box H-97, Spokane, WA 99202. Perennials, daylilies, irises. Catalog free.

*Lounsberry Gardens:* PO Box 135, Oakford, IL 62673. Rock garden plants. Catalog 25¢

*Mincemoyer's:* Rt. 526-HP, Jackson, NJ 08527. Wildflowers, ferns. Catalog 25¢.

*Miniature Gardens:* Box 757, Stony Plain, Alberta TOE 2G0. Dwarf plants. Catalog $1.00.

*Nature's Farm:* Box 252, Morrison, TN 37357. Wildflowers. Catalog free.

*Nature's Garden:* Rt. 1, Box 488, Beaverton, OR 97005. Sedums and sempervivums. Catalog 50¢.

*Peekskill Nurseries:* Box HP, Shrub Oak, NY 10588. Ground cover. Catalog free.

*The Rock Garden:* Dept. HP, Hallowell Rd., Litchfield, ME 04350. Flowers and shrubs. Catalog 40¢.

*Savage Gardens:* PO Box 163, McMinnville, TN 37110. Wildflowers, ferns. Send stamp for list.

## HERBS

*A.B.C. Herb Nursery:* B313-HP, Rt. 1, Lecoma, MO 65540. Catalog 25¢.

*Capriland's:* Dept. PG, Coventry, CT 06238. List 15¢.

*Herb Shop:* PO Box 362-PHP, Fairfield, CT 06430. Catalog 50¢.

*Herbs 'n Honey Nursery:* Rt. 2, Box 205, Monmouth, OR 97361. Catalog $1.00.

*Howe Hill Herbs:* Dept. HP, Howe Hill Rd., Camden, ME 04843. Catalog 35¢.

*Le Jardin Du Gourmet:* Box 404-HP, West Danville, VT 05873. Catalog 25¢.

*Taylor's Herb Garden, Inc.:* 1535-HP Lone Oak Rd., Vista, CA 92083. Catalog 25¢.

*Well-Sweep Herb Farm:* 317 Mt. Bethel Rd., Port Murray, NJ 07865. Catalog 25¢.

*Yankee Peddler Herb Farm:* Box 76-HP, Rt. 4, Hwy. 36N, Brenham, TX 77833. Catalog $1.00.

# Mail order seed houses and nurseries APPENDIX V

**Abundant Life Seed Foundation,** PO Box 772, Port Townsend, WA 98368. Nonhybrid seed for vegetables, herbs, wildflowers, and trees. A seed-collecting, nonprofit organization that offers a newsletter as well as their informative catalogue. $2 ($3, Canada).

**Ahrens Nursery,** RR 1, Huntingburg, IN 47542. Berry, asparagus, and rhubarb plants. Catalogue free.

**Alberta Nurseries and Seeds, Ltd.,** Box 20, Bowden, Alberta TOM OKO, Canada. Garden seed and landscape plants especially selected for short growing seasons. Catalogue free.

**Alexander's Nurseries,** 1225 Wareham Street, Middleboro, MA 02346. Organically grown seed. Blueberry bushes and lilacs. Price list free; enclose self-addressed stamped envelope.

**W. F. Allen Co.,** 2179 Strawberry Lane, Salisbury, MD 21801. Asparagus, berries, grapes. Catalogue free.

**Allen, Sterling and Lothrop,** 191 Route 1, Falmouth, ME 04105. Vegetable, flower, and grass seed. Catalogue free.

**Applewood Seed Co.,** 333 Parfet Street, Lakewood, CO 80215. Herbs and teas, widlflowers, seeds for sprouting, vegetable seed. Catalogue free.

**Armstrong Nurseries, Inc.,** PO Box 4060, Ontario, CA 91761. Fruit trees, small fruits, roses. Catalogue free.

**Aubin Nurseries, Ltd.,** Box 268, Carman, Man. ROG OJO, Canada. Fruit trees. Catalogue free.

**The Banana Tree,** 715 Northampton Street, Easton, PA 18042. Chinese vegetable seed. Import distribution of cashew and cocoa seed. Tropical fruit and nut tree seed. Catalogue 50¢.

**Baum's Nursery,** RD 2, New Fairfield, CT 06810. Old and new varieties of fruit trees. Price list free; enclose self-addressed stamped envelope.

**Borchelt Herb Gardens,** 474 Carriage Shop Road, East Falmouth, MA 02536. Herb plants and seeds. Price list free; send self-addressed stamped envelope.

**Boston Mountain Nurseries,** Route 2, Mountainberg, AR 72946. Berries for Zones 10 to 5. Price list free.

**Bountiful Ridge Nurseries,** Box 250, Princess Anne, MD 21853. Berry plants, fruit trees. Catalogue free.

**Brittingham Plant Farms,** PO Box 2538, Salisbury, MD 21801. Asparagus and berry plants. Catalogue free.

**Broom Seed Co.,** PO Box 236 Q, Rion, SC 29132. Vegetable, flower, and herb seed. Catalogue free.

**Burgess Seed and Plant Co.,** 67 E. Battle Creek St., PO Box 5000, Galesburg, MI 49053. Vegetable seed, fruit trees, house plants, roses, and flower seed. Catalogue free.

**Burpee Seed Co.,** Warminster, PA 18991; or Clinton, IA 52732; or Riverside, CA 92502. Complete selection of vegetable seed, flower seed, house plants, bulbs, fruit trees, shrubs, vines. Catalogue free.

**D. V. Burrell Seed Growers Co.,** P.O. Box 150, Rocky Ford, CA 81067. Vegetable and flower seed. Catalogue free.

**California Nursery Co.,** PO Box 2278, Fremont, CA 94536. Fruit and nut trees; grape vines. Price list free.

**Caprilands Herb Farm,** Silver Street, Coventry, CT 06238. Seed, herb plants, gardening books. Price list free; enclose self-addressed stamped envelope.

**Carroll Gardens,** 444 E. Main Street, PO Box 310, Westminster, MD 21157. Perennial and annual plants, herbs, shrubs, and vines. Catalogue free.

**Casa Yerba,** Star Route 2, Box 21, Days Creek, OR 97429. Unusual plants and herb seed. Catalogue $1. Seasonal plant list for spring and fall available.

**Cascade Forestry Service, Inc.,** RR 1, Cascade, IA 52033. Pine seedlings, nut trees, windbreak trees. Specializing in American nut trees. Price list free.

**Catnip Acres Farm.** PO Box 142-FA, Seymour, CT 06483. Herb plants and seeds. Geraniums. Catalogue $1.

**Comstock, Ferre and Co.,** 263 Main St., Wethersfield, CT 06109. Vegetable, herb, and flower seed. Catalogue free.

**Converse Nursery,** Amherst, NH 03031. Apple trees. Price list free.

**C. A. Cruickshank, Ltd.,** 1015 Mount Pleasant Road, Toronto, Ont. M4P2M1, Canada. Vegetable and flower seed. Catalogue $1.

**DeGiorgi Co.,** 1411 Third Street, Council Bluffs, IA 51502. Flower and vegetable seed from around the world. Catalogue 92¢.

**J. A. Demonchaux Co.,** 827 North Kansas Avenue, Topeka, KS 66608. Imported seed and gourmet products. Catalogue 50¢.

**Epicure Seeds,** Box 23568, Rochester, NY 14692. Gourmet vegetable and herb seed. Catalogue free.

**Exotica Seed Co.,** 1742 Laurel Canyon Blvd., Los Angeles, CA 90046. Unusual vegetable seed, flowering trees, tropical vine, and fruit seed. Catalogue $2.

**Farmer Seed and Nursery Co.,** 818 Northwest 4th St., Faribault, MN 55021. Vegetable and flower seed, hardy northern nursery stock, and bulbs. Catalogue free.

**Faubus Berry Nursery,** Star Route 4, Elkins AR 72727. Berries. Price list free.

**Henry Field Seed and Nursery,** 407 Sycamore, Shenandoah, IA 51602. Vegetable and flower seed, house plants, bulbs, perennials, roses, trees, nuts, and berries. Catalogue free.

**Dean Foster Nurseries,** Route 2, Hartford, MI 49057. Fruit bushes. Catalogue free.

**Garden of Eden Nurseries,** Box 1086, Route 2, Spruce Pine, NC 28777. Vegetable and flower seed, fruit trees. Price list $1; enclose self-addressed stamped envelope.

**Louis Gerardi Nursery,** RR 1, O'Fallon, IL 62269. Nut and persimmon trees. Price list free.

**Gleckiers Seedman,** Metamora, OH 43540. Vegetable seed, many unusual varieties. Brochure 50¢.

**Green Herb Gardens,** Greene, RI 02827. Herb seed. Price list 25¢.

**R. Guidi Tree Nursery,** 193 Curtis Ave., Dalton, MA 01226. Nut trees. Price list free.

**Gurney Seed and Nursery Co.,** Yankton, SD 57079. Vegetable and flower seed, bulbs, shrubs, fruit and nut trees, house plants. Catalogue free.

**Joseph Harris Co.,** Moreton Farm, 3760 Buffalo Road, Rochester, NY 14624. Vegetable, flower, and herb seed. Catalogue free.

**H. G. Hastings Co.,** PO Box 42-74, Atlanta, GA 30302. Vegetable, flower, and herb seed; trees, berries, shrubs, bulbs for southern gardens. Catalogue free.

**Hemlock Hill Herb Farm,** Hemlock Hill Road, Litchfield, CT 06759. Herb plants. Catalogue 50¢.

**Herbst Brothers Seedsmen,** 1000 N. Main Street, Brewster, NY 10509. Vegetable, flower, and tree seed. Catalogue free.

**Herb Shop,** Box 362Y, Fairfield, CT 06430. Large variety of herb seed. Catalogue 50¢.

**Hickory Hollow,** Route 1, Box 52, Peterstown, WV 24963. Herb seed. Price list 25¢.

**Hilltop Herb Farm,** PO Box 1734, Cleveland, TX 77327. Herb seed, plants, and geraniums. Catalogue $2.

**Horticultural Enterprises,** Box 340082, Dallas, TX 75234. Large selection of pepper plants.

**Howe Hill Herbs,** Camden, ME 04843. Excellent selection of herb plants. Price list 50¢.

**J. L. Hudson Seedsman,** World Seed Service, Box 1058, Redwood City, CA 94064. Vegetable, flower, tree, and shrub seed. Rare seed a specialty. Limited catalogue free; 128-page catalogue $1.

**Le Jardin du Gourmet,** Box 48, West Danville, VT 05873. Imported herb, vegetable, and flower seed. French shallots and leeks. Price list 50¢.

**Jersey Chestnut Farm,** 58 Van Duyne Ave., Wayne, NJ 07470. Chinese chestnut and American persimmon trees. Price list free.

**Johnny's Selected Seeds,** Albion, ME 04910. Vegetable seed, hardy, short-season, and standard varieties. Catalogue free.

**J. W. Jung Seed Co.,** 339 South High Street, Randolph, WI 53956. Vegetable and flower seed, bulbs, trees, shrubs, berries, house plants. Catalogue free.

**Kelly Brothers Nurseries,** 23 Maple Street, Dansville, NY 14437. Vegetable and flower seed bulbs, trees, shrubs, berries, grapes. Catalogue free.

**Kilgore Seed Co.,** 1400 West First Street, Sanford, FL 32771. Tropical climate vegetable, herb, and flower seed.

**Kitazawa Seed Co.,** 356 West Taylor Street, San Jose, CA 95110. Asian vegetable seed. Price list free.

**Lakeland Nurseries,** 340 Poplar St., Hanover, PA 17331. Vegetable seed and fruit trees. Catalogue 50¢.

**D. Landreth Seed Co.,** 2700 Wilmarco Avenue, Baltimore, MD 21223. 195-year-old seed house offering seeds of old-time vegetables. Catalogue $2.

**Lawson's Nursery,** Route 1, Box 294, Ballground, GA 30107. Fruit trees. Price list free.

**Letherman's Inc.,** 1203 East Tuscarawas, Canton, OH 44707. Vegetable and flower seed, bulbs. Catalogue free.

**Henry Leuthardt Nurseries, Inc.,** Montauk Highway, East Moriches, NY 11940. Fruit trees, berries, grapes. Price list free. Catalogue $1.

**Makielski Berry Farm and Nursery,** 7130 Platt Road, Ypsilanti, MI 48197. Berry and asparagus plants. Catalogue free.

**Maple Hill Herb Farm,** Route 1, Anna, IL 62906. Herb seed and plants. Catalogue 50¢.

**Margrave Plant Co.,** Box 100, Gleason, TN 38229. Sweet potato plants. Price list free.

**Earl May Seed and Nursery Co.,** 100 North Elm, Shenandoah, IA 51603. Vegetable and flower seed, plants, fruit trees, roses, trees. Catalogue free.

**Meadowbrook Herbs and Things,** Wispering Pines, Wyoming, RI 02898. Herb seed. Catalogue 50¢.

**Mellinger's,** 2310 West South Range Rd., North Lima, OH 44452. Vegetable, tree, and flower seed, vines, fruit, and nut trees, berries. Catalogue free.

**The Meyer Seed Co.,** 600 South Caroline Street, Baltimore, MD 21231. Vegetable and flower seed, bulbs. Catalogue free.

**Midwest Seed Growers, Inc.,** 505 Walnut, Kansas City, MO 64106. Vegetable and flower seed. Catalogue free.

**J. E. Miller Nurseries, Inc.,** 5060 West Lake Road, Canandaigua, NY 14424. Small fruit trees, old-time apple trees, berries. Catalogue free.

**Mountain Seed and Nursery,** Route 1, Box 271, Moscow, ID 83843. Vegetable seed, extensive tomato seed selection, perennial plants. Price list 25¢.

**Natural Development Co.,** Box 215, Bainbridge, PA 17502. Vegetable and flower seed for organic gardens. Catalogue 25¢.

**New York State Fruit Testing Cooperative Association,** Geneva, NY 14456. Large selection of berries, grapes, and fruit trees. Catalogue free.

**Nichol's Garden Nursery,** 1190 North Pacific Highway, Albany, OR 97321. Unusual and rare vegetable and herb seed. Catalogue free.

**L. L. Olds Seed Co.,** PO Box 7790, Madison, WI 53707. Vegetable and flower seed, bulbs, berries, grapes, fruit trees. Catalogue free.

**Ozark National Seed Order,** Drury, MO 65638. Vegetable, herb and flower seed (untreated, open-pollinated). Price list 25¢; send self-addressed stamped envelope.

**George W. Park Seed Co.,** Box 31, Greenwood, SC 29647. Vegetable and flower seed, plants, bulbs. Catalogue free.

**Piedmont Plant Co.,** 807 North Washington Street, PO Box 424, Albany, GA 31703. Vegetable plants. Catalogue free.

**Porter and Son Seedsmen,** 1510 East Washington, Stephenville, TX 76401. Vegetable and flower seed. Catalogue free.

**Redwood City Seed Co.,** PO Box 361, Redwood City, CA 94064. Vegetable, herb, fruit, nut and wildflower seed. Dye plants. Catalogue 50¢.

**The Rosemary House,** 120 South Market Street, Mechanicsburg, PA 17055. Herbs. Catalogue 50¢.

**Rutland of Kentucky,** Route 1, Box 17, Maysville, KY 41056. Herb seed, plants, scented geraniums. Catalogue $2.

**Sanctuary Seeds,** 1913 Yew Street, Vancouver, BC V6K 3G3, Canada. Vegetable and medicinal seed (untreated). Catalogue free.

**Seedway, Inc.,** Railroad Place, Hall, NY 14463. Vegetable seed. Catalogue free.

**R. H. Shumway Seedsman, Inc.,** 628 Cedar Street, Rockford, IL 61101. Vegetable and flower seed, trees, shrubs, bulbs, house plants. Catalogue free.

**Southmeadow Fruit Gardens,** 2363 Tilbury Place, Birmingham, MI 48009. Large selection of hard-to-find fruit trees. Price list free. Catalogue $6.

**Spring Hill Nurseries,** 6523 North Galena Road, Peoria, IL 61632. Berries, fruit trees, perennials, bulbs. Catalogue free.

**Stark Brothers Nurseries and Orchards,** Box A3411A, Louisiana, MO 63353. Standard, dwarf, and semi-dwarf fruit trees, nut trees, shrubs, and ornamentals. Catalogue free.

**Steele Plant Co.,** Box 191, Gleason, TN 38229. Vegetable plants, specializing in sweet potatoes

**Stokes Seeds, Inc.,** Box 548, 737 Main Street, Buffalo, NY 14240. Vegetable and flower seed. Catalogue free.

**Sunrise Enterprises,** PO Box 10058, Elmwood, CT 06110. Oriental vegetable and flower seed; oriental gardening and cooking books. Catalogue free.

**Taylor's Herb Garden,** 1535 Lone Oak Road, Vista, CA 92083. Herb plants. Catalogue $1.

**Thompson and Morgan,** Box 100, Farmingdale, NJ 07727. Vegetable and flower seed. Catalogue free.

**Tsang and Ma International,** 1306 Old County Road, Belmont, CA 94002. Chinese vegetable seed. Price list free.

**Otis S. Twilley Seed Co.,** PO Box 65, Trevose, PA 19047. Vegetable and flower seed. Catalogue free.

**Unwins,** PO Box 9, Farmingdale, NJ 07727. Excellent selection of sweet pea seed. European flower and vegetable seed. Herb, tree, shrub, and house plant seed. Catalogue free.

**Vermont Bean Seed Co.,** Garden Lane, Box 308, Bomoseen, VT 05732. Wide selection of bean and pea seed. Catalogue free.

**Vesey's Seeds,** York, P.E.I. COA 1PO, Canada. Vegetable and flower seed for short season areas. Catalogue free.

**Waynesboro Nurseries, Inc.,** PO Box 987, Waynesboro, VA 22980. Berry plants, fruit, nut, and shade trees. Catalogue free.

**Willhite Melon Seed Farms,** PO Box 23, Poolville, TX 76076. Vegetable, watermelon, and cantaloupe seed. Catalogue free.

**Dave Wilson Nursery,** Box OF81, Hughson, CA 95326. Fruit and nut trees, grapevines. Catalogue free.

**Wilton's Organic Potatoes,** Box 28, Aspen, CO 81611. Red skin and russet seed potatoes. Price list free.

**Wyatt-Quarles Seed Co.,** PO Box 2131, Raleigh, NC 27602. Vegetable seed for the South, herbs, strawberry plants. Catalogue free.

## APPENDIX VI Garden tools for horticultural therapy

*Lightweight Long-handled Tools,* Wilkinson Sword-manufacturers, distributed by Brookstone Company, 127 Vose Farm Rd., Peterborough, NH 03458.

*Trig-A Grip Trowels and Fork,* distributed by Brookstone and many seed companies; Stokes, Burpees.

*Rachet Cut Pruner,* distributed by Brookstone Company.

*Anita Garden Grab,* distributed by Brookstone Company.

*Baronet Cut and Hold Flower Gatherer,* A. Wright and Son, Midland Works, 16-18 Sidney St., Sheffield S1 4 RH, England.

*Baronet Firm Grip Weed Puller,* A. Wright and Son, address above.

*Cori Easi-Kneeler Stool,* J. B. Corrie & Co. Ltd., Frenchman's Road, Petersfield, Hants GU32 3AP, England.

*Smith and Hawken Tool Co.,* 68 Homer, Palo Alto, CA 94301

**Suggested Reading:**

*Equipment for the Disabled, Leisure and Gardening,* order from 2 Foredown Drive, Portslade, Sussex BN4 2BB, England.

*Horticulture for the Disabled and Disadvantaged,* D. Olszowy, 1978, Charles C Thomas Publishers, Springfield, IL.

*Gardening for the Handicapped,* Betty Massingham, Shire Publications London.

*Barrier Free Site Design,* US Dept of Housing and Urban Development, July 1977, Superintendant of Documents, US Govt. Printing Office, Washington, DC 20402, Stock No. 023-000-00291-4.

# Glossary

***Abscisic acid*** One of the major plant hormones which is associated with dormancy in seeds and buds.

***Abscission*** The process of separation of leaves, fruits, petioles, or other plant parts from a plant shoot. It is triggered by low temperature, frost, drought, wind, ethylene, and often by herbicides.

***Accent plant*** In landscaping, a plant used in a situation where it can create a point of interest in keeping with the overall unity of the design.

***Achene*** A small, dry, nonopening (indehiscent), one-seeded fruit with a thin pericarp.

***Adventitious*** Said of a plant root, stem, or leaf that arises in an abnormal position on the plant; for example, adventitious roots on a corn stem or adventitious buds on a stem other than at the normal node site.

***Aerosol*** A fine mist spray produced by the nozzle of a sprayer.

***After-ripening*** See stratification.

***Aggregate fruit*** A fruit formed by the coherence of several mature pistils that were distinct in a single flower, as in blackberry.

***Agronomic*** Pertaining to the management of farm land and crop production.

***Air layering*** Process of scraping the bark or outer surface of a stem, surrounding it with wet sphagnum moss, and wrapping the sphagnum in plastic to induce rooting. After cutting off the rooted shoot just below the newly formed roots, the original "mother plant" may be saved to produce new lateral shoots.

***Alkaloid*** An organic substance that is alkaline, bitter, and contains nitrogen; plant alkaloids may have physiological or toxic effects on animals including man.

***Amo 1618*** A growth retardant used to dwarf shoots. It acts by inhibiting the synthesis of native gibberellins in the plant.

***Angiosperm*** Flowering plant; plant with a reproductive organ called a *flower* and possessing seeds enclosed in a fruit (mature ovary). This is the most advanced category of plants and consists of two classes, the monocots and the dicots.

***Annual*** A plants which grows, flowers, sets seed, and dies within a single growing season.

***Anther*** The pollen sac (microsporangium) of a stamen in a flower.

***Antibiotic*** A chemical substance produced by certain microorganisms that inhibit the growth of or destroy other microorganisms.

***Apical meristem*** The apex or growing tip of a root or shoot.

***Aquarium*** Tank of water with aquatic plants and fish; it may have inanimate objects present to simulate a marine or freshwater aquatic "landscape."

***Aril*** A fleshy or leathery outer covering of the seed in addition to the seed coat.

***Aroid*** A member of the aroid family (Araceae). Aroids include *Monstera, Philodendron, Anthurium,* and *Arisaema* (jack-in-the-pulpit).

***Availability*** Refers to a form(s), for example, as ions, of an element which in soluble form can be absorbed by plant roots. Unavailable forms may be insoluble precipitates of such elements, as occurs at some pHs in the soil.

***Awn*** A stiff, needlelike appendage at the tip of one of the bracts of a grass flower (the lemma). It aids in the dispersal of the seeds by various means.

***Bacteriocide*** A chemical that kills bacteria.

***Balanced light spectrum*** Lighting system which emits the white light spectrum from the violet to the far-red. This is achieved with full-spectrum lamps or with a combination of fluorescent and incandescent lamps.

***Bark*** All tissues outside the vascular cambium; in old trees it is divided into inner, living bark (functional secondary phloem) and dead, outer bark.

***Benzyl adenine*** A synthetic cytokinin that stimulates cell division in plants and is used to induce bud formation or to release buds from apical dominance.

***Berry*** Fleshy, indehiscent (nonopening) few- or many-seeded fruit, originating from a single pistil. Examples: the fruits of tomato and grape.

**Biennial** A plant that completes its life cycle in two years, producing a rosette of leaves the first year, then flowering, and setting seed in the second year.

**Binomial** Refers to the generic name and specific epithet, in combination, for a given plant. Thus, the genus and specific epithet for a plant is its binomial. An example is *Coleus blumei*, where *Coleus* is the genus, and *blumei* is the specific epithet.

**Biodegradable** Materials decomposed by microorganisms such as fungi and bacteria.

**Biological clock** A time-keeping process which internally may regulate in plants, for example, their time of flowering (as in short-day plants) or their leaf movements day and night (as in legumes such as beans).

**Biomass** Total weight of living plant or animal material.

**Blade** The broad, expanded part of a leaf; also called the lamina.

**Bog** An aquatic ecosystem that has a highly organic substrate, often with acid pH, and no inlet or outlet for water flow. It may have a characteristic flora which includes orchids, members of the heath family *(Ericaceae)*, and carnivorous plants.

**Bonsai** A miniature tree in a pot with the tree made to look old.

**Bonsai pot** A ceramic container, usually made of clay, in which one grows a bonsai tree.

**Bonsai triangle** Refers to the form of a bonsai tree's branches after appropriate pruning; the entire system should form a triangle with the points representing man, earth, and heaven.

**Botany** The basic science of the study of plants. It is also called the science of plant biology or simply plant science. It covers such subareas as plant ecology, plant taxonomy (or systematics), plant anatomy and morphology, plant physiology and development, plant cytology and genetics, and paleobotany.

**Bract** A modified leaf usually associated with a flower or inflorescence.

**Bud** An embryonic shoot with a short axis surrounded by protective scale leaves; for example, the lateral and terminal buds on shoots of woody plants, the quiescent lateral tiller shoots of grasses, or the "eyes" of a potato tuber.

**Budding** The process of inserting a bud from a scion shoot into the bark of a rootstock.

**Bulb** A swollen, underground shoot made up of a basal plate (stem), shoot tip, and surrounding fleshy, scalelike leaves.

**Bulblet** A small bulb produced as a lateral bud from a "mother" bulb or in some plants, for example, the bunching onion, from the inflorescence; also, some lilies produce bulblets along the stems at the nodes.

***Caffeine*** An alkaloid stimulant found in coffee beans, tea leaves, the cola nut, maté, and chocolate.

***Callus*** A mass of undifferentiated tissue growing over a wound or as the result of cellular multiplication in plant tissue cultures.

***Calorie*** A unit of food value or energy equivalent.

***Calyx*** The collective term for the sepals of a flower.

***Capillary water*** Refers to soil water which is held fairly loosely to soil particles and is therefore readily available for uptake by plant roots.

***Capsule*** A simple, dry, dehiscent, many-seeded fruit as, for example, in lily and iris.

***Carbohydrate*** A polymer made up of sugar subunits. It occurs, for example, in very high amounts as starch in seeds of cereals and in tubers of potatoes.

***Carcinogen*** A substance which causes cancer in mammals.

***Carnivorous*** A plant that traps and digests insects or other small animals.

***Caudiciform*** A type of succulent plant with a disproportionately enlarged basal portion, or caudex. The caudex may be a root or stem.

***CCC*** A growth retardant [(2-chloroethyl) trimethylammonium chloride] used to dwarf shoots. It acts by inhibiting the synthesis of native gibberellin hormones in the plant.

***Chlorosis*** A light green or greenish yellow condition in plants due to a lack of chlorophyll, resulting from a mineral deficiency or a disease.

***Clay*** One of the major inorganic components of the soil with particles of small, colloidal size, less than 0.002 millimeters in diameter.

***Climax community*** The last and most stable plant community to occupy a given habitat in a successional sequence.

***Cloning*** The process of mass propagating genetically identical individuals from an original stock by vegetative means.

***Cold frame*** A kind of miniature unheated greenhouse. It is basically a frame covered with one to two layers of glass and/or plastic film (glazing) in which plants can be grown and at the same time protected from frost and freezing.

***Cold-hardiness*** Condition in which portions of a plant or entire plants are able to withstand cold winter temperatures that prevail at given latitudes.

***Collenchyma*** A tissue composed of living cells with irregularly thickened primary walls. Such tissue primarily provides support for plant structures such as leaf petioles or the pulvini (swollen leaf-sheath bases) of grasses.

***Column*** That central portion of an orchid flower made up of fused stamens and styles.

***Commensualism*** A type of symbiosis in which one partner is benefited from the association and the other is neither hurt nor helped.

***Competition*** Interactions among organisms in the same habitat as each obtains a share of the limiting resources necessary for survival, such as food, light, nutrients, and water.

***Compost*** Decomposition product (also called humus) of microbial breakdown of organic material, as occurs in a compost pile.

***Convection*** Natural circulation of warm air from cooler, lower levels to warmer upper levels, as occurs in a greenhouse or south window of a house.

***Cordon*** A technique in which a free-growing specimen is pruned to grow in only one plane, with only selected branches allowed to grow. (See espalier.)

***Cork cambium*** The lateral, meristematic layer that produces cork cells of the outer bark (periderm).

***Corm*** An enlarged, shortened, underground stem which stores food. Plants with corms include gladiolus and crocus.

***Corolla*** The collective term for the petals of a flower.

***Cortex*** Tissue of stem or root lying immediately inside the epidermis. It is usually made up of parenchyma cells but may also include sclerenchyma (fibers or sclereids), collenchyma, and vascular bundles (in monocots).

***Cotyledon*** A seed leaf or first leaf of a fully developed embryo. It is evident below ground in hypogeous seedlings and above ground in epigeous seedlings.

***Cover crop*** Same as green manure crop.

***Crop rotation*** Growing sequence in which one crop follows another crop in successive years as, for example, rice-wheat-rice or wheat-corn-alfalfa.

***Cross*** Process of transferring pollen from an anther to a stigma.

***Cross-pollination*** Pollination resulting from the transfer of pollen from the anther of one flower to the stigma of another flower.

***Crown*** A large clump of dormant shoots, above the root portions, as occurs in herbaceous perennials. A crown also refers to the uppermost or canopy level of a tree.

***Culinary*** Of or pertaining to cooking and eating.

***Culm*** Refers to a single shoot of a grass plant. Lateral culms are called *tiller shoots.*

***Cultivar*** Term derived from cultivated variety; the cultivar of a plant that may be propagated sexually or asexually, in cultivation, to maintain its distinctive characteristics.

***Cutting*** A plant piece (stem, leaf, or root) removed from a parent plant and placed in a growing medium to produce new roots and shoots.

***Cycocel*** A commercial name for CCC, a growth retardant used to dwarf shoots.

***Cytokinin*** One of the types of native plant hormones produced in relatively high amounts in developing fruits and in roots of plants. Cytokinins promote cell division and are responsible for the release of lateral buds from apical dominance.

***Damping off*** A disease of seedlings caused by certain fungi which attack plants at the soil level and kill the delicate seedlings. It is promoted by unsterile soil and humid conditions.

***Day-neutral*** Refers to a plant which will flower under any day-length condition, long or short days. A good example is tomato.

***Deciduous*** Condition in which plants shed their leaves annually.

***Decomposer*** An organism (a fungus or bacterium) which secretes enzymes which break down organic materials to humus as occurs in compost piles or in the forest floor with decomposing logs and leaves.

***Dichogamy*** Condition in which either the anthers or the stigmas mature before the other in the same flower.

***Dicots (dicotyledons)*** A large group of flowering plants characterized by floral parts in four's or five's, two cotyledons, net-veined leaves, and a distinct ring of primary vascular bundles in the stem.

***Dimorphic*** Having two different growth forms; for example, the fertile and sterile leaves of some ferns.

***Dioecious*** (of two houses) Condition in which a plant has unisexual male or female flowers. Thus, pollen must be transferred from "male" plants to "female" plants.

***Diploid*** Cells which carry two sets of chromosomes in their nuclei; they are termed 2N.

***Disease resistance*** Genetically determined condition in which an organism, such as a plant or animal, resists infection and disease by a given pathogen (a pathogen is a disease-causing organism such as a parasitic fungus or bacterium).

***Dispersal*** Process of dissemination; for example, pollen during pollination by wind, water, or insects or wind-blown or animal-disseminated seeds.

***Division*** Process of propagating a plant vegetatively by dividing it.

***Dolomitic limestone*** Soil amendment which is a good source of calcium; it is often added to compost piles along with other nutrient sources to increase the supply of essential nutrients and to raise the pH.

***Dormancy*** A period in a plant's life cycle during which time it is not in active growth. May relate to bulbs, buds, seeds, or an entire plant.

***Double cross hybrid*** Production of a hybrid plant by means of two sets of crosses to produce inbred parental lines, which in turn are crossed to produce a hybrid that shows heterosis or hybrid vigor, as with hybrid maize (corn). (See heterosis; hybrid vigor.)

***Double-digging*** One of the essential steps in preparing a raised bed for growing crop plants, in which one digs down two trench shovel lengths and throws the soil forward in the process of preparing the bed.

***Double fertilization*** The fusion of the egg and sperm, resulting in formation of a diploid (2N) fertilized egg (zygote) and the simultaneous fusion of a second sperm with the two polar nuclei in the embryo sac, resulting in the formation of a triploid (3N) primary endosperm nucleus. It occurs only in flowering plants (angiosperms).

***Double glazing*** Two layers of transparent material (separated by an air space) on a greenhouse, cold frame, or in a window. It may be a combination of glass and polyethylene, or glass alone, or polyethylene alone.

***DNA*** Deoxyribonucleic acid, the primary genetic information found in the nuclei of plant and animal cells. DNA makes up the genes of the cell.

***Drupe*** A simple fleshy fruit in which the inner layer is hard and stony as, for example, in olive, peach, and plum. Normally, such fruits are one seeded.

***Ecosystem*** The interacting living and nonliving components of the environment.

***Egg nucleus*** A 1N (haploid) nucleus in the embryo sac (female gametophyte tissue) which represents a female gamete. It fuses with a sperm nucleus to form a 2N (diploid) zygote.

***Emasculation*** In plant breeding, the process of removing anther sacs from stamens to prevent self-pollination of a flower from occurring when one is making a cross.

***Embryo*** The multicellular structure derived from the zygote. It develops within the embryo sac in flowering plants or within the female gametophyte tissue in mosses, scouring rushes, ferns, fern allies, cycads, and conifers.

***Embryo sac*** The female gametophyte (gamete = sex cell; phyte = plant) of angiosperms (flowering plants). It contains 1N (haploid) nuclei, including egg, synergids, polar nuclei, and antipodals.

***Emergent aquatic plant*** Aquatic plant whose shoots grow above the water level in an aquatic plant community.

***Endosperm*** Nutritive tissue adjacent to the embryo in seed plants.

***Entomology*** The study of insects.

***Epicotyl*** That part of a seedling shoot axis which occurs above the cotyledon(s).

**Epidermis** The outermost layer(s) of tissue of a root, stem, leaf, or fruit.

**Epigeous** Refers to a seedling whose cotyledons appear above ground following seed germination.

**Epiphyte** Mostly tropical plants that grow upon other plants (epi = upon; phyte = plant), for example, orchids on the trunks and branches of large trees.

**Erosion** A geologic process by which soil is lost from a given area through the action of wind and/or water. It may take the form of wind-blown soil erosion or water-induced gully or sheet erosion.

**Espalier** A technique whereby a specimen is pruned to grow flat against a wall or fence.

**Ethephon** Commercial name for 2-chloroethylphosphonic acid, which produces ethylene, a product used to induce flowering in bromeliads.

**Ethylene** One of the plant hormones which is involved in causing leaf fall (abscission) and is used to induce flowering in bromeliads, including pineapple, and to hasten ripening in apples and bananas.

**Evaporative cooling** Cooling of air due to evaporation of water, where fans pull air over and through a large surface to drier atmosphere, as with water flowing through a porous surface saturated with water. It is used commonly in greenhouses to cool the air within.

**Evergreen** Condition in which a plant holds its functional leaves for one full year or longer.

**F-1** The first felial generation, or the immediate progeny of a cross between two parent plants.

**F-2** The second felial generation, or the generation produced by crossing F-1 parents.

**Family** Category of classification usually containing several genera with some definite characters in common.

**Fen** An aquatic ecosystem which has soil and water of alkaline pH and an inlet and outlet for water flow. It may have a characteristic flora which includes *Potentilla* (cinquefoil), *Chara,* and/or *Nitella* (both of the latter are macroscopic stoneworts).

**Fertile** Term in plant breeding referring to the ability of a plant to produce viable seed and thus to reproduce sexually.

**Fertilization** The fusion of two gametes (sperm with egg), resulting in the formation of a zygote (the first stage of embryo formation). Also, the act of applying fertilizer to improve plant growth.

**Fertilizer** A term referring to a nutrient source such as inorganic fertilizer containing N (nitrogen), P (phosphorus), and K (potassium) or an organic fertilizer such as manure containing high amounts of N.

**Fiber** A long, narrow, generally thick-walled sclerenchyma cell whose primary function is that of providing support in roots, stems, and leaves.

**Fiberglass** Lightweight, semitransparent covering used on roofs of greenhouses. It is also used to make boats and is used on the sides of buildings.

**Filament** The stalk of a stamen to which the anther is attached.

**Floating aquatic plant** An aquatic plant that floats and grows at the surface of the water.

**Florigen** A generic name given to the as yet undiscovered plant hormone(s) which causes flowering in long-day and short-day plants.

**Flower** The reproductive structure of an angiosperm. The flower is a shoot axis bearing specialized leaves modified as sepals, petals, stamens, and pistils.

**Flowering** The reproductive phase in an angiosperm's development, resulting in the production of flowers.

**Fluorescent lamp** A type of lamp which emits in the visible region up to and including red light (600 to 650 nanometers), but not including the physiologically active and essential far-red light (700 to 800 nanometers).

**Foliage houseplant** Houseplant which is grown primarily because of its foliage and is especially showy because of its size, form, texture, and/or color.

**Follicle** Dry, dehiscent fruit, opening only on the dorsal (front) suture, and the product of a single pistil.

**Forcing** The process of inducing bulbs to flower after they have been chilled; or inducing flower buds on twigs to open after they have been subjected to a period of low temperatures.

**Foundation plant** In landscaping, a plant used next to a building structure to soften the appearance of the base of the structure.

**Friable** Refers to soil whose texture is such that it is easily cultivated because the particles are of relatively large aggregates, in contrast to clay soils, which are hard to cultivate due to the very small, compacted particles.

**Frond** The whole leaf of a fern plant.

**Fructan** A storage polysaccharide of dahlia tubers and Jerusalem artichoke tubers made up of fructose subunits, much as starch is made up of glucose subunits. Fructose is also the primary sugar in honey.

**Fruit** In flowering plants, the mature ovary or ovaries of one or more flowers and sometimes adjacent parts.

**Fungicide** A chemical that kills fungi.

**Fungus** (pl. ***fungi***) Refers to one of a large group of nonphotosynthetic organisms, whose bodies are organized into strands called *hyphae,* aggregates of which are termed *mycelia*. Fungi include both parasites and saprophytes. Examples are slime molds, yeasts, mushrooms, smuts, and rusts.

***Gall*** A tumor formed on a plant, produced by a fungus, insect, or bacterium.

***Genetic engineering*** A technology of transferring genes from one organism to another in order to produce new strains of organisms with new traits or new combinations of traits.

***Genetics*** The science of heredity (gene transmission) and variation.

***Genus*** Subdivision of a family. Grouping in which certain gross characters make the members of a genus easily recognizable as such; for example, all oaks, maples, and elms are easily recognizable at the genus level. It is the first part of the Latin name (binomial) of an organism.

***Geodesic dome*** A hemispheric-shaped structure supported by struts put together in such a way that the outer surface is covered with many triangular or hexagonal faces.

***Germinate*** The sprouting of a seed, resulting in the emergence of the first root (radicle) from the seed.

***Gibberellin*** One of the types of native plant hormones produced by vascular plants and by some fungi (for example, *Gibberella fujikuroi*, a pathogenic fungus which causes foolish seedling disease in rice by secreting copious amounts of gibberellin in the host rice plants). In vascular plants gibberellins promote cell elongation and in some plants cell division; also, flowering in long-day plants and cone initiation in pines, cedars, cypress, and redwood.

***Gibbing*** Process of removing one flower bud of a close pair, then applying gibberellin to the wound; the hormone is absorbed and causes the remaining flower bud to open into a much larger flower.

***Glazing*** Any transparent material, such as glass, plastic, or fiberglass used to let light enter a structure such as a greenhouse or cold frame.

***Glycoside*** An organic substance containing sugar, often having deleterious physiological effects on animals, including man. Many plant poisons are glycosides.

***Grafting*** The process of "annealing" one part of a plant to another, with union occurring because of compatible cambial activity by both partners. The shoot portion of a graft is called the *scion*, and the root portion is called the *stock*.

***Grain (caryopsis)*** A type of fruit, also called an achene; a grain is usually restricted to the fruits of grasses (Poaceae family), as distinguished from the achenes of members of the aster family (Asteraceae) by being derived from a superior ovary in the former, and an inferior ovary in the latter.

***Gravitational water*** Water which moves freely in the soil; thus, it is water which is not tightly held by soil particles.

***Gravitropism*** A growth response of a plant to gravity. Roots, when placed horizontally, show a positive gravitropic response by growing downward; shoots, when placed horizontally (lodged), show a negative gravitropic response by growing upward.

***Greenhouse*** A glass- or plastic-covered structure in which plants are grown.

***Greenhouse effect*** The trapping of infrared light (heat) when light passes through glass into an enclosed structure such as a greenhouse or a passive solar collector.

***Green manure crop*** A crop which is grown for a short time, then plowed or tilled into the soil to increase its humus (organic matter) content. Annual ryegrass is one such green manure or cover crop.

***Greensand*** Glauconite (a green mineral made up of iron potassium silicate) often mixed with clay or sand, a potassium source to enrich soil or compost piles.

***Ground cover*** Refers to a plant which is low growing and effectively covers the soil. Examples include pachysandra, lily-of-the-valley, English ivy, ice plant *(Mesembryanthemum)*, and myrtle *(Vinca)*.

***Growth chamber*** A controlled environment in which plants are grown. Controls include humidity, day length, night length, and temperature.

***Gully erosion*** Type of soil erosion caused by floods or excess rainfall, resulting in the formation of gullies due to soil loss.

***Gypsum*** A term used for calcium sulfate. Gypsum is a soil conditioner used to break up hard clay soils into tillable form as, for example, with adobe clay. It also adds calcium to the soil without raising the pH.

***Habitat*** The place where a plant or animal naturally lives.

***Haploid*** Cells which carry one set of chromosomes in their nuclei; they are termed 1N.

***Hardwood cutting*** A shoot cutting made from woody portions of a vine, tree, or shrub.

***Heading back*** A pruning technique whereby branches or twigs are cut back by about one-half to one-fourth, just beyond a node, to stimulate lateral branching and to produce a bushier growth habit.

***Herb*** A nonwoody (herbaceous) or soft-stemmed plant which is annual, biennial, or perennial.

***Herbaceous*** Plants that do not produce hard, woody tissue.

***Herbicide*** A chemical that kills plants.

***Hesperidium*** Technically, a berry with a leathery rind. It is the fruit of all members of the citrus family, which includes orange, lemon, and grapefruit.

***Heterosis*** Hybrid vigor. The hybrid is more vigorous, in some ways, than either parent.

***Heterostyly*** Condition in which the style of a flower is significantly longer or shorter than the stamens in different individuals of a given species and thus is not positioned favorably for self-pollination.

***Histamine*** An amine derived from histidine, an amino acid, found in all organic matter. It is a powerful constrictor of blood vessels and a stimulator of gastric secretions.

***Hormone*** An organic substance produced by plants in small amounts; it is synthesized at one site(s) and translocated to another site (the target site) where it causes a physiological or growth response. Hormones in plants include gibberellins, auxins, cytokinins, ethylene, and abscisic acid.

***Horticulture*** The art and science of cultivating fruit, flowers, and vegetables. It includes the fields of pomology (fruit growing), olericulture (vegetable growing), viticulture (grape culture), floriculture (flower culture), nursery management, and landscape architecture.

***Humus*** Well-decayed organic matter, making up the major end-product of compost piles. Also, it is the nonliving, organic component of the soil.

***Hybrid*** The product of a cross between two parental lines which differ genetically.

***Hybrid vigor*** Also called heterosis, it is a condition of increased size (usually) and crop yield, manifested when two inbred parental lines are crossed with each other, as exemplified by hybrid corn (maize).

***Hybridization*** Process of forming a hybrid by cross-pollination of plants of different types.

***Hydathode*** A small pore structure, usually on leaves at their tips, through which water escapes; it is not a stomate.

***Hydrocarbon*** Any compound containing hydrogen and carbon.

***Hydroponics*** Soilless culture; growing plants in aerated, nutrient solutions.

***Hygroscopic water*** Refers to soil water which is held to soil particles fairly tightly and is therefore not available for uptake by plant roots.

***Hypha*** Filamentous strand of a fungus.

***Hypocotyl*** That stem part of a seedling shoot axis which occurs below the cotyledon(s).

***Hypogeous*** Refers to a seedling whose cotyledons remain below ground during seed germination and seedling development.

***IBA*** Indole-3-butyric acid, one of the synthetic auxin-type plant hormones. It is used to stimulate rooting in cuttings.

***Inbred line*** A genetic line or selection which is maintained by repeated back-crossing of a hybrid to one of its parents in plant breeding.

***Inbreeding*** Process in plant breeding of crossing a parental line of a given selection with itself repeatedly in successive crosses. For example, it is used to produce different inbred parental lines of short height; when these inbred parental lines are crossed, they may produce tall hybrids, as in hybrid corn (an example of hybrid vigor or heterosis).

***Incandescent lamp*** A type of lamp which emits light in the full visible range of the white light spectrum [from violet (around 400 nanometers) to the far-red (around 800 nanometers)], thus providing plants with all colors (qualities) of light necessary for normal development.

***Indusium*** A flaplike tissue that covers and protects the sorus in ferns. The sorus is a cluster of spore cases (sporangia).

***Inflorescence*** The arrangement of flowers on a stem or axis.

***Infrared light*** That portion of the white light spectrum which represents the far-red region (between 700 and 800 nanometers). It is important for normal plant growth and development in addition to the blue and red regions which are used in photosynthesis.

***Inoculum*** Dried bacterial preparation used to inject soil with nitrogen-fixing bacteria, such as *Rhizobium*, which will form nodules on the roots of peas, beans, clovers, vetch, and alfalfa. It is usually dusted on the seeds just before planting them.

***Insulation*** Material used in a wall or ceiling to prevent heat loss from a building or from the north wall of a greenhouse or cold frame.

***Intercalary meristem*** Meristematic tissue, originally derived from the apical meristem of a shoot, which continues its meristematic activity some distance below the apical meristem, as at the base or top of an internode or at the base of a leaf. Thus, it is "sandwiched" between parts that are more mature.

***Internode*** That portion of a stem which lies between two nodes. (See node.)

***Insecticide*** Chemical used to kill insects.

***Integrated pest management*** A method of limiting pest populations, utilizing many different means of control, but with the least possible damage to the environment.

***IPA*** Indole-3-propionic acid, one of the synthetic auxin-type of plant hormones which is used to stimulate rooting in cuttings.

***Kinetin*** A synthetic cytokinin type of plant hormone used to induce bud development in plants.

***Lateral bud*** A bud arising from a node in the axil of a leaf and which is lateral to the terminal bud of the shoot axis.

***Lath shades*** Pieces of wood slats strung together by wire or attached to frames and put over greenhouse glass to provide shade or over a lath-house for the same purpose.

***Layering*** A method of rooting shoots of shrubs or woody vines which are still attached to the parent plant. It is accomplished by bending a lower shoot or twig to the ground and covering it with earth until it is rooted. It is then severed from the parent plant and grown independently on its own roots.

***Leaf*** A foliar appendage of the shoot system in plants whose primary function is to harvest the sun's energy in the process of photosynthesis; it consists of blade and petiole or blade and sheathing leaf base.

***Leaf cutting*** A cutting made from the leaf of a plant.

***Leaf primordium*** A young, embryonic leaf as seen in the young, undifferentiated leaves associated with the shoot apex.

***Leaf stalk*** A supporting stem (petiole) by which a leaf is attached to a twig or larger branch.

***Lean-to greenhouse*** A greenhouse attached to a building; for example, the south side of a house.

***Legume*** Any of a large group of plants belonging to the pea family *(Fabaceae)* and producing a simple, dry, dehiscent fruit or pod, as in peas and beans.

***Lenticel*** An opening through which gases pass; occurs most often on woody stems.

***Lime*** Term used for calcium oxide, $CaO_2$. Lime is used to make soils more alkaline (more basic) in pH.

***Lip*** The lowermost petal in an orchid flower; usually, it is the most elaborately colored and enlarged petal.

***Loam*** Ideal soil mixture made up of varying parts of sand, silt, and clay.

***Lodging*** The falling over of a plant due to the action of wind and/or rain. It is especially a problem in tall cultivars of cereal grains such as wheat and rice, and it can result in significant reductions in yield of cereal grain.

***Long-day plant*** A plant which will flower only when grown under long days (or short nights). Good examples are spinach and barley.

***Longevity*** Condition describing the length of time during which a structure, such as a seed, is still alive (or viable).

***Louvre*** A window which can be opened, as in a greenhouse.

***Macronutrient*** Refers to an inorganic element essential for normal plant growth and development, which is required in relatively large amounts. Examples include nitrogen (N), phosphorus (P), potassium (K), calcium (Ca), magnesium (Mg), and sulfur (S).

***Mercury vapor lamp*** Special high intensity light source used to illuminate plants in greenhouses.

**Mericloning** A technique of plant tissue culture in which the apical meristem of a plant is vegetatively propagated in sterile, nutrient media to obtain hundreds of genetically identical specimens.

**Meristem** A region of dividing cells from which mature cells are derived. Examples are shoot and root apical meristems, vascular and cork cambia, and intercalary meristems.

**Mesophyll** The photosynthetic tissue of leaves.

**Methane generator** A digester of manure or plant materials involving anaerobic decomposition brought about by methane bacteria with consequent release of methane (natural gas), carbon dioxide, and small amounts of nitrogen and hydrogen sulfide.

**Microorganisms** Microscopic organisms; for example, mycoplasmas, bacteria, some fungi, and blue-green algae.

**Micronutrient** Refers to an inorganic element essential for normal plant growth and development and which is required in relatively small amounts. Examples include boron (B), manganese (Mn), copper (Cu), zinc (Zn), iron (Fe), molybdenum (Mo), and chlorine (Cl).

**Mineral nutrient** An inorganic element essential for normal plant growth and development. Over thirteen such elements are considered to be essential for plants: N, P, K, Ca, Mg, S, Fe, Zn, Cu, Mo, Mn, Cl, and B.

**Miticide** A chemical used to kill mites which attack plants.

**Molluscicide** A chemical that kills molluscs such as slugs and snails.

**Monocots (monocotyledons)** A class of flowering plants characterized by floral parts in threes, parallel-veined leaves, a single cotyledon, vascular bundles usually not arranged in a ring, and the absence of a vascular cambium.

**Monoculture** Growing of a single crop.

**Monoecious** (of one house) Condition in which a plant has separate male-only and female-only flowers on the same plant. Thus, pollen must be transferred from male (staminate) flowers to female (pistillate) flowers. (See dioecious.)

**Morel** Special edible mushroom with convoluted fruiting body. It is a fungus in the group called ascomycetes (sac fungi). It is collected as a wild, edible food in the spring.

**Mottled** Condition in which leaves have patches of yellow and green. It can be caused by genetic mutation, physiological disorders such as nutrient deficiencies, and disease-causing organisms.

**Muck** A type of soil, black in color, and very rich in organic matter. It is derived from the decay of plants in once aquatic environments such as bogs or old lake bottoms.

**Mulches** Material placed on top of the soil to conserve water, prevent the growth of weeds, and increase soil organic matter. They include, for example, bark chips, leaves, sawdust, straw, rice hulls, and cocoa bean hulls.

***Multiple fruit*** A fruit formed from several flowers united into a single structure having a common axis, as in mulberry and pineapple fruits.

***Mushroom*** Another name for the fruiting body of a fungus.

***Mutualism*** A type of symbiosis in which both organisms involved benefit from the relationship.

***Mycelium*** A mass of hyphae making up the fungus body.

***Mycoplasmas*** Small, intracellular, parasitic organisms of plants and animals, including humans. They lie between viruses and bacteria in size.

***Mycorrhiza*** (pl. ***mycorrhizae***) A structure composed of a filamentous body of a fungus and the root of a vascular plant growing together in a symbiotic relationship.

***NAA*** β-naphthaleneacetic acid, one of the synthetic auxin-type plant hormones which is used to stimulate rooting in cuttings.

***Necrosis*** Death or decay of tissue.

***Nectar*** A sugar-containing secretion found in many flowers, which attracts certain pollinators because it serves as a food source for them.

***Nectary*** The floral or extra-floral structure that produces a sugary liquid capable of attracting insects and other small animals.

***Nematicide*** A chemical that kills nematodes.

***Nematode*** A microscopic, nonsegmented roundworm found in soils, many of which attack plants and animals.

***Nitrogen fixation*** Process by which atmospheric nitrogen is converted to ammonia by blue-green algae (for example, *Anabaena, Nostoc*) or nitrogen-fixing bacteria (for example, *Rhizobium, Azospirillum*) growing in association with plant roots.

***Node*** That locus on a stem where leaves and/or buds are produced.

***N-P-K*** Refers to a nitrogen-phosphorus-potassium ratio used for inorganic fertilizers such as a 10-10-10 fertilizer (10 parts N, 10 parts P, and 10 parts K).

***Nut*** A large, indehiscent, one-seeded, hard or "bony" fruit.

***Occult balance*** Uneven balance, a term often used in landscaping in informal design schemes.

***Offset*** A lateral shoot which originates from the crown of the "mother plant," as in hen and chickens, sedum, agave, and yucca.

***Orchard*** A term referring to fruit trees grown in cultivation in condensed rows in a given area.

***Organic gardening*** A system of gardening which relies on building up soil organic matter with cover crops and compost; uses alternative biological means of pest control; and utilizes extensive cropping systems which rely on mixed crops (polyculture), low tillage (or none), mulches, and often, trickle irrigation.

***Osmunda fiber*** The root mass of the cinnamon fern; it is used as a growing medium for epiphytes such as orchids and bromeliads.

***Ovary*** The enlarged, basal part of a pistil. It contains the ovules and eventually develops into the fruit.

***Ovule*** A reproductive structure in seed plants which contains the embryo sac (with egg cell), surrounding nucellus tissue (a nutritive jacket), and one or two integuments. A fertilized mature ovule is a seed.

***Parasite*** An organism living off another plant or animal.

***Parenchyma*** A tissue composed of living cells which are more or less isodiametric (same size in all directions), and generally, though not always, thin walled.

***Parent*** In plant breeding, an individual member of a selection capable of producing progeny either by crossing with another parent (cross-pollination) or by selfing (self-pollination) if that is genetically feasible.

***Parthenocarpy*** Development of seedless fruits, either occurring naturally in some fruit selections or as the result of hormone treatment (auxins or gibberellins) used to enlarge the fruits.

***Particulate matter*** Air pollutants consisting of solid particles and liquid droplets.

***Pasteurization*** The use of moist heat or steam to kill certain pathogenic microorganisms in the soil before use in greenhouse potting mixes. The material is usually heated to 60° to 77°C (140° to 170°F) for 30 minutes.

***Pathogen*** Any disease-causing organism.

***PBB*** Polybrominated biphenyls, which are toxic to humans and an industrial contaminant in our environment.

***PCB*** Refers to polychlorinated biphenyls, industrial contaminants in our environment which are toxic to humans.

***PCP*** Pentachlorophenol, a wood preservative used to prevent rotting of the wood caused by fungi. It is toxic to humans and thus considered to be an industrial contaminant in the environment.

***Peat*** The breakdown product of sphagnum moss and/or sedges found in bogs. It is used to acidify soil and to add humus to the soil. Peat has high water-holding capacity and is quite acidic and sterile.

***Peat pot*** Small container made from pressed and shaped peat moss. It is used to grow plants before directly planting them in the garden (the peat pots can be planted with the plants and break down in the soil).

***Peduncle*** Stalk of a flower cluster or of one flower borne singly.

***Perennial*** A woody or herbaceous plant that grows and potentially flowers and sets seed year after year. The plant does not die after flowering.

***Perfect flower*** Flower having both functional stamens and pistils.

***Perlite*** A pumicelike white, fluffy material used in combination with soil or fir bark or vermiculite to improve aeration in potting mixes.

***Permanent wilting percentage (PWP)*** This refers to the percentage of water in the soil at which plants permanently wilt, that is, will not recover even when the soil is then watered.

***Pesticide*** A chemical substance used to control a plant pest such as weeds (herbicide), insects (insecticide), fungi (fungicide), and mites (miticide).

***Petal*** One of the units of the corolla of a flower; it is a leaflike part of the flower which may be brightly colored.

***Petiole*** The stalk of a leaf.

***pH*** A measure of acidity (hydrogen ion concentration) or alkalinity (hydroxyl ion concentration). pH values range from 0 to 14 with pH 7 being neutral, pH's below 7 being acid, and pH's above 7 being alkaline.

***Pheromones*** Chemical sex attractants. They may be naturally occurring or synthetically produced. They are used as one of the biological controls of insect pests.

***Phloem*** Food-conducting tissue in vascular plants.

***Phosphon D*** A growth retardant (2, 4-dichlorobenzyl-tributyl phosphonium chloride) used to dwarf shoots. It acts by inhibiting the synthesis of native gibberellin hormones in the plant.

***Photoperiod*** The relative length of day/night exposure that triggers a developmental response in plants, such as flowering or no flowering, onset of bud dormancy, or breaking of bud dormancy. The length of the dark period is critical in determining the developmental response.

***Photoperiodism*** The phenomenon in which the relative length of day and night has an effect on the growth response of a plant. (See photoperiod.)

***Photosynthesis*** The process by which light energy from the sun is harvested by plants and converted into foods (sugars) and high energy-containing ATP (adenosine triphosphate). Basically, the overall photosynthetic reaction is: $CO_2 + H_2O \xrightarrow[\text{chlorophyll}]{\text{light}} O_2 + CH_2O + \text{ATP}$. $CH_2O$ refers to the various sugars produced.

***Phototropism*** Growth curvature response of a plant toward light as seen, for example, with a plant bending toward light of greater intensity coming in a window.

***Phytochrome*** The photoreceptor pigment (a protein) which absorbs light in the far-red (peak around 730 nanometers) and in the red (peak around 660 nanometers); thus, it has two forms: $P_{fr}$ and $P_r$. Phytochrome is responsible for controlling many developmental responses in plants such as hook opening in seedlings, leaf expansion, internodal extension, seed germination, and flowering in long-day and short-day plants.

***Pip*** A dormant bud along a section of rhizome as seen, for example, with lily-of-the-valley *(Convallaria majalis).*

***Pistil*** The female reproductive structure of a flower. It consists of stigma, style, and ovary.

***Pistillate*** Refers to flowers with female parts (pistils) only, as occurs in monoecious or dioecious plants.

***Pith*** Innermost tissue of a stem or root made up of parenchyma cells.

***Pit-type greenhouse*** A greenhouse in which the lower one-third to one-half portion is below ground level to provide better insulation.

***Plant breeding*** The process of making selected crosses and propagating plants through sexual reproduction to produce new desirable plants useful to man.

***Plant pathology*** The study of plant diseases.

***Plumule*** The shoot portion of a developing seedling.

***Polar nucleus*** One of the two 1N (haploid) nuclei in the embryo sac. It is also one of three (two polar nuclei and one sperm nucleus) involved in forming the 3N (triploid) nuclei of endosperm cells in ovules which are developing into seeds.

***Pollarding*** A pruning technique in which the main limbs of a large tree are drastically cut back to reduce the mass and volume of the crown of the tree. It is also called topping.

***Pollen*** Microscopic grains borne by the anthers. They contain vegetative and generative nuclei; the latter forms two 1N (haploid) sperm nuclei before or during pollen germination but prior to the pollen tube entering the ovule and bringing about fertilization in the embryo sac of the ovule.

***Pollen tube*** A tube formed after the germination of the pollen grain on the stigma. It carries the two male sex cells (sperm) into the embryo sac of the ovule.

***Pollination*** The process of transfer of pollen from an anther of the stamen to a stigma of the pistil.

***Pollinator*** A vector (for example, insect, bird, bat, moth, butterfly) which effects transfer of pollen from one flower to another (cross-pollination) or from anther to stigma of the same flower (self-pollination).

***Polyethylene film*** Sheets made from a plastic polymer (polyethylene) that are more or less transparent and of varying thickness. They are used to cover greenhouses instead of glass.

***Polyploid*** Condition in which cells have more than two sets of chromosomes.

***Pomander*** A ball of aromatic substances such as an apple studded with cloves.

***Pome*** The fruit of apple, pear, quince, crabapple, and hawthorn. It is an inferior ovary made up of several carpels having a fleshy pericarp.

***Potherb*** Another term for a plant that can be used as a cooked vegetable.

***Potpourri*** Literally means rotten pot; refers to a fragrant mixture of flowers, herbs, and spices contained in a jar.

***Predator*** An organism that catches and feeds on another organism (a parasite also destroys).

***Primary phloem*** That phloem (food-conducting tissue) produced by the procambium during primary growth of a root or shoot; in contrast, secondary phloem is produced by the vascular cambium during secondary growth in a root or shoot.

***Primary plant body*** Refers to the part of a plant body arising from the root and shoot apical meristems and their derivative meristematic tissues. It is composed entirely of primary tissues. (For comparison, see secondary plant body.)

***Primary xylem*** That xylem (water-conducting and nutrient-conducting tissue) produced by the procambium during primary growth of a root or shoot; in contrast, secondary xylem is produced by the vascular cambium during secondary growth in a root or shoot.

***Propagation*** To increase the numbers of a plant by such methods as division, cuttings, seeds, mericloning, and layering.

***Protein*** Polymer made up of amino acid subunits. Seeds contain many storage proteins that are important in the human diet (especially those found in the cereal grains and in legumes such as soybeans, peas, and beans) and in germinating seeds, where they are used as reserve or storage food substances.

***Protocorm*** A spherical mass of cells without roots or shoots, representing an early stage in the development of an orchid embryo.

***Pruning*** The process of cutting off superfluous or dead branches from a plant. It is a trimming process.

***Pseudobulb*** In orchids, an enlarged or swollen food-storage stem near the base of the plant.

***Quonset greenhouse*** A greenhouse that is hemispherical in cross section and much longer than wide. It is supported by riblike pipes and covered with plastic.

***Radicle*** The root of a developing seedling, appearing during and just following seed germination.

***Rag-doll*** A wet, folded paper device used to test seed germination.

***Receptacle*** The enlarged, terminal portion of the axis of a flower stalk that bears the floral organs (sepals, petals, stamens, and pistils).

***Relative humidity*** An expression of the percentage of moisture actually in the air as compared to the amount of moisture the air could hold if saturated at a given temperature. As the temperature goes up, relative humidity goes down unless more moisture is added to the air.

***Respiration*** The process by which food materials, such as sugars, are oxidized to simpler compounds, such as carbon dioxide ($CO_2$) and water ($H_2O$), and in which the energy released is transformed to do work in cells.

***Rhizome*** The horizontal, creeping, enlarged stem of a fern or flowering plant; it is usually produced underground and may act as a means of vegetative propagation.

***RNA*** Ribonucleic acid, the primary product of transcription from DNA in the cell. It codes for the formation of proteins in the cell by the process of translation.

***Rodenticide*** A chemical that kills rodents.

***Root*** The descending (positively gravitropic) portion of a plant which serves to anchor the plant and to absorb and transport water and mineral nutrients.

***Root apex*** The apical meristem of a root, responsible for generating new cells and tissues of the root.

***Root cutting*** A cutting made from the roots of a plant for the purpose of propagation.

***Rootstock*** The lower, or root-producing, portion of a graft. Also called stock.

***Rosette*** A dense cluster of overlapping leaves growing in the form of a shallow cup or saucer, as in dandelions, the bromeliads and many leaf succulents. It is also the first-year growth stage in biennials.

***Runner*** Also called stolon. An elongate, prostrate, above-ground stem which, as it grows along the soil, produces roots and lateral shoots at the nodes.

***Rust*** Term applied to a type of parasitic fungus which causes rustlike, reddish or yellowish pustules to form on the undersides of leaves of one of its host plants. Rust diseases caused by rust fungi include white pine blister rust, barberry-wheat rust, hollyhock rust, and cedar-apple rust.

***Samara*** An indehiscent, winged fruit as seen, for example, in the maples and ashes.

***Sand*** One of the major inorganic components of the soil with particles of large size, ranging between 0.2 and 0.02 millimeter in diameter.

***Saprophyte*** An organism living off the dead remains of another plant or animal as, for example, mushrooms growing on decayed wood or straw.

***Scaly bulb*** A bulb whose fleshy, scalelike leaves overlap one another loosely as, for example, in the lily bulb.

***Scarification*** An abrasion process in which one scratches a nick in the seed coat, rendering it permeable to gases and water. It is used to break dormancy in seeds with hard seed coats.

***Schizocarp*** A dry, dehiscent fruit that splits into two halves at maturity. Each half is called a mericarp. This type of fruit, and the process of its splitting into two mericarps, occurs in most members of the carrot family *(Apiaceae)*.

***Scion*** The upper or shoot-producing portion of a graft.

***Sclereid*** A sclerenchyma cell with a thick, lignified secondary wall having many pits. One type of sclereid is the stone cell; stone cells occur in clumps in the fleshy part of the pear fruit, rendering it somewhat gritty in texture.

***Sclerenchyma*** Supporting and protective tissue whose cells have thick, lignified walls. Sclerenchyma tissue is composed of either sclereids or fibers.

***Secondary body*** That part of the plant body produced by the vascular cambium and the cork cambium. It consists of secondary phloem, secondary xylem, and periderm (bark). (For comparison, see primary plant body.)

***Secondary phloem*** That phloem (food-conducting tissue) which is produced by the vascular cambium. It is a primary component of living bark. Its production contributes to axis thickening during secondary growth.

***Secondary xylem*** That xylem (water- and nutrient-conducting tissue) which is produced by the vascular cambium. Its production results in axis thickening or secondary growth. The bulk of the wood portion of a tree is made up of secondary xylem.

***Seed*** A mature, fertilized ovule containing the embryo and borne in a mature ovary, the fruit.

***Seed germination*** The resumption of growth of the embryo, resulting in the protrusion of a root (also called the radicle) and a shoot (also called the plumule).

***Seed-set*** Condition in which the ovary of a pistil enlarges and produces a mature fruit with seeds.

***Self-compatible*** A term in pollination biology and plant breeding referring to the ability of the pollen of a given plant to be able to germinate, form a pollen tube, and its sperm to fertilize the egg in the ovary of the same plant.

***Self-incompatible*** Condition in which the pollen of a given plant will not germinate or whose pollen tubes will not develop normally (abort) in the style of the pistil of the same plant. It can also arise as a result of failure of self-pollination due to structural or maturation timing problems.

***Self-pollination*** Pollination resulting from the transfer of pollen from the anther to the stigma of the same flower.

***Sepal*** The outermost, leaflike appendage of a flower and one of the units of the calyx.

***Sexual reproduction*** In seed plants, the entire process of getting the sperm nucleus to the egg nucleus, resulting in viable seeds.

***Shading cloth*** Plastic, netlike cloth with mesh size that allows for 25 to 75 percent light transmittance. It is used over a greenhouse glass surface or over plants in a greenhouse to provide shade.

***Shading compound*** Slaked lime or other similar compound applied to glass of a greenhouse to cut down on light transmittance and hence to afford shading during warm summer weather.

***Sheet erosion*** Type of soil erosion caused by rain in relatively large amounts, resulting in soil loss to more or less the same extent over large areas.

***Shoot apex*** The apical meristem of a shoot responsible for generating new cells and tissues of the shoot.

***Short-day plant*** A plant which will flower only when grown under short days (or long nights). Good examples are chrysanthemum, Christmas cactus, and poinsettia.

***Side layering*** A type of asexual reproduction in which the shoot of a plant is buried for some distance along its length to induce rooting and new shoot formation.

***Sieve-tube element*** A cell of the phloem concerned with long-distance transport of organic solutes (food substances, such as sugars).

***Silique*** The long, dry, dehiscent fruit of certain members of the mustard family (Brassicaceae), which open by two valve pieces separating the seeds by a partition in between. A rounded form of this fruit type is termed a *silicle.*

***Silt*** One of the major inorganic components of the soil with particles of intermediate size, ranging between 0.02 and 0.002 millimeter in diameter.

***Sod*** Grass- or turf-covered soil. Sod can thus come from a lawn, pasture, or field where grasses (and sometimes weeds) predominate.

***Softwood cutting*** A shoot cutting made from nonwoody or herbaceous plants or from herbaceous (young) shoots of a woody plant, which, when rooted, form new plants.

***Soil*** The composite material on the earth's land surface which is made up of varying amounts of sand, silt, and clay, all derived from the weathering of rocks initially. Soil can be modified by the addition of organic matter (humus) from decaying plants, fallen leaves, or animal remains.

***Soil binder*** Refers to a plant which effectively prevents soil erosion from occurring on slopes. Examples include crown vetch, ivy, honeysuckle, and St. Augustine grass.

***Soiless mix*** Plant-growing medium without soil; for example, a mixture of vermiculite, perlite, and peat.

***Solar heated*** Heated by the sun.

***Solar panel*** A solar collector device which converts solar energy to heat via heated air or water, which can be stored and used to heat a home or greenhouse. The collector is a closed system with glass or plexiglass cover and a black backing separated by an air space through which air passes or water flows in copper or aluminum pipes.

***Solar radiation*** The emission of heat energy from the sun.

***Sorus*** (pl. ***sori***) A small cluster of sporangia (spore cases) on fern leaves. The shape and arrangement of sori are useful identification features for many ferns.

***Species*** Basic taxonomic group of plants, being a subdivision of a genus. Species are usually recognizable as a certain kind of organism. All members of a given species will normally interbreed with one another. It is the second part of the Latin name (binomial).

***Specimen plant*** In landscaping, an outstanding individual plant given a position of attention because of attractive and characteristic form, color, and/or size.

***Sperm*** A haploid (1N) male gamete (sex cell). It is usually motile and smaller than the female sex cell (gamete), the egg.

***Sphagnum*** A moss characteristic of acid bogs, which when dried is called *peat moss.*

***Spice*** Aromatic flavoring product derived from some part of a woody perennial shrub or tree that usually grows in the tropics.

***Spine*** A stiff, sharp protuberance from a plant, especially one that is a leaf modification.

***Sporangium*** (pl. ***sporangia***) A small, usually microscopic structure that produces spores in ferns. Sporangia are often arranged into characteristic sori which may be covered by an indusium.

***Stamen*** The male reproductive structure of a flower, consisting of filament and anther sac(s).

***Staminate*** Refers to flowers with male parts (stamens) only as occur on monoecious and dioecious plants. (See monoecious, dioecious.)

***Starch*** A storage polysaccharide made up of glucose subunits found, for example, in potato tubers.

***Stem*** The main, upward-growing (negatively gravitropic) portion of a shoot. In some plants, it is also a modified underground structure to which roots are attached (as in bulbs, corms, rhizomes, and tubers). Stems possess nodes and internodes; at the nodes, leaves or scalelike leaves are attached.

***Stem cutting*** A section removed from a stem of a plant, which when rooted forms a new plant. Making cuttings is a form of asexual propagation in plants.

***Sterile*** Term in plant breeding referring to the inability of a plant to produce viable seeds, and thus to reproduce sexually. Sterility is often observed with interspecific hybrids.

***Stick-tight*** Term applied to those fruits which have barbed awns or spines—structures that aid in their adherence to clothing of humans or fur of animals, thus facilitating their dissemination (dispersal).

***Stigma*** The uppermost region of a pistil, serving as a receptive surface for pollen grains. It is also the site where pollen grains germinate to produce pollen tubes.

***Stock*** (or ***rootstock***) The lower or stem portion of a plant to which scions are grafted.

***Stoma*** (pl. ***stomata***) A small opening in the epidermis of leaves and stems through which gases pass.

***Stratification*** Also called after-ripening. Refers to moist, low temperature storage of seeds, usually done for four to six weeks at 4° C in order to break dormancy resulting from the presence of chemical inhibitors in the seed coat or other parts of the seeds.

***Style*** An elongation of tissue between the stigma and the ovary in the pistil of a flower. Pollen tubes grow within the style from the stigma (where pollen grains germinate) to the ovules in the ovary.

***Styrofoam*** Porous, plastic beads or sheets used for insulation.

***Submerged aquatic plant*** A water plant which grows primarily below the water surface level.

***Succession*** The phenomenon in plant ecology in which a sequence of plant communities occupies a given habitat over a period of time, each new community changing the habitat somewhat in character.

***Succulent*** A plant which stores water in specialized roots, stems, or leaves; these organs may appear enlarged and juicy and may bear thorns, spines, or dense hairs.

***Superphospate*** A high phosphate inorganic fertilizer.

***Symbiosis*** Two or more different organisms living in close association with one another for the mutual benefit of both.

***Temperate zone*** The climatic regions north and south of the tropical latitudes, characterized by strong seasonal fluctuations in daylength and temperature. The seasons in the north temperate zone are just the opposite of those in the south temperate zone.

***Terrarium*** (pl. ***terraria***) A glass or plastic container in which one grows plants; a miniature greenhouse. (See wardian case.)

***Testa*** Refers to the seed coat (derived from the integuments of an ovule).

***Thermoperiod*** Refers to temperature period; for example, a period of low temperature necessary to break bud or seed dormancy or to induce flowering in biennial long-day plants or temperate zone (hardy) bulbs.

***Thinning out*** A pruning technique whereby whole branches or twigs are removed from a dense growth. Used to rejuvenate an overgrown shrub.

***Thorn*** A stiff, sharp modified stem of a plant; may be branched.

***Tip layering*** Type of asexual propagation in which the shoot tip of a plant is buried in the soil where it roots and produces a new shoot. It may occur naturally as, for example, with black raspberries.

***Top-dressing*** Term referring to the application of fertilizer to the soil around growing plants or to soil in which seeds have been planted.

***Topiary*** A technique in which evergreen shrubs are pruned into geometric shapes.

***Trace element*** Element essential for plant growth which is required only in small amounts. Includes iron (Fe), copper (Cu), molybdenum (Mo), boron (B), manganese (Mn), zinc (Zn), and chlorine (Cl).

***Tracheid*** An elongated thick-walled conducting cell of the xylem (water- and nutrient-conducting tissue) found in nearly all vascular plants. It has tapering ends; its walls are pitted; and, there are no perforations at the ends of the cell as one finds with vessel elements. Tracheids are dead (without protoplasts) at maturity.

***Transpiration*** Evaporation of water from plants through the stomata.

***Trickle irrigation*** Watering system for plants in which water is delivered to the soil through perforated, usually plastic, pipes.

***Triploid*** Cells which carry three sets of chromosomes in their nuclei; they are termed 3N.

***Trophic level*** The particular way in which an organism obtains its nutrition in a given biological community. For example, plants are producers of food; animals are consumers; and fungi are decomposers.

***Tropical zone*** The climatic regions extending about 20° north and south of the equator. It is the zone in which daylengths and temperatures remain about the same year-round and where there are no pronounced cold winter seasons causing the plants to go dormant; however, there may be brief dry seasons.

***TTC*** Refers to 2,3,5-triphenyltetrazolium chloride, a chemical used to test for seed viability.

***Tuber*** A fleshy, enlarged, underground stem as seen, for example, in Irish potato or Jerusalem artichoke. Propagation from "eyes".

***Tuberous root*** An underground root which is enlarged and fleshy, as occurs in dahlia, sweet potato, and tuberous-rooted begonia.

***Tungsten lamp*** Another name for an incandescent lamp. It emits light over the entire white light spectrum as well as considerable heat.

***Tunicate bulb*** A bulb whose fleshy, scalelike leaves overlap one another tightly, one inside the other, as in the onion or tulip bulb.

***2,4-D*** 2,4-dichlorophenoxyacetic acid, a synthetic auxin (growth hormone) which is used as an herbicide to control broad-leaved dicot weeds.

***2,4,5-T*** 2,4-5-trichlorophenoxyacetic acid, a synthetic auxin (growth hormone) which is used as an herbicide to control woody plants. 2,4,5-T is currently banned by the EPA because of the presence of the very toxic substance, dioxin.

***Ultraviolet light*** Portion of the light spectrum below the violet/blue region, and which is harmful to both humans and plants. It is surrounded by the ozone layer in the atmosphere, thus protecting us from excess UV radiation at the earth's surface because the ozone layer prevents much of the sun's UV rays from reaching us.

***Vascular bundle*** A conducting strand ("vein") in a leaf or stem. It is composed of both primary xylem (water- and nutrient-conducting tissue) and primary phloem (food-conducting tissue) and is frequently enclosed by a bundle sheath of parenchyma or fibers.

***Vascular cambium*** The cylindrical meristem of roots and stems which gives rise to secondary xylem and phloem.

***Vascular system*** The conducting system of plants. It is composed of xylem (nutrient- and water-conducting tissue) and phloem (food-conducting tissue).

***Vector*** Any organism that is a carrier of a pathogen.

***Vegetative reproduction*** An asexual means of producing new plants using roots, stems, or leaves. It is a cloning process of propagating genetically identical plants.

***Vermiculite*** A seed germination and rooting medium composed of heat-expanded mica. It has high water-holding capacity and little or no nutritive value.

***Vernalization*** Exposing a plant to chilling or low temperature (such as winter outdoors in temperate regions) in order to satisfy a requirement for flowering as seen, for example, in winter wheat and winter rye. (See stratification.) Many woody plants require vernalization.

***Vessel element*** (or ***vessel member***) A xylem cell that has perforated end-walls and is one of many such cells connected in series to make up a vessel. Vessel elements are dead (without a protoplast) at maturity.

***Viable*** Alive, such as viable seeds.

***Virus*** A living particle so small that it can only be seen by the use of a transmission electron microscope. It consists of a core of nucleic acid enclosed by a sheath of protein. Viruses may infect plants and animals and cause serious diseases in them.

***Visible light*** Light visible to the human eye between the ultraviolet and the far-red.

***Vitamin*** An organic substance required in the human diet, often acting as a cofactor in enzyme reactions in metabolism.

***Wardian case*** An old-fashioned terrarium or mini-greenhouse of various designs, made up of glass plates sealed together like pieces of stained glass window. It was first used in the early ninteenth century to ship delicate tropical plants back to Europe by providing high humidity and requiring little care.

***Wilt*** Term used variously to refer to (1) condition where a plant suffers from lack of water; (2) a type of fungus disease which causes the leaves to droop and slowly die as seen in Dutch elm disease, oak wilt, and *Fusarium* wilt of tomato; and (3) a type of bacterial disease symptom as seen in the bacterial wilt of cucumber, potato, and tomato.

***Woody plant*** A plant which produces a significant amount of secondary xylem or "wood" in its plant body.

***Xylem*** Water-conducting tissue of vascular plants; it conducts water and minerals in tracheids and/or vessels upwards from the roots.

***Zygote*** The product of fusion of a sperm cell (1N gamete) with an egg cell (1N gamete); this product is a fertilized egg (2N).

# Index

Plants are listed by common name, or the generic name used as a common name. Text entries are indicated by regular type; figures and photos by bold face type; and charts and tables by italics.